T0155664

NEUROMETHODS

Series Editor
Wolfgang Walz
University of Saskatchewan
Saskatoon, SK, Canada

For further volumes:
http://www.springer.com/series/7657

Techniques to Investigate Mitochondrial Function in Neurons

Edited by

Stefan Strack

Department of Pharmacology, University of Iowa Carver College of Medicine, Iowa City, IA, USA

Yuriy M. Usachev

Department of Pharmacology, University of Iowa Carver College of Medicine, Iowa City, IA, USA

Editors
Stefan Strack
Department of Pharmacology
University of Iowa Carver College of Medicine
Iowa City, IA, USA

Yuriy M. Usachev
Department of Pharmacology
University of Iowa Carver College of Medicine
Iowa City, IA, USA

ISSN 0893-2336 ISSN 1940-6045 (electronic)
Neuromethods
ISBN 978-1-4939-8329-2 ISBN 978-1-4939-6890-9 (eBook)
DOI 10.1007/978-1-4939-6890-9

Printed on acid-free paper

This Humana Press imprint is published by Springer Nature
The registered company is Springer Science+Business Media LLC
The registered company address is: 233 Spring Street, New York, NY 10013, U.S.A.

Dedication

"To our families"

Preface to the Series

Experimental life sciences have two basic foundations: concepts and tools. The *Neuromethods* series focuses on the tools and techniques unique to the investigation of the nervous system and excitable cells. It will not, however, shortchange the concept side of things as care has been taken to integrate these tools within the context of the concepts and questions under investigation. In this way, the series is unique in that it not only collects protocols but also includes theoretical background information and critiques which led to the methods and their development. Thus it gives the reader a better understanding of the origin of the techniques and their potential future development. The *Neuromethods* publishing program strikes a balance between recent and exciting developments like those concerning new animal models of disease, imaging, in vivo methods, and more established techniques, including, for example, immunocytochemistry and electrophysiological technologies. New trainees in neurosciences still need a sound footing in these older methods in order to apply a critical approach to their results.

Under the guidance of its founders, Alan Boulton and Glen Baker, the *Neuromethods* series has been a success since its first volume published through Humana Press in 1985. The series continues to flourish through many changes over the years. It is now published under the umbrella of Springer Protocols. While methods involving brain research have changed a lot since the series started, the publishing environment and technology have changed even more radically. Neuromethods has the distinct layout and style of the Springer Protocols program, designed specifically for readability and ease of reference in a laboratory setting.

The careful application of methods is potentially the most important step in the process of scientific inquiry. In the past, new methodologies led the way in developing new disciplines in the biological and medical sciences. For example, Physiology emerged out of Anatomy in the nineteenth century by harnessing new methods based on the newly discovered phenomenon of electricity. Nowadays, the relationships between disciplines and methods are more complex. Methods are now widely shared between disciplines and research areas. New developments in electronic publishing make it possible for scientists that encounter new methods to quickly find sources of information electronically. The design of individual volumes and chapters in this series takes this new access technology into account. Springer Protocols makes it possible to download single protocols separately. In addition, Springer makes its print-on-demand technology available globally. A print copy can therefore be acquired quickly and for a competitive price anywhere in the world.

Saskatoon, Canada *Wolfgang Walz*

Preface

Mitochondria are double membrane-bound organelles present in nearly every eukaryotic cell. Thought to originate from a primitive bacterial endosymbiont, mitochondria generate most of the cellular supply of ATP and are therefore often referred to as the power plant of the cell. In addition to their central role in metabolism, mitochondria generate and sequester reactive oxygen species, serve as high capacity buffers of intracellular calcium, and mediate programmed cell death (apoptosis).

Neurons are particularly dependent on proper mitochondrial function, because of the high energy demands associated with the maintenance of ionic gradients during action potential firing and synaptic transmission, the principal components of neuronal computation. As well, adult neurons are difficult to replace and have a complex cytoarchitecture with dendrites and axons extending over considerable distances. For these reasons, mitochondrial quality control and transport are of unique importance in the nervous system.

Mounting evidence indicates that mitochondrial dysfunction is an essential contributing factor in several common neurodegenerative disorders, including Alzheimer disease, amyotrophic lateral sclerosis, Huntington disease, and Parkinson disease. Moreover, impaired mitochondrial dynamics and energetics have been implicated in peripheral diabetic neuropathy, chemotherapy-induced peripheral neuropathy, Charcot-Marie-Tooth disease, and several other types of acquired and inherited peripheral neuropathies. A better understanding of mitochondrial structure and function is therefore paramount to the development of effective therapies for many major CNS and PNS diseases.

This book is intended as a laboratory manual for a wide range of researchers who study mitochondria in the nervous system. The 16 chapters of this volume combine contributions of leading investigators in the field and describe a broad spectrum of experimental approaches for investigating structure, function, and transport of neuronal mitochondria in health and disease. Many of these approaches were only recently developed and, to the best of our knowledge, have never been assembled in book form. The state-of-the-art techniques compiled in this volume range from electron tomography-based 3D reconstruction of mitochondrial cristae to patch clamp recording from mitochondria in intact neurons. Several chapters describe optical approaches based on the use of genetically engineered fluorescent sensors for monitoring synaptic ATP and axonal ROS generation, mitochondrial Ca^{2+} cycling and pH changes, and mitochondrial dynamics and axonal trafficking in live neurons in real time. With recent advancements in mass spectrometry, this book also includes a chapter that details the use of mass spectrometry for mitochondrial proteomics analysis in neurons. Additional chapters in this volume describe respirometry, NADH imaging, and methods for studying pyruvate transport and mitophagy.

Each chapter focuses on a specific method for studying neuronal mitochondria and is prefaced with an introduction to its underlying scientific principles and areas of implementation. The main sections of each chapter describe specific materials, reagents, tools, and equipment required for a given technique, which is followed by a step-by-step protocol and practical details for employing the method. Each chapter also contains a section that discusses potential problems and their solutions associated with a specific method as well as provides examples of results and outcomes for the described method.

We hope that this book will be a valuable practical resource for a broad range of investigators interested in the function of neuronal mitochondria in health and disease states and would like to thank all the authors for their contribution to this book.

Iowa City, IA, USA

Stefan Strack
Yuriy M. Usachev

Contents

Contributors

THOMAS S. BLACKER • *Department of Cell and Developmental Biology, University College London, London, UK; Department of Physics and Astronomy, University College London, London, UK*

NICKOLAY BRUSTOVETSKY • *Department of Pharmacology and Toxicology, Indiana University School of Medicine, Indianapolis, IN, USA*

TATIANA BRUSTOVETSKY • *Department of Pharmacology and Toxicology, Indiana University School of Medicine, Indianapolis, IN, USA*

KAROLINA CAN • *Zentrum Physiologie und Pathophysiologie, Institut für Neuro-und Sinnesphysiologie, Universitätsmedizin, Göttingen, Germany; Cluster of Excellence and Research Center Nanomicroscopy and Molecular Physiology of the Brain (CNMPB), Göttingen, Germany*

JENNIFER A. CODDING-BUI • *Department of Neurosurgery, Stanford University School of Medicine, Stanford, CA, USA*

MEREDITH M. COURSE • *Department of Neurosurgery, Stanford University School of Medicine, Stanford, CA, USA; Neurosciences Graduate Program, Stanford University School of Medicine, Stanford, CA, USA*

RUBEN K. DAGDA • *Department of Pharmacology, Reno School of Medicine, University of Nevada, Reno, NV, USA*

MICHAEL R. DUCHEN • *Department of Cell and Developmental Biology, University College London, London, UK*

KYLE H. FLIPPO • *Department of Pharmacology, University of Iowa Carver College of Medicine, Iowa City, IA, USA*

HOWARD S. FOX • *Department of Pharmacology and Experimental Neuroscience, University of Nebraska Medical Center, Omaha, NE, USA*

LAWRENCE R. GRAY • *Department of Biochemistry, Carver College of Medicine, University of Iowa, Iowa City, IA, USA*

CHUNG-HAN HSIEH • *Department of Neurosurgery, Stanford University School of Medicine, Stanford, CA, USA*

MAXIM V. IVANNIKOV • *Department of Neuroscience and Physiology, NYU School of Medicine, New York, NY, USA*

ELIZABETH A. JONAS • *Department of Internal Medicine, Yale University School of Medicine, New Haven, CT, USA; Department of Neuroscience, Yale University School of Medicine, New Haven, CT, USA*

SEBASTIAN KÜGLER • *Cluster of Excellence and Research Center Nanomicroscopy and Molecular Physiology of the Brain (CNMPB), Göttingen, Germany; Klinik für Neurologie, Universitätsmedizin, Göttingen, Germany*

ZHIHONG LIN • *Department of Pharmacology, University of Iowa Carver College of Medicine, Iowa City, IA, USA*

GREGORY T. MACLEOD • *Department of Biological Sciences & Wilkes Honors College, Florida Atlantic University, Jupiter, FL, USA*

BRYCE A. MENDELSOHN • *Gladstone Institute of Neurological Disease, San Francisco, CA, USA; Department of Pediatrics, University of California – San Francisco, San Francisco, CA, USA*

RONALD A. MERRILL • *Department of Pharmacology, University of Iowa Carver College of Medicine, Iowa City, IA, USA*

NELLI MNATSAKANYAN • *Department of Internal Medicine, Yale University School of Medicine, New Haven, CT, USA*

MICHAEL MÜLLER • *Zentrum Physiologie und Pathophysiologie, Institut für Neuro-und Sinnesphysiologie, Universitätsmedizin, Göttingen, Germany; Cluster of Excellence and Research Center Nanomicroscopy and Molecular Physiology of the Brain (CNMPB), Göttingen, Germany*

KEN NAKAMURA • *Gladstone Institute of Neurological Disease, San Francisco, CA, USA; Department of Neurology and Graduate Programs in Neuroscience and Biomedical Sciences, University of California – San Francisco, San Francisco, CA, USA*

LALITA OONTHONPAN • *Department of Biochemistry, Carver College of Medicine, University of Iowa, Iowa City, IA, USA*

GUY PERKINS • *National Center for Microscopy and Imaging Research, Center for Research in Biological Systems, School of Medicine, University of California – San Diego, La Jolla, CA, USA*

ADAM J. RAUCKHORST • *Department of Biochemistry, Carver College of Medicine, University of Iowa, Iowa City, IA, USA*

MONICA RICE • *Department of Pharmacology, Reno School of Medicine, University of Nevada, Reno, NV, USA*

ALIX A.J. ROUAULT • *Department of Physiology and Molecular Biophysics, Carver College of Medicine, University of Iowa, Iowa City, IA, USA*

JACOB E. RYSTED • *Department of Pharmacology, University of Iowa Carver College of Medicine, Iowa City, IA, USA*

JULIEN A. SEBAG • *Department of Physiology and Molecular Biophysics, Carver College of Medicine, University of Iowa, Iowa City, IA, USA; Fraternal Order of the Eagles Diabetes Research Center, University of Iowa Carver College of Medicine, Iowa City, IA, USA; Abboud Cardiovascular Research Center, University of Iowa Carver College of Medicine, Iowa City, IA, USA; Pappajohn Biomedical Institute, University of Iowa Carver College of Medicine, Iowa City, IA, USA*

ATOSSA SHALTOUKI • *Department of Neurosurgery, Stanford University School of Medicine, Stanford, CA, USA*

LAUREN Y. SHIELDS • *Gladstone Institute of Neurological Disease, San Francisco, CA, USA; Biomedical Sciences, University of California – San Francisco, San Francisco, CA, USA*

WILLIAM I. SIVITZ • *Division of Endocrinology and Metabolism, Department of Internal Medicine, The University of Iowa Hospitals and Clinics, Iowa City, IA, USA; Iowa City VAMC, Iowa City, IA, USA*

KELLY L. STAUCH • *Department of Pharmacology and Experimental Neuroscience, University of Nebraska Medical Center, Omaha, NE, USA*

STEFAN STRACK • *Department of Pharmacology, University of Iowa Carver College of Medicine, Iowa City, IA, USA*

ERIC B. TAYLOR • *Department of Biochemistry, University of Iowa Carver College of Medicine, Iowa City, IA, USA; Fraternal Order of the Eagles Diabetes Research Center, University of Iowa Carver College of Medicine, Iowa City, IA, USA; Abboud Cardiovascular Research Center, University of Iowa Carver College of Medicine, Iowa City, IA, USA; Holden Comprehensive Cancer, University of Iowa Carver College of Medicine, Iowa City, IA, USA; Pappajohn Biomedical Institute, University of Iowa Carver College of Medicine, Iowa City, IA, USA*

EUGENIA TRUSHINA • *Department of Neurology, Mayo Clinic, Rochester, MN, USA; Department of Molecular Pharmacology and Experimental Therapeutics, Mayo Clinic, Rochester, MN, USA*

PEI-I TSAI • *Department of Neurosurgery, Stanford University School of Medicine, Stanford, CA, USA*

YURIY M. USACHEV • *Department of Pharmacology, University of Iowa Carver College of Medicine, Iowa City, IA, USA*

XINNAN WANG • *Department of Neurosurgery, Stanford University School of Medicine, Stanford, CA, USA*

LIANG ZHANG • *Department of Neurology, Mayo Clinic, Rochester, MN, USA*

Chapter 1

Three-Dimensional Reconstruction of Neuronal Mitochondria by Electron Tomography

Guy Perkins

Abstract

Electron tomography (ET) is a method that uses higher voltage electron microscopes and computer image processing to generate reconstructed volumes of cells, organelles and macromolecules. A renaissance in the study of neuronal mitochondrial structure and function is being fueled by improved techniques for ET with the ability now to generate volumes with nanometer resolution at relatively high throughput. High-quality ET requires knowledge and skill and even though generally performed by well-trained scientists, its use can be extended to those with no training in electron microscopy. This premise is the goal of this chapter. Detailed instructions for performing ET on neuronal mitochondria in situ are presented here. Modern microscopes and the software used to collect tilt series and perform the various image processing tasks have become sufficiently "user friendly" that scientists trained in the biological sciences can learn them relatively quickly, especially with guidance from personnel at ET facilities.

Key words Electron tomography, Image processing, Mitochondria, Movie, Reconstruction, Segmentation, Specimen preparation, Tilt series

1 Introduction

A goal of neurobiology is to generate a theoretical framework that merges structural, physiological, and molecular knowledge of neuronal mitochondrial function. These realms do not advance in synchrony; advances in one realm define new experiments in other realms. For example, in the last few decades, our understanding of mitochondrial function has been driven by physiological and molecular techniques. These advances demand higher resolution structural images of mitochondria in situ. During the 1990s and 2000s, a renaissance in mitochondrial structure–function research was fueled by improved techniques for electron tomography (ET) with the ability now to generate volumes with nanometer resolution. Over the last 18 years, ET has been applied to neuronal mitochondria with special attention to the crista and crista junction architectures.

Stefan Strack and Yuriy M. Usachev (eds.), *Techniques to Investigate Mitochondrial Function in Neurons*, Neuromethods, vol. 123, DOI 10.1007/978-1-4939-6890-9_1, © Springer Science+Business Media LLC 2017

What is ET? ET is a method that uses higher voltage electron microscopes and computer image processing to generate 3D images from multiple 2D projection images of an object, obtained over a wide range of viewing directions, i.e., tilt angles. The 3D reconstruction is generated by using a computer algorithm to back-project each 2D image with appropriate weighting [1, 2] and may include iterations of refinement [3, 4]. As such, ET is comparable to the medical imaging method of X-ray computerized axial tomography (CAT) in that it generates projection images to provide a 3D view of an object, yet with nanometer resolution. The 3D perspective it provides has revised our understanding of cellular and organellar organizations, their dynamics in normal development, and their perturbations due to disease.

The quality of the tomographic reconstruction depends on each step of the process and care must be exercised to perform each step in the best way to preserve the original structure. The steps that require the most attention to detail are, in order,

1. Specimen preparation, including the application of fiducial markers.
2. Tilt series collection, paying attention to optimal imaging parameters.
3. Image processing, with consideration of the strengths and limitations of common methods.

Other steps that follow are for presenting the 3D data in ways that highlight the rich structural information and focus on interrelationships between structural components and include,

1. Segmentation of mitochondrial compartments for display and measurements.
2. Figure and movie making for analysis and presentation.

2 Methods

2.1 Specimen Preparation for ET

Excellent ultrastructural preservation lays the foundation for a high-quality ET reconstruction and starts with attention to detail. The process is relatively long and the accurate execution of nearly every step is important. Although neurons may be grown in culture, either with or without supporting astrocytes, the focus here is tissues and applies to both tissues in the central nervous system and peripheral nerves. The preparation will address mice and rats only (Note 1). An alternative preparation technique is cryopreservation (Note 2).

2.1.1 Preparing Primary Fixative

For 50 ml of 2% paraformaldehyde and 2.5% glutaraldehyde in 0.1 M sodium cacodylate buffer:

1. Heat 20 ml of double-distilled (dd)water in a glass beaker to above 60 ° C, but less then 100 ° C (i.e., not boiling).
2. In a fume hood, add 1 g paraformaldehyde (Prills EM quality). Stir with a stir bar.
3. Add 2 drops 1 N NaOH solution into the stirring solution.
4. When the solution clears, take the beaker off the hotplate (and after cooling close to RT) filter through a #1 Whatman filter, using a funnel into a clean beaker.
5. Add 5 ml of 25% glutaraldehyde (EM grade).
6. Add 16 ml of a 0.3 M cacodylate stock solution pH 7.4.
7. Bring total volume to 50 ml with dd-water.
8. Warm to 37 ° C in a water bath before perfusion fixation.

2.1.2 Preparing Ringer's Solution

250 ml is needed to perfuse 4–6 mice.

1. Add the following components together from stock solutions:
   2.5 ml Na2HPO4 (18 g/L).
   24.8 ml NaCl (80 g/L).
   2.5 ml KCl (38 g/L).
   2.5 ml MgCl2. 6 H_2O (20 g/L).
   6.3 ml NaHCO3 (50 g/L).
   0.5 g dextrose.
2. Bring the total volume to 250 ml with dd-water.
3. Bubble 95% air/5% CO_2 through the solution at 37 °C for 5 min.
4. Add 2.5 ml $CaCl_2$•2 H_2O (30.0 g/l).
5. Continue to bubble the gas until switching the line to the fixative.
6. Because high quality ultrastructure is desired, add 2.5 ml xylocaine (anesthetizes smooth muscle) and 5 ml heparin (prevents blood clotting).

2.1.3 Perfusing the Animal (Example Given Is a Mouse)

For mice, use a 25-G needle.
For medium rats, use a 22-G needle.
For large rats, use an 18-G needle

1. Set out the necessary surgical instruments: large and small scissors, hemostat, tweezers, and iris scissors.
2. Anesthetize the mouse by giving an intraperitoneal injection of nembutol (0.1 cc/100 gm body weight). Ensure the needle is in the peritoneal cavity by drawing back needle and checking for air (no blood).

3. Allow the anesthetic to take effect and test the consciousness of the mouse by using a tail pinch and a paw prick. Ensure that breathing does not stop because mitochondrial structure often changes after even only 1 min of ischemia.
4. Pin the anesthetized mouse to a board.
5. Using small scissors and tweezers, lift up the skin below the rib cage and cut until the liver is visible. Cut upward along the sides of the body cavity until the diaphragm is visible. Cut the diaphragm and then quickly cut up along the sides of the body, being very careful not to damage the heart.
6. Once the heart is sufficiently exposed, use a hemostat to pull the skin up over the mouse's head. Place the needle into the left ventricle and then cut the right atrium with the iris scissors.
7. Turn on the flow of the Ringer's solution to flush the blood via cardiac perfusion. The heart should continue to beat and the liver should quickly turn pale in color.
8. Perfuse with the Ringer's solution for about 1 min. Switch the valve on the pump to the fixative solution. The heart should stop beating and the body should begin to stiffen.
9. Perfuse with the fresh fixative solution (made the same day) for 3–5 min. Take the needle out of the heart. Remove the brain or peripheral nerve desired and place in a "20 ml" scintillator vial containing ice-cold fixative for 1 h. It is important to keep the samples cold at all times until the 100% ethanol step below.

2.1.4 Preparing Brain or Peripheral Nerve Slices

1. Make up the cacodylate and osmium solutions (next three steps). Get them cooling on ice.
2. Using a vibrating microtome or Vibratome, cut the portion of the brain desired into 80–100 μm thick sections in ice-cold 0.1 M cacodylate buffer. Peripheral nerve tissue may be cut into small (<2 mm × 2 mm) pieces with a razor blade. Usually, four slices or four pieces of tissue per desired region is sufficient. Place back in the scintillator vial containing the fixative solution.
3. Wash the slices with ice-cold 0.1 M sodium cacodylate buffer (from a stock solution of 0.3 M cacodylate buffer) for three washes, each lasting 3 min (3 × 3 min) to remove excess aldehydes before postfixing with osmium tetroxide. The other component is dd-water. This is best done by pipetting. Do not use glass pipets because tiny shards get in the sample and can ruin diamond knives used for sectioning below; plastic pipets are suitable, however). It is important to precool all solutions on ice before adding to the sample. It is also important that the sample be uncovered by solution for not more than a few seconds. So, when washing or replacing solutions, add the new solution quickly after removing the previous and do this for each sample before going to the next sample. Keep on ice between washes.

4. Postfix the tissue in ice-cold 1% osmium tetroxide (from a stock solution of 4%) diluted with dd-water + 0.8% potassium ferrocyanide (1% = 1 mg/100 ml) in 0.1 M sodium cacodylate for 2 h on ice. Work in the fume hood because osmium tetroxide is volatile and toxic. Make sure that the ferrocyanide is dissolved before adding; this might take 15 min; mix by shaking with hand; solution should turn dark. The tissue slices should turn black during this step. Get the dd-water cold for the next step.
5. Get the 2% uranyl acetate (UA) in dd-water solution cold that will be used in the next step. Note that UA is light sensitive and will precipitate if not stored in a covered or brown glass container. Rinse the tissue in dd-water 3 × 3 min to remove the excess osmium tetroxide.
6. Get the four vials of 20, 50, 70, 90% ethanol in dd-water cold that will be used in the next step. Stain and stabilize the tissue in ice-cold 2% uranyl acetate (in DDW) for 60 min on ice. If getting late, this step can go overnight.
7. Dehydrate in cold ethanol series: at least 4 ml each of ice-cold 20, 50, 70, 90% ethanol for 10 min each on ice. For the 70 and 90% steps do not incubate longer than 10 min otherwise lipids will be lost. Keep on ice.
8. At room temperature (RT), dehydrate in 100% ethanol for 3 × 10 min. Keep the bottle and vial caps on when not removing liquid to minimize the absorption of water vapor. Remove as much of the liquid as possible from each wash by pipet. Make sure that no water gets inside the vial during these washes, and, of course in subsequent steps. Try not to breathe into the vials to minimize water vapor. Remove and empty the ice bucket. From here to the end, water is the enemy. While incubating, make up the ethanol/Durcupan solutions.
9. Infiltrate in well-mixed 67% ethanol/33% Durcupan, followed by 50% ethanol/50% Durcupan, followed by 33% ethanol/67% Durcupan ACM resin for at least 6 h each (can be overnight if running late) at RT in the scintillation vial. The ethanol is used as a vehicle to penetrate into the tissue higher concentrations of the resin with time. Estimate how much total solution is needed. About 2–4 ml of solution per vial is needed. Four different bottles labeled A (or A/M), B, C, and D will be used. The ratios of the Durcupan components should be 11.4 ml A: 10 ml B: 0.3 ml C: 0.1 ml D. Because certain of the components are viscous, measure 10 g of A and B on a scale. For C and D, use 1 ml syringes to suck up the solution. Component C is smelly, so work in the fume hood. Make sure that the Durcupan is well mixed with a wooden applicator stick before adding in the 100% ethanol. The solution should be a reddish brown color

and viscous. Mix well the Durcupan/ethanol solutions. Because the solutions are viscous, agitate on a rotator or teeter-totter for the full time. Even though the second and third incubations will occur later, make them up at this time.

10. Infiltrate tissue in 2–4 ml of 100% Durcupan ACM for 3 × 8 h. Agitate on a rotator or teeter-totter for the full time. Try to extract as much of the ethanol/Durcupan solution as possible with a pipet with each exchange as ethanol disrupts the resin polymerization. For the second and third 100% Durcupan incubations, transfer the slices to another vial with a minimal amount of Durcupan using a wooden applicator stick. Place in the oven for 5 min before transferring and the resin will soften to make the transfer easier. If it is getting late, an overnight incubation is fine (with agitation). A 12-h incubation is fine, but a 24-h incubation is too long because polymerization (and hardening) of the resin occurs. Each incubation must be done with freshly made Durcupan for good embedment.
11. Using a wooden applicator stick, transfer the slices or tissue pieces into either an aluminum dish, rubber mold, or sandwich between two Mould release slides. Secure the slides and apply pressure to drive out unneeded Durcupan volume using binder clips. It is important that the slides be Mould release otherwise it may be nearly impossible to separate the slides after resin polymerization. Place in an oven. Polymerize at 60–80 °C for 48 h. After removing from the oven, the resin should be "rock" hard.

2.2 Sectioning, Poststaining, and Gold Application

Because ET uses higher voltage electron microscopes, the sections can be thicker.

Note that Sato lead should be made at least 1 day in advance of poststaining.

Sato Lead Preparation.

Weight the following out and add together:

1.0 g—lead nitrate.

1.0 g—lead acetate.

1.0 g—lead citrate.

2.0 g—sodium citrate.

Boil 150 ml of dd-water for 20 min. Pour off 40.0 ml into two 20 ml scintillation vials and cap them. Pour off another 82 ml of dd-water in a graduated cylinder. Pour the 5.0 g of lead and sodium citrate components into a clean Erlenmeyer flask. Pour 82 ml of boiled water in the flask. Cover the flask with Parafilm. Sonicate for 3 h. Make 20.0 ml of 1 N NaOH by adding 0.8 g of NaOH pellets with 20 ml double distilled water that was boiled. Dissolve completely. Pipette out 18 ml of the 1 N NaOH and add to the lead mixture. Mix gently. The 1 N NaOH will dissolve the solute.

Cover the flask opening and then cover around the flask with aluminum foil. Let sit overnight. Next morning pour off into five 20 ml scintillation vials. Cover the opening of the scintillation vials with Parafilm and then screw back on the cap. Wrap the glass portion of the vials with aluminum foil because Sato lead will precipitate upon exposure to light. Sato lead will also precipitate upon exposure to carbon dioxide that is in the atmosphere and so should be quickly capped with Parafilm after each use.

1. After prying off the two slides or removing the aluminum dish or removing from the rubber mold, make blocks for sectioning. The slice that was between slides is thin enough so that a razor blade can be used to cut out the region of interest, which is then mounted on a dummy block with superglue. Embedded tissue coming from a dish or mold will need to be cut out using a hacksaw after the resin block is secured with a vise clamp. Take the dummy block and on the side write the sample information using a paint pen.
2. Place a drop of super glue on top of the dummy block. Carefully place the cutout sample piece onto the superglue. Let dry for several hours before sectioning.
3. Cut 300–500 nm thick sections. Thinner for heavily stained samples and thicker for lightly stained samples, including samples labeled with genetic tags [5, 6]. The sections should be placed inside a clamshell grid (size 100:100 or 50:50 work best) or on a coated slot grid (Fig. 1). The coat is a support film and is typically formvar or luxol.

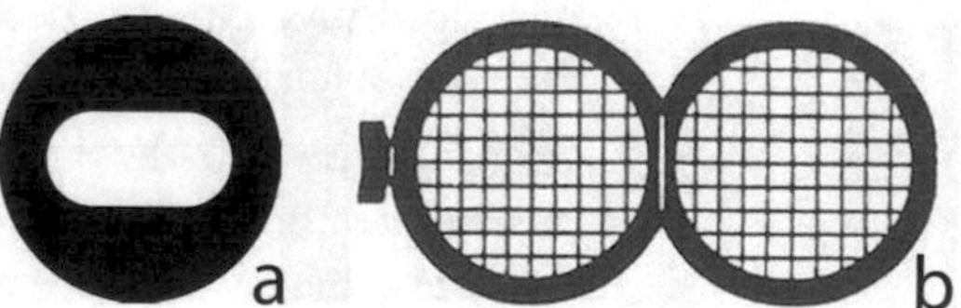

Fig. 1 The best types of grids to use for ET are (**a**) slot grids, which have a large oval opening in the middle to place sections, even serial sections and (**b**) clamshell grids, also called folding grids because the sections are placed on one half of the grid, then the other half is folded over and secured with the metal tab shown on the *left*. Slot grids require a support film that is commonly made from formvar or luxol. Slot grids are more difficult to work with because the support film is fragile and can be easily broken during poststaining and the applying of colloidal gold, and including when subjected to the vacuum of the microscope. They have one advantage, though, that often supersedes the danger of breaking the support film and this is that there is no grid bar to obscure parts of the section(s). Slot grids are almost a necessity when doing serial ET where it is essential to track the same region of interest across serial sections. The advantage of a clamshell grid is that the section is securely sandwiched between two grids and thus will not float off during poststaining and the applying of colloidal gold. There is no worry about handling the grid

4. For poststaining: obtain two small petri dishes (25 mm diameter): one for the uranyl acetate (UA) poststaining, one for the Sato lead poststaining. Also obtain 3 5-ml syringes, and 3 0.22-μm filters, one filter paper (tear into two halves), and two weigh boats.
5. Cut out a square of Parafilm to fit into the center of each petri dish. For UA the size does not matter so long as it fits within the petri dish. For Sato lead, make sure that the size is big enough for the droplet of lead, but small enough so that it does not touch the sodium hydroxide pellets. Tap down the edge of the Parafilm to secure in place.
6. First poststain with UA. Attach a 5-ml syringe to a 0.22-μm filter. Take out the plunger portion of the syringe. Pour UA into the syringe. Add the plunger to the top of the syringe. Let the first few drops fall back into the UA bottle as these may have impurities from the filter. Then put about three drops of UA onto the Parafilm in the petri dish to form one large droplet.
7. Place the grid with section side down on top of the droplet and poststain for 10 min. Note that samples with genetic tags will require little or no poststain.
8. While the grid is on the UA, prepare the Sato lead petri dish. Place sodium hydroxide pellets around the periphery of the petri dish; make sure that the pellets are not touching any part of the Parafilm in the center. The sodium hydroxide will absorb carbon dioxide from the atmosphere in the small volume of the petri dish and thus reduce precipitation of the lead. It is important to minimize lead precipitation because it shows up as dark particles on the section surface and can obscure features of interest.
9. Examine the Sato lead and make sure that there is no white precipitate that has formed inside the vial (on the bottom or the sides) because even with filtering (next step) smaller particles will pass through.
10. Attach a 5-ml syringe to a filter and take out the plunger. Peel back the Parafilm covering the top of the vial. Pour the Sato lead into the syringe, and as with the UA solution, let the first few drops fall back into the bottle. Then put about three drops of solution onto the Parafilm in the petri dish to form one large droplet. Re-cover the top of the Sato lead vial with Parafilm and cap. All this should be done as quickly as possible to minimize the formation of precipitate.
11. Prepare the wash by attaching a 5-ml syringe to a filter and take out the plunger. Pour dd-water into the syringe. With the plunger, push 5 ml of dd-water into a small weigh boat. Do this again for the second weigh boat. One water bath will be used to wash the grids after UA incubation and the other will be used after the lead incubation.

12. After the UA incubation has been completed, use forceps to remove the grid from the droplet. Flip the grid so that the section side is facing up. This is important for sections on slot grids because when facing down, the section can fall off during the agitation of the wash. Gently insert the grid under the water in the weight boat and wash away the UA droplet adhering to the grid by gently swirling in the water bath.
13. Take a piece of filter paper and blot dry from the side of the grid. Place the grid onto another piece of filter paper with section facing up and let dry.
14. Poststain for 10 min in Sato lead by following the same procedure described for the UA incubation.
15. Remove the grid from the lead and wash as described in step 12, but using the second water bath.
16. Apply colloidal gold of defined size to the grid to be used as fiducial marks. Useful gold particle sizes for ET of neuronal mitochondria are 10, 15 and 20 nm (Fig. 2). The right size to use depends on the magnification of the tilt series, the section thickness and how electron dense the cells are. With higher magnifications (0.8 nm per pixel or smaller) 10 nm gold is used for 300-nm thick sections. With medium-high magnifications (0.9–1.5 nm per pixel) 15 nm gold is used for 300-nm thick sections. For lower magnifications (1.6 nm per pixel or larger) 20 nm gold is used for 300-nm thick sections. Because of more material the electron beam has to pass through to form an image with thicker sections, the gold size should be shifted higher to visualize it. For example, for 500-nm thick sections and a pixel size of 0.8 nm or smaller, 15 nm gold is better. Likewise, with heavily stained material, switch to a larger gold size so that the gold is not "lost" in the heavy metal stain, thus making it difficult for the auto-tracking employed by the image-processing software. Conversely, with the lighter stained genetically tagged samples, switch to a smaller size gold. In summary, the problem to avoid with gold too small is not being able to visualize the particles for tracking purposes and the problem to avoid with gold too large is streaks extending through the reconstructed volume that are artifacts of the backprojection algorithm used in image processing.
17. To have an even spread of gold particles across the grid without clumping, use BSA (bovine serum albumin) as a surfactant (Note 3). Prepare a gold–BSA mixture as follows.
    (a) Make up a 10× stock solution of BSA (0.5% in dd-water).
    (b) Dilute the stock solution 10× with dd-water (0.05%) and filter through a syringe.
    (c) Mix 1:3 gold–BSA to make a volume of 4 ml and place in a scintillator vial. Can store this solution in a fridge for several weeks.

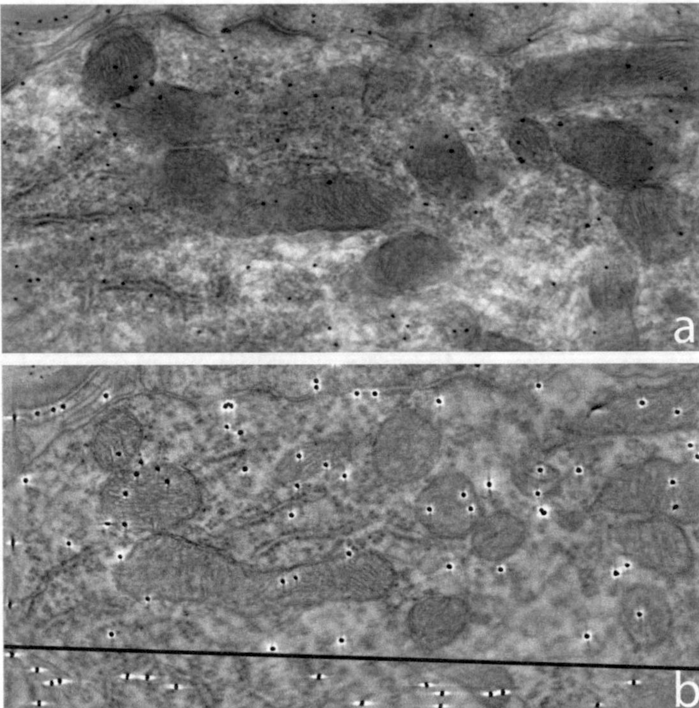

Fig. 2 A good size and distribution of gold particles are shown in (**a**). This image shows part of the 0° image as seen on the computer screen of the electron microscope. The 20-nm sized gold particles appear as *small black* "dots." They are large enough to be seen (although in the darker areas, just barely) and thus tracked, but not so large that they dominate or obscure the biological material. More than 40 gold particles can be seen and are spread out nicely over the area, which aids the triangulation used for image alignment. (**b**) View of the surface of the section of the tomographic reconstruction of the area in (**a**) showing the gold particles on that side of the section. This slice is from a 2-tilt reconstruction and is a useful diagnostic to verify that the gold particles are indeed sharp and round with symmetrical white "wings." Note the shape of the white wings for the gold particles under the black bar. They are elongated in the horizontal direction because the overlap of the two halves of the 2-tilt reconstruction was not perfect after 90° rotation and so the region under the bar comes from only one of the halves, hence the elongation. Above the black bar, the "wings" appear as "halos" because both halves of the reconstruction are present. The symmetry of the wing or halo is thus a useful diagnostic for reconstruction quality

18. Holding the edge of the grid with forceps, dip the grid in this solution for 10 s. Longer times will give a higher density of gold and conversely, shorter times a lower density of gold. The optimal incubation time will be determined by trial and error after observing the number of gold particles in the area of the image collected in a tilt series. For optimal image processing, it is recommended to have between 50–150 gold particles in the

zero-tilt image of the series. Too few gold particles and the TxBR program will not optimally correct for beam distortions and local sample warping. Too many gold particles and the tracking of the fiducial (gold) markers may jump to nearby gold.

19. Let air-dry for 10 min still held in the tweezers.

2.3 Tilt Series Collection

Currently, double-tilt (also called 2-axis or 2-tilt) tomography is the most commonly practiced version of ET (Fig. 3) because it provides more homogeneous resolution. However, single-tilt tomography is still practiced in three situations: (1) when the region of interest is close to a grid bar or (2) when a survey reconstruction suffices or (3) when a simple reconstruction answers the biological question. Instructions are as follows:

1. For stability in the beam and to minimize buildup of static charge, the section may be coated with carbon.
2. Insert the grid into the 300 or 400 kV microscope.
3. Spread the beam to well beyond the imaging area for good optics. During data collection, the illumination should be held to near parallel beam conditions. With microscopes where the main viewing is a computer screen (essentially all modern microscopes) that uses a video-rate camera mounted inside the microscope, adjust the contrast histogram to adequately see the features of interest.
4. Pick the spot size and adjust the exposure time to give a good dynamic range for the CCD camera used. For example, with

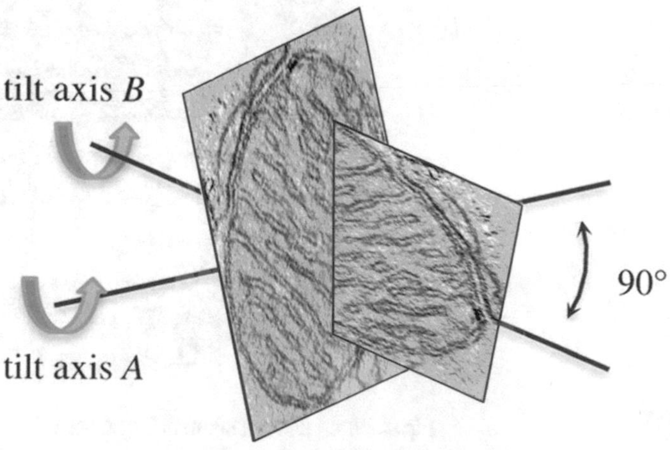

Fig. 3 Double-tilt (also called 2-axis or 2-tilt) tomography provides more homogeneous resolution than single-tilt tomography and thus is the most commonly practiced version of ET. It is performed by rotating the grid 90° after acquiring the first tilt series (tilt axis A) and collecting a second tilt series (tilt axis B). It is best to use a rotation holder for this purpose, but a regular holder can be used by manipulating the grid with forceps to rotate it

the most commonly used tilt series software, serialEM, and CCD camera, Gatan, an electron count of 8000–12,000 is good. Be cautious about having image recording times be longer than 4–5 s, depending on the stability of the microscope, because even slight drift of the specimen during the image recording will produce streaked and blurry images.

5. Survey/mark positions of interest on the grid. Surveying may be done at lower magnifications. For sections in a clamshell grid, test whether the tilt range can go ±60° (Fig. 4) for each region of interest.
6. Set up the tilt series collection to be done automatically at the desired magnification:
   (a) Flat-field over an area with no section or support film, i.e., empty space.
   (b) Set up the automated beam "blank" for 1–2 s for each image recorded in the tilt series, which actually exposes the area to the beam for 1–2 s before recording each tilt image. This is to avoid beam-induced specimen movement that is usually a quick "jump" upon exposing the area for each image.

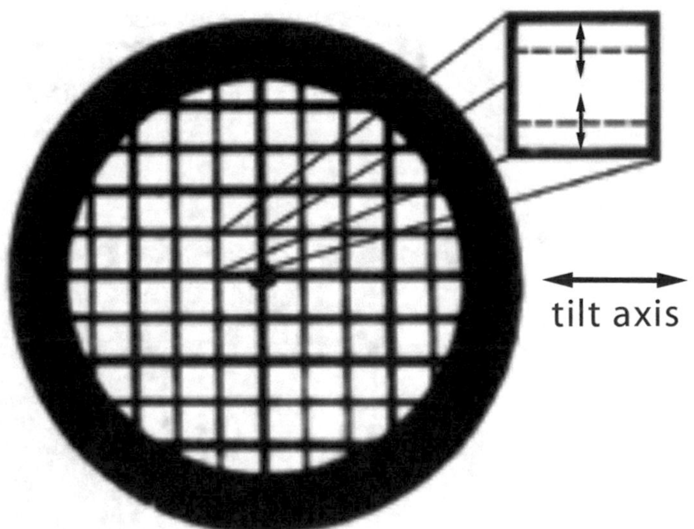

Fig. 4 When using a clamshell grid for ET, it is advised to start the survey of good regions in the center of the window because when tilting the grid to high angles, the grid bars parallel to the tilt axis will move towards the center (represented by the *double-sided arrows*) as represented by the *dotted lines* in the enlargement of a window shown in the *upper right*. Thus, the usable area for a full tilt range is not the entire window, but actually the *rectangular portion* between the *dotted lines*. Before selecting a surveyed region of interest for tilt series collection, it is a good idea to test whether the tilt range can go ±60°

(c) Set up the autofocus parameters. A long and thin rectangle in the center of the image with long axis along the tilt axis is ideal because the full image size is not needed and the focal gradient perpendicular to the tilt axis at higher tilts may cause the autofocusing to be off (hence the thin dimension perpendicular to the tilt axis). Also, the full dynamic range is not needed, so bin the pixels 4×. This will also allow 16× (4 × 4) less exposure time, thus speeding up this step.

(d) Insert the optimal condenser and objective apertures. Insert the largest condenser aperture because it provides the strongest beam, which is important for the thicker sections used in ET. Generally, insert the smallest objective aperture because it will maximize the contrast, which is important because more highly energetic electrons interact less strongly with the sample and so this is a way to increase the contrast. If the objective aperture is so small that its edge interferes with the wobbler used for autofocusing, then insert the next larger aperture.

(e) Set up the "trial" image parameter as a binned by two image. The trial image is used in automatic ET to cross-correlate to re-center the region of interest for the next tilt image in the series. This will also allow 4× (2 × 2) less exposure time, thus speeding up this step. Choose the option to center on the region currently displayed, as this will be important for step 9 below.

(f) Set the target defocus to zero because the contrast in the image will be "amplitude" contrast, as opposed to "phase" contrast, important for single-particle EM and cryo-ET, where a negative defocus (underfocus) will greatly enhance the contrast. Also, choose the option to keep the beam size constant and not vary with tilt angle.

(g) Make the region of interest "eucentric" (Fig. 5), which means that the region of interest is at the same height in the microscope as the tilt axis. Thus, upon tilting, this region does a true rotation and not a precession, which would distort the tomographic reconstruction. Note that this step must be done on the region of interest; making eucentric any other region will not apply to the one of interest. On modern microscopes, this step is straightforward and should take less than a minute to perform:

- Press the eucentric height button.
- Use the "alpha" wobbler to minimize specimen movement.

(h) Set the tilt angle range to ±60°, generally starting at −60° and finishing at +60° and the tilt angle increment to 1°,

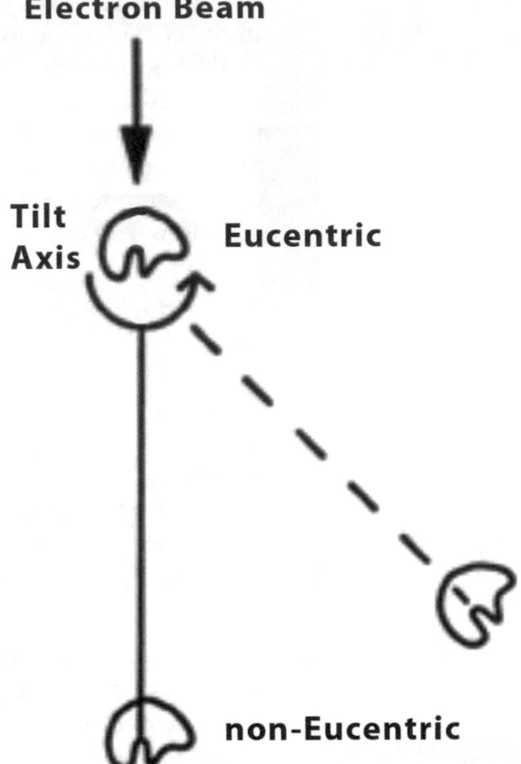

Fig. 5 Eucentricity is defined as positioning the height of the object of interest, shown as a "blob" here, at the same height in the microscope as the tilt axis. Thus, upon tilting, this region does a true rotation and not a precession, as indicated at the bottom with the non-eucentric situation. The change in height of the object during a tilt series collection for the non-eucentric situation would cause a magnification change and thus would introduce a distortion in the ET reconstruction

producing 121 images in the tilt series. If the specimen cannot take a lot of electron dose without being degraded, make the tilt angle increment 2° (61 images). For high-resolution tilt series, make the tilt angle increment 0.5° (241 images) because within the limits of electron dose a specimen can handle, the resolution is roughly proportional to the number of images in a given tilt angle range.

7. "Cook" the region of interest. This means irradiate the region for about 30 min before initiating a tilt series to limit anisotropic specimen thinning and warping during image collection. Modern microscopes will have a program (also part of serialEM) to irradiate at each tilt angle for a user-defined duration, e.g., 15 s per tilt for a 1° tilt schema (30 min total time).

8. Check the eucentricity again because cooking can change the "height" of the region of interest.
9. Position the region of interest precisely as desired in the image area. Record an image of this region to be used for tracing and start the tilt series. Instead of walking away at this point and coming back when the tilt series is finished, it is a good idea to watch the autoET track from 0° to −60° to make sure that the region of interest does not drift, which would only happen if the auto-tracking malfunctions. Also watch the collection of the first couple images to make sure that everything is proceeding as desired.
10. Multi-tilt Series.

    For increased resolution in all dimensions, 2-, 4-, and even 8-tilt series are performed (Fig. 6).

    (a) After the first tilt series has finished, rotate the grid the required angle (90° for 2-tilt, 45° for 4-tilt, 22.5° for 8-tilt). It is best to use a rotation holder for this purpose so that the holder does not have to be removed from the microscope.

    (b) Before rotating, assure that the region of interest can be found again after rotating. Some microscopes are set up so that the region can be watched with a video-rate camera while rotating and be brought back into view.

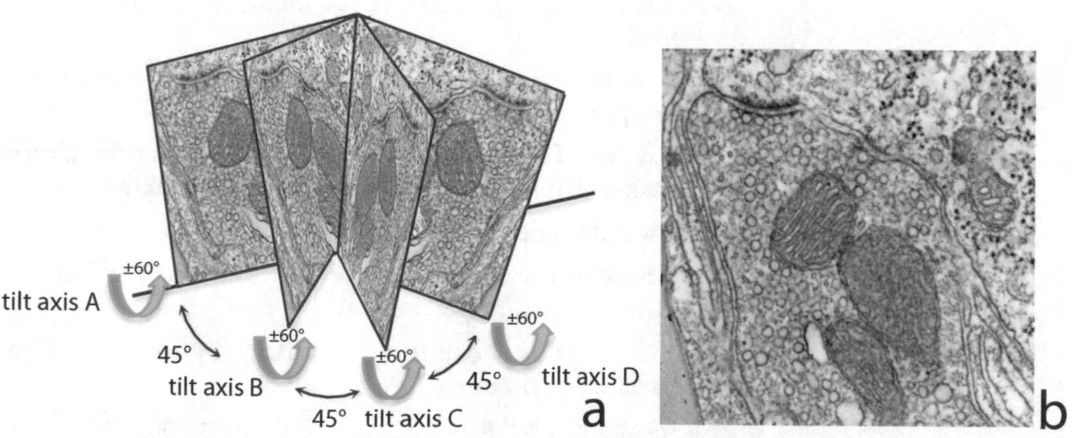

Fig. 6 (**a**) Example of a 4-tilt series in which the grid is rotated 45° after acquiring the first tilt series (tilt axis A) and collecting a second tilt series (tilt axis B), followed by another 45° rotation, collection of a third tilt series (tilt axis C), followed by another 45° rotation and collection of a fourth tilt series (tilt axis D). If the region of interest can handle the increased electron dose, then a 4-tilt reconstruction has better quality than a 2-tilt reconstruction because the sampling is greater and there is less missing information in Fourier space. (**b**) A slice through an ET reconstruction generated from a 4-tilt series of a region of the brain that shows the high-quality structural details. Notice how well defined the mitochondrial, and other, membranes are and how clear the synaptic vesicles, ribosomes, and presynaptic and postsynaptic densities are

It often helps to go down in magnification to use a larger viewing field.

(c) Redo the eucentricity correction.

(d) Center the region of interest to correspond to the first tilt series. It is helpful to have the central (0°) image open for comparison. The industry-standard program IMOD can be used for this purpose.

(e) Autofocus the region and take an image.

(f) Name this new series with the same "basename," but with a "b" to distinguish it from the "a" tilt series.

(g) Do not cook the region. It is important to perform the last three steps as quickly as possible to minimize the beam causing further mass loss or warping. In other words, for the second, third, etc. tilt series taken, it is desired to have as much of the electron dose as possible for the collection of the tilt series proper.

(h) Repeat steps (c) through (g) if 4- or 8-tilt series is desired remembering to name subsequent files "c," "d," etc.

2.4 Image Processing

1. Instruction for computer processing of tilt series has been well documented including excellent online tutorials and practice datasets of real tilt series at http://bio3d.colorado.edu/imod/. IMOD is the industry standard and has a strong support group. IMOD has these advantages:

   (a) It is free and open source for those interested in adding plugins.

   (b) It works on any platform because of the etomo java wrapper.

   (c) It is been around for many years with frequently released version that have bug fixes and added capabilities.

   (d) It was designed specifically with ET in mind.

   (e) It handles the whole ET process, except for the final part of movie-making (see below).

   (f) It is versatile because of its many options, including "advanced" options.

2. IMOD and its display window, 3dmod, open image files in the MRC format, which typically have the extension *.rec* or *.mrc* and are a stack of 2D images in a tilt series or a stack of 2D images making up slices of a 3D tomogram. IMOD can also open tiff files. The command line programs tif2mrc and mrc2tif are used to convert back and forth between formats. Segmentation or "modeling" as used in IMOD parlance creates a file in "IMOD Binary Model" file format, which should be given a .mod extension.

3. Within IMOD, each model contains layers. Objects contain contours and contours contain points. For modeling mitochondria, contours are usually closed.
4. An alternative approach is to use the program TxBR, which uses a "transformed-based reconstruction" approach and uses IMOD as a front-end loader. This program is available upon request at the link https://confluence.crbs.ucsd.edu/display/ncmir/NCMIR+Software;jsessionid=08725A15FAAB10D5E6019490D1182BF7.
5. Comparison of IMOD with TxBR; advantages and limitations of each:
   (a) The documentation for IMOD is better than for TxBR, but this may change as more users external to the National Center for Microscopy and Imaging Research (NCMIR) use TxBR.
   (b) TxBR does not align the images in a tilt series well if the gold particles used for fiducial marks are only on one side of the section. Which side a gold particle is on can be determined using the 3Dmod display GUI of IMOD or etomo, the IMOD java wrapper. In movie mode, watch the movement of a gold particle from image to image. Gold particles on one side of the section will move in one direction and those on the other side of the section will move in the opposite direction. For TxBR to work well, a minimum of three gold particles must be on each side of the section for triangulation purposes. Although fiducial-less alignment may become more promising (Note 4).
   (c) In contrast, IMOD will align well if all the gold particles are on one side. For simple alignment, about a dozen gold particles are best. To use the "local" alignment option, it is recommended that 40 or more gold particles be used. Both IMOD and TxBR now have the option to automatically select and track fiducial marks; it is the default in TxBR and is a simple set of buttons and input boxes in etomo.
   (d) IMOD is limited to 1- and 2-tilt series reconstructions. In contrast, TxBR performs a reconstruction from any scheme of data acquisition, notably multiple tilt series (any number).
   (e) TxBR has an option that is simple to invoke that adds more sophisticated alignment schemes that represent first, second and third order nonlinear corrections. Adding a level of sophistication, however, requires more gold particles. The rule of thumb is that first order requires a minimum of 40 gold particles, second order requires a minimum of 60 gold particles and third order requires a minimum of 80 gold particles to over determine the alignment equations.

(f) TxBR has the advantage that it accounts for the curvilinear nature of electron trajectories, making it suited for the processing of large-field ET [7]. This is important because the pixel array of CCD cameras has been increasing. Many modern microscopes now have 4 k × 4 k and even 8 k × 8 k cameras. Moreover, serialEM allows for montaged images. It is common now to have 2 × 2 to 4 × 4 montaged images. The downside to large images is the computer storage memory required. For example, with the simplest tilt series scheme: single-tilt, 1° tilt interval, ±60° tilt range, 4 k × 4 k camera—the file is 4 Gb in size. This grows to 512 Gb with an 8-tilt, 4 × 4 montage! Moreover, several intermediate files are generated during the processing that are just as large.

(g) Both IMOD and TxBR have the option to run SIRT (simultaneous iterative reconstruction technique) instead of the usual R-weighted backprojection algorithm, which may improve the reconstruction quality.

(h) In their current versions, TxBR is easier (more automated) to run than IMOD. Yet, IMOD can be more robust. Thus, the default at the NCMIR is to run TxBR, but when something does not look right with the reconstruction, including the quality, then IMOD is also run and a comparison made between the two reconstructions. In addition to looking at the reconstructed cell and organelles to judge the quality, looking at the gold particles is advisable. They should be sharp and round with the white "wings" (reconstruction artifact, but always present to some extent) symmetrical (Fig. 2).

6. To improve the signal-to-noise ratio, the images in the tilt series can be binned down 2× (or even 4×) by averaging adjacent pixels. The IMOD command, newstack –bin two infile outfile, will do this. This pixel averaging is advised based on the modulation transfer function (MTF) of the CCD camera and is a good idea for Gatan cameras. This averaging, of course, affects the pixel resolution, so if the original pixel resolution was 1 nm, the pixel resolution will now be 2 nm, which is certainly sufficient to see the mitochondrial membranes with clarity.

7. Because the reconstructed volume generated is greater than the actual volume by default, it is a good idea to trim the volume to remove empty space and to reduce the file size. This is easily done with the trimvol command in IMOD or with the trimming option in etomo after noting which slices are empty after loading the volume in 3dmod.

8. Trimming is necessary for serial-section ET to efficiently stack the volumes [8]. Excellent documentation exists under the IMOD webpage. Use the "Join Serial Tomograms" option in etomo.

2.5 Segmentation

Although examination of a 3D reconstruction in combination with simple image processing to enhance contrast and remove background noise may be informative for some simple structures, it is more often the case, especially with the complex membrane topology found in neuronal mitochondria, that segmentation techniques and methods for defining and dissecting components of the structure are required to facilitate interpretation and measurement (Fig. 7). The capability to segment and classify features of the tomographic volume was essential for the interpretation of

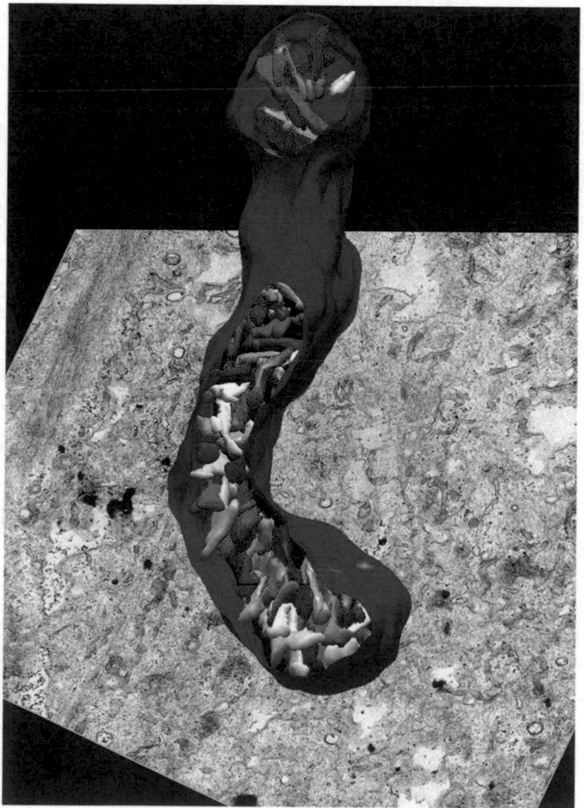

Fig. 7 Segmentation, what other groups call modeling, of the mitochondria in an ET reconstruction not only helps to visualize the various mitochondrial compartments, but is also useful for defining and dissecting components of the structure in 3D. Segmentation is often required to facilitate interpretation and measurement. Shown here is the 3D segmentation of the membranes of a neuronal mitochondrion overlaid on a tilted slice of the ET volume. This visualization is called a "surface-rendered volume" because the volume surfaces can be rotated in any direction and displayed in any color or transparency or lighting. Here, the mitochondrial outer membrane is shown in *blue* and the many cristae are shown in various *colors*. After segmentation, measurements of mitochondrial and cristae volumes, surface areas, connectivity, shapes, sizes, constrictions, and orientations can be performed. In summary, with segmentation, the mitochondrial architecture can be quantified

mitochondrial architecture and led to the recognition of a new mitochondrial structure, the crista junction [9]. Currently, manual tracing is necessary to segment mitochondria accurately, although effort is ongoing to develop automatic segmentation of mitochondria (Note 5).

1. Open a tomographic reconstruction in IMOD. Switch to "model" mode in the main 3D window before tracing contours in the "Zap window." The main 3D window shows the object, contour and point that are currently selected. Under the "help" menubar are more extensive instructions on segmenting (what the authors of IMOD call "modeling") using 3dmod.
2. Adjust the contrast using the slider.
3. Add a new object with Menubar > Edit > Object > New. It is wise to place the mitochondrial outer membrane, inner boundary membrane and each crista membrane in a separate object so that each can be manipulate and displayed separately when desired. Select "closed" contour.
4. Open the plugins "DrawingTools" and "Interpolator" found under Menubar > Special > DrawingTools. These two plugins accelerate manual segmentation. Instructional videos are found in http://www.andrewnoske.com/student/imod.php > then click the "Drawing Tools and Interpolator - Overview" video
5. The Drawing Tools plugin enables quick tracing, editing and erasing of contours. Start with the "Normal" drawing mode and using the left mouse button trace the membrane of interest. If the trace goes off the membrane, then use the "Sculpt" mode to push contours in or out. On the mouse use the scroll wheel to resize the Sculpt circle to a size that works best for the correction desired. The most useful shortcut keys are:

   [page up] / [page down] to scroll up and down slices.

   [+] / [−] to zoom in and out

   [n] to start a new contour

   [shift+d] to delete the selected contour

   [v] to open the "Model View" window

   A complete list of shortcut keys is under Menubar > Help > Shortcut keys.
6. The Interpolator plugin reduces the number of contours that are manually tracing by automatically generating interpolated contours between the traced contours. Most ET volumes are hundreds of slices, so tracing mitochondrial features on every slice is time-consuming. Previously, it was suggested to trace every fourth or so slice because mitochondrial features do not change substantially from slice to slice. A better solution

is to trace every fourth or fifth slice and use the "Linear" interpolation option.

(a) Set up a "Z Bridge" of ten slices.

(b) Trace the first contour and hit [Enter]. This will close the contour. Note that it is not required to be precise with closing the contour because the Interpolator will connect the first point traced to the last point.

(c) Move up four or five slices and trace the same feature in this slice. By hitting [Enter] again, not only will this close the contour, but it will also add a series of interpolated contours in between the traced slices and will do so on both sides of the contour just added or edited. These will be shown as dotted contours, but will become solid upon "sculpting" or otherwise editing them. Think of a dotted contour as one that has not yet been checked for accuracy.

(d) Continue tracing every fourth or fifth slice until the feature of interest has been fully traced. Remember to hit [Enter] every time a new contour is traced or corrected. With mitochondria, editing is minimal and is usually needed only when the manual tracing goes off the membrane; it is hard for humans to be perfect tracers. The precision is helped by zooming in as much as the screen allows. The other time editing is needed is when the mitochondrion or one or more cristae inside it are branched. The slice showing the branching point is where the Interpolator can become confused.

7. ModelView meshing and 3D viewing. It is a good idea to check the segmentation/modeling from time to time.

(a) Open the ModelView window by using the shortcut key [v]. The contours in the ZAP window are not automatically meshed, so to do this in the ModelView window go to Menubar > Edit > Objects... [O]. A new window will pop up.

(b) Click on the "Meshing" option (bottom left) and set the iterations to "100" to improve the surface rendering.

(c) Hit "Mesh one" or "Mesh all" depending on if only the selected object or all objects are to be meshed, respectively. Often, the cristae look better when the meshing option "Cap" (to close the top and bottom of each surface) is used. The surface-rendered 3D objects will now appear in the ModelView window.

(d) In the ModelView window, the mouse can be used to rotate and zoom in/out the surface-rendered volume.

(e) The "Edit Objects" window also allows turning objects on or off and changing their display, including color and transparency. Changing the transparency of the mitochondrial outer membrane and inner boundary membrane is often needed to visualize the cristae inside and makes for a nice publication figure. Make sure to turn objects back on afterwards to see them in the Zap Window.

2.6 Capturing Images for Presentations

To capture a slice of the ET volume or a portion of it:

1. Make the ZAP window as large as possible on the computer screen. Use the entire slice area or zoom in on the desired portion of the area.
2. Adjust the black and white levels using the scroll bar.
3. Go to File > Movie/Montage. Usually, the ET slice is too big to fit on the computer screen at 1× zoom. To grab the entire slice:
   (a) Click [−] until the entire area fits in the window.
   (b) Press [Ctrl]+[R] to shrink the window to fit the image perfectly. For each captured image pixel to represent one tomogram pixel:
      - Adjust the ZAP zoom × and the montage number. For example, if the zoom is 0.25 (displayed on the ZAP bar) set the montage value to 4. The output will then be the original size.
4. Select Snapshot as a "TIFF" file.
5. Take a single TIFF snapshot by using [Ctrl]+[S] or [Apple]+[S] (Macintosh). The image zap000.tif will be saved to the IMOD home directory. Note what the slice number is (*z* value). It is wise to change the file name to include the slice number, e.g., basename_slice#.tif.

To capture a view of the segmented volume or a portion of it:

1. Go to the *ModelView window*:
2. Set up the desired view, including objects on/off, colors, transparencies and zoom in/out.
3. Save the view by using Edit > Views. Click "New View" and "Store" (allowing the return to this view later).
4. Go to File > Movie/Montage
   (a) Set Make to "Montage" and Write to "TIFFs".
   (b) Set the # of montage frames to "4" and turn on "Write files".
   (c) Click "Make".
   (d) The model image modv0000.tif will be saved to the IMOD home directory.
5. Rename the file with a name that is more descriptive.

6. Can further adjust either slice or model image
   (a) Open the image file in Adobe Photoshop.
   (b) Use the crop tool to remove unneeded area.
   (c) Adjust the contrast using "Levels" and the slider bar.
   (d) Go: Image > Image Size
      - Change the resolution as specified by the desired journal and click okay, e.g., 300 pixels per inch.
   (e) Save as a new file.

2.7 Movie Making

There are several different programs that can be used to make movies from ET volumes. Presented here is perhaps the simplest approach and uses IMOD. IMOD does not actually produce a movie file. It only writes a series of snapshots, and thus a moving-making program such as QuickTime Pro will be needed to turn the image stack into a movie file (one image per frame). First is instruction to make a movie of the slices of the reconstructed volume.

1. Open the ET reconstruction in 3dmod.
2. Set the ZAP window to the desired size and slice to the desired zoom in (use [+] and [−]).
3. Make sure 3dmod is set to "movie mode".
4. Set the black and white scroll bar to the desired contrast, and type [shift]+[r] to resize to the image.
5. Click File > Movie/Montage to display the movie dialog. Note that a specific version of IMOD may be needed to see the Movie option.
6. Set the start, end and increment to the range and spacing of slices desired.
7. Set "snapshot" to "TIFF". Note that if "None" is selected images will not be written. Tip: Use the "Zap montage" option to make an image or movie larger than the size of the computer screen.
8. Middle and/or right-click once anywhere inside the ZAP window and the movie files will be written as a series of images "zap001.tif, zap002.tif,. ." will be written in the directory 3dmod was opened from, as shown in the main 3dmod window.
9. Once the movie files have been written, open QuickTime Pro (Note 6), then click File > Open Image Sequence and open the first image in the sequence; it is recommended to use 10–20 frames per second and save as a .mov file using compression in the Export feature. QuickTime player is a free movie player for Mac and Windows. QuickTime comes preinstalled on Mac computers, or can be downloaded free from the Apple webpage.

For a small fee, QuickTime 7 Pro features will be available; the "Pro" is a simple program for compressing movies and making edits. Unlock QuickTime 7 Pro by clicking menubar > Help > Buy QuickTime Pro.

10. Export the movie to one of many different movie formats by going: File > Export. The recommended option is "Movie to QuickTime Movie"—especially if using a MAC. By clicking on Options, the desired Size and Settings of the movie can be changed. By default "Movie to QuickTimeMovie" will save the .mov file using "H.264 compression"—a compression highly recommended as there is minimal quality loss. The compression can be further adjusted with "kbits/sec" under "Options > Settings" to dictate the size of the file and movie quality. Another option is to make AVI files using "Movie to. aVI". This has the advantage that .avi format plays on essentially every computer and is the only movie format able to play in Microsoft PowerPoint on Windows.
11. To cut or copy movie frames, use the selection markers at the bottom of the timeline to make a selection. Then use the cut and copy commands (under Edit). If no selection is made a single frame is cut.
12. Multiple movies can be edited by opening the movies and copying frames from one to another. One movie can be appended on the end of another.

To make a movie from a ModelView in IMOD:

1. Open the model in 3dmod and press [v] to display the ModelView window.
2. Use the ModelView window for all the following commands:
   (a) Under "Edit" use the object list under the "Objects" panel to adjust the display if needed, such as objects on/off, colors or transparency of objects.
   (b) Make the ModelView window the desired size, using the "Controls" panel to change the camera settings.
   (c) With the ModelView window selected, click File > Movie (m).
   (d) Use the mouse keys to set the desired starting position then click "Set Start".
   (e) Use the mouse keys to change the desired ending position then click "Set End".
   (f) Check "write files" and then click "Make". As long as "write files" is clicked, the frames will be saved out as TIFFs—e.g., modv0000.tif, modv0001.tif,. Note that unfortunately IMOD currently only supports "point-to-point" linear animation, and thus to make longer movies,

click "Set Start" at the end of each movie segment and set up the next segment of the movie.

(g) Once finished, open QuickTime Pro, then click File > Open Image Sequence and follow the instructions above.

Options for making more sophisticated movies are available (Note 7). For compressing movies down to the file size often required by journals, download the free program "handbrake," which is simple to use and will compress movies even as large as 100 Mb down to 10 Mb without much loss in quality.

3 Conclusions

High-quality ET requires knowledge and skill and has been generally performed by well-trained scientists. This chapter provides detailed instructions for performing ET on neuronal mitochondria in situ with the goal to broadening its application to those who have an interest and access to the required equipment, but may not work in a dedicated ET facility. The interested scientist would still need knowledge of "wet lab" techniques, sectioning, microscope operation, and have sufficient knowledge of computers to be able to learn new, although not difficult, software. What makes this list of skills less daunting is that most ET facilities have arrangements to either hire dedicated personnel to do one or more of the tasks or have the personnel teach the interested scientist. For example, sectioning still is a skill that can take weeks to learn properly and so is often "contracted out." On the other hand, modern microscopes and the software used to collect tilt series and perform the various image processing tasks have become so "user friendly" that scientists trained in the biological sciences can learn them relatively quickly, especially with guidance from personnel at ET facilities. A summary of the basic stages in ET is shown in Fig. 8 and highlights the beautiful end product of a rendered volume of a neuronal mitochondrion.

4 Notes

1. Obviously, perfusion fixation is not possible with human biopsied tissue. However, if carefully performed, immersion fixation of human biopsied tissue can provide good ultrastructural preservation. It is important that the human tissue be extracted as quickly and possible after the blood supply to the region is cut off, e.g., by a scalpel, and placed in ice-cold fixative because mitochondrial structure can start to change even after as little as 1 min ischemia. The same fixative recipe used for perfusion fixation can be used for immersion fixation. Once in the fixative,

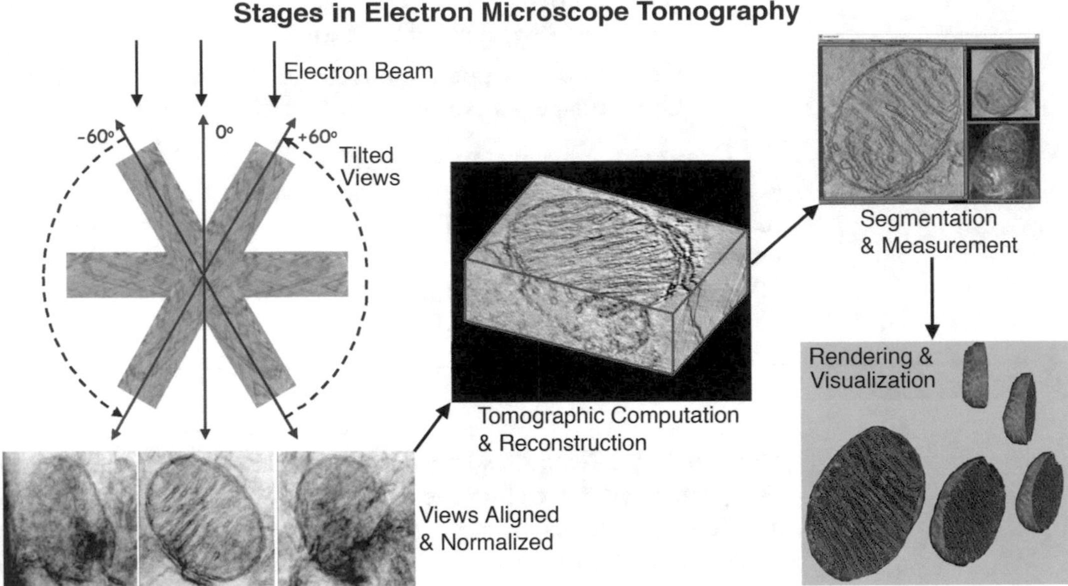

Fig. 8 The basic stages in ET. From *left* to *right*: a section is tilted over a range of tilt angles and images of the region of interest are collected into a tilt series. At high tilt, the view of the mitochondrion may be quite different from at 0° tilt because of the orientation of the cristae membranes and the fact that each image is a projection through the section volume. Each view is aligned and the density normalized in an automatic procedure. After tomographic computation and reconstruction, the volume is a stack of image slices 1-pixel thick. This stack can be oriented in any way and displayed in various ways. Segmentation of the mitochondrial volume offers added visualization and measurement capabilities. Surface rendering adds another layer of visualization and is helpful for publication figures and movie presentations

the biopsied tissue should be diced into smaller pieces about 1 mm cubed for improved penetration of the fixative. After this step, the same procedures detailed for perfusion fixation sample preparation should be followed.

2. Cryo-ET uses biological structures preserved in vitreous ice (also referred to as being frozen-hydrated) or preserved via freeze substitution. Two advantages of cryo-ET are that [1] rapid freezing can immobilize all molecules in a sample within milliseconds, thus allowing the quick physical fixation of all molecules in their current position, and [2] frozen-hydrated samples have contrast from the protein density itself and not from a heavy metal stain or replica and this allows specimen preservation to atomic detail. The development of high-pressure freezing (HPF) of frozen-hydrated samples followed by cryo-sectioning [10] has extended this cryo-technique to generate high-quality preservation of cellular architecture, in situ. However, investigators quickly discovered that frozen-hydrated samples had their own challenges, in particular low contrast and susceptibility to electron beam induced damage. Image processing strategies and low-dose modes of electron

microscopy have made progress in overcoming these challenges. The cryo-sectioning of frozen-hydrated samples has provided molecular details of intracellular structures imaged with high clarity. Cryo-sectioning is still far from a routine procedure. The sectioning itself and handling of frozen-hydrated sections is tricky and requires substantial training. Once frozen, the specimens cannot be further manipulated, e.g., by staining or immunolabeling. Cryo-sections cannot be as thick as plastic sections commonly used for ET and should be kept thinner than about 100 nm to avoid formation of crevasses. They currently do not produce the volume of plastic sections because they are not suited for serial sectioning as substantial material is lost between each cut. Freeze substitution is a cryo-technique complementary to vitrification in that rapid freezing is first performed. Freeze substitution then diverges by replacing the vitrified ice with chemical fixatives slowly perfused into the sample at low temperature [11]. Chemical cross-linking happens as the sample slowly is raised to room temperature. At this point, dehydration and plastic embedment follow similarly to the conventional procedure presented in this chapter. HPF followed by freeze substitution is currently the ideal method for examining neuronal mitochondria at high-resolution in a cellular environment because it reduces fixation artifacts and avoids the challenges still facing cryo-sectioning.

3. An alternative to the “dipping” method is to apply a droplet of colloidal gold, either with or without BSA on one side of the grid and then after letting air-dry, applying a droplet on the other side of the grid. The edge of the grid is held in place with forceps. Use a “20 microliter” pipetman and set the volume to 8 μl. Apply this volume of gold onto one side of the grid taking care not to touch the grid surface. Let air-dry for 20 min. Apply the same volume of either the same size gold particle or a different size to the other side of the grid. The advantage of different size gold particles, e.g., 15 and 20 nm in diameter, is that it is easy to see the size difference in the electron microscope and thus determine whether there is enough gold on each side of the region of interest for TxBR image processing. The pitfall is that this approach tends to produce more clumped gold particles and less even distribution, especially if no BSA is used.
4. ET processing took a step forward with marker-free alignment, which has the advantage of not requiring fiducial gold particles on or in the samples [12, 13]; it is an option in IMOD. However, using gold particles as fiducial marks is still the “gold” standard.
5. Promising current efforts are underway for developing automatic recognition and segmentation of mitochondria in tomograms [14, 15].

6. Other attractive options for making ET movies are:
   (a) ImageJ because it is free and uses Java to be platform independent (Windows, Mac OS X, and Unix). ImageJ writes only one type of movie format—AVI. Instructions are as follows:
      - Copy all the images/frames to a single directory with nothing else in the directory, and make sure they are in alphabetical order. For example: *"frame000.jpg"*, *"frame001.jpg"*, etc.
      - Open ImageJ.
      - Click File > Import > Image Sequence, navigate to the desired directory and click Select. Scaling and other options are here.
      - Click File > Save As > AVI and select frames per second (10–20 provides good movie quality) and save. A drawback is that there are no options for AVI quality.
   (b) Most ET facilities have Adobe Photoshop. Instructions are as follows:
      - Go to File > Scripts > Load Files into Stacks. This option will add all the images into one file with separate layers.
      - Open the Animation Palette under: Window > Animation.
      - Click on the Flyout Menu at the top right corner of the Animation window and select Make Frames from Layers. If the frames are backwards click the bottom left icon of the animation window to get into "frame animation mode" and select all the frames, then click the Flyout again and reverse frames.
      - Adjust the timing of each frame by selecting the 0 s. on the frame and inputting the desired timing.
      - Go to File > Export > Render Video. Among the many options the Quick Time or MPEG4 options are most useful. Compression type "H.264" is recommended.
7. To make more sophisticated movies, specialized programs are available, such as Amira, Cinema4D, or Chimera. Chimera is one of the few programs that will open an IMOD model file (.mod) directly. For most other programs it is advised to export the model to a VRML version 1 or 2 format using the IMOD command line program imod2obj or imod2vrml.

References

1. Frank J (1992) Electron tomography: three dimensional imaging with the transmission electron microscope. Plenum, New York, NY
2. Perkins G, Renken C, Song JY et al (1997b) Electron tomography of large, multicomponent biological structures. J Struct Biol 120:219–227
3. Wan X, Zhang F, Chu Q et al (2011) Three-dimensional reconstruction using an adaptive simultaneous algebraic reconstruction technique in electron tomography. J Struct Biol 175:277–287
4. Wolf D, Lubk A, Lichte H (2014) Weighted simultaneous iterative reconstruction technique for single-axis tomography. Ultramicroscopy 136:15–25
5. Shu X, Lev-Ram V, Deerinck TJ et al (2011) A genetically encoded tag for correlated light and electron microscopy of intact cells, tissues, and organisms. PLoS Biol 9:e1001041
6. Martell JD, Deerinck TJ, Sancak Y et al (2012) Engineered ascorbate peroxidase as a genetically encoded reporter for electron microscopy. Nat Biotechnol 30:1143–1148
7. Phan S, Lawrence A, Molina T et al (2012) TxBR montage reconstruction for large field electron tomography. J Struct Biol 180:154–164
8. Vranceanu F, Perkins G, Terada M et al (2012) Three-dimensional structure of the striated organelle in vestibular type I hair cells as revealed by electron microscope tomography. Proc Natl Acad Sci U S A 109:4473–4478
9. Perkins G, Renken C, Martone M et al (1997) Electron tomography of neuronal mitochondria: 3-D structure and organization of cristae and membrane contacts. J Struct Biol 119:260–272
10. Al-Amoudi A, Chang JJ, Leforestier A et al (2004) Cryo-electron microscopy of vitreous sections. EMBO J 23:3583–3588
11. McDonald KL, Auer M (2006) High-pressure freezing, cellular tomography, and structural cell biology. Biotechniques 41:137–143
12. Kim HW, Oh SH, Kim N et al (2013) Rapid method for electron tomographic reconstruction and three-dimensional modeling of the murine synapse using an automated fiducial marker-free system. Microsc Microanal 19(Suppl 5):182–187
13. Winkler H, Taylor KA (2013) Marker-free dual-axis tilt series alignment. J Struct Biol 182:117–124
14. Mumcuoglu EU, Hassanpour R, Tasel SF et al (2012) Computerized detection and segmentation of mitochondria on electron microscope images. J Microsc 246:248–265
15. Tasel S.F., Hassanpour R., Mumcuoglu E.U. et al. (2014) Automatic detection of mitochondria from electron microscope tomography images: a curve fitting approach. SPIE conference, Feb 2014, San Diego, CA

Chapter 2

Measuring Mitochondrial Shape with ImageJ

Ronald A. Merrill, Kyle H. Flippo, and Stefan Strack

Abstract

Mitochondria are shaped by opposing fission (division) and fusion events. Mounting evidence indicates that mitochondrial shape influences numerous aspects of mitochondrial function, including ATP production, Ca^{2+} buffering, and quality control. Despite the recognized importance of mitochondrial dynamics, the literature is rife with subjective, categorical estimates of mitochondrial morphology, preventing reliable comparison of results between groups. This chapter describes stringent, but easily implemented methods for quantification of mitochondrial shape changes using the open-source software package ImageJ. While we provide examples for analysis of epifluorescence images of cultured primary neurons, these methods are easily generalized to other cell types and imaging techniques.

Key words ImageJ, Fiji, Microscopy, Mitochondrial fission/fusion, Morphometry, Digital image analysis

1 Introduction

The word mitochondrion derives from the Greek words for thread (mitos) and grain (khondros), hinting at the ability of mitochondria to assume diverse morphologies. Mitochondria can be spherical, rod-shaped, or even branched and highly interconnected, forming a "mitochondrial reticulum." Reflecting their complex cytoarchitecture and the need for long-distance transport of mitochondria, neurons typically harbor mitochondria than span the entire shape spectrum. For instance, a serial section electron microscopy study documented branched mitochondria tens of microns in length in primary dendrites of hippocampal pyramidal neurons, while mitochondria in axon terminals were near spherical in shape [1].

Mitochondrial shape is determined by precisely orchestrated fission and fusion events. Fission, or division, followed by growth is how cells generate new mitochondria in response to increased energy demand or to replace damaged mitochondria. Fission is also intimately involved in programmed cell death, or apoptosis, facilitating cytochrome C release from the mitochondrial intermembrane space. In highly polarized cells, such as neurons, fission

Stefan Strack and Yuriy M. Usachev (eds.), *Techniques to Investigate Mitochondrial Function in Neurons*, Neuromethods, vol. 123, DOI 10.1007/978-1-4939-6890-9_2, © Springer Science+Business Media LLC 2017

is prerequisite for transport of mitochondria. Fusion, on the other hand, allows for mixing of mitochondrial content and the formation of reticular mitochondrial networks, which are generally associated with increased calcium buffering and bioenergetic capacity [2, 3].

Not easily replaceable, torturous in shape, highly demanding for energy, and relying largely on oxidative phosphorylation for ATP production, neurons are particularly susceptible to an imbalance of mitochondrial fission and fusion. For instance, evidence is mounting that dysregulated mitochondrial dynamics may be causative for common neurodegenerative disorders, including Alzheimer's and Parkinson's disease [3]. Moreover, mutations in mitochondrial fusion enzymes can lead to hereditary neurological disorders. Mutations in optic atrophy 1 (Opa1), an enzyme involved in fusion of the inner mitochondrial membrane, causes autosomal dominant optic atrophy, the most common inherited cause of blindness [4]. Mutations in the outer-membrane fusion protein mitofusin-2 (Mfn2) are responsible for Charcot–Marie–Tooth disease type 2A, a common peripheral sensory and motor neuropathy [5]. On the fission side, the mitochondria-dividing enzyme dynamin-related protein 1 (Drp1) and mitochondrial fission factor (Mff), the receptor for Drp1 at the outer mitochondrial membrane, have been found mutated in several cases of severe birth defects, childhood epileptic encephalopathy, and early infant death [6–10].

The development and adoption of rigorous analytical methods has lagged behind the growing interest in the field of mitochondrial dynamics. Indeed, primary research articles that describe mitochondrial shape in subjective, categorical terms (e.g., percent of cells with normal, fragmented, or elongated mitochondria) continue to be published. Subjective shape scores may be difficult to compare between experimenters, can introduce bias, often lack sensitivity, and reduce the power of statistical analyses, all of which may lead to erroneous conclusions. Subjective scores do, however, have a place as benchmarks, i.e., when validating new image analysis algorithms. In this chapter, we briefly touch on methods for labeling mitochondria before introducing the reader to mitochondrial shape analysis (morphometry) using ImageJ. Available bundled with various addons ("plugins") for life scientists as the Fiji distribution, ImageJ is an open-source, cross-platform application with powerful scripting capabilities that is continually improved by its user community [11]. ImageJ often outperforms commercial software solutions and facilitates increased reproducibility by providing transparent (open-source) data analysis for free. Our methods focus on single time point, "snapshots" of mitochondrial shape rather than counting individual fission and fusion events over time, as this allows the sampling of a larger population of neurons. Limited programming experience is helpful, but is not required to follow the code samples provided in this chapter. A simple, but fully functional macro for mitochondrial morphometry is listed in Appendix and is available for download at http://imagejdocu.tudor.lu/.

2 Materials and Equipment

1. Primary neuronal cultures on 4-chamber cover glasses (No. 1 or 1.5, Nunc Lab-Tek or In Vitro Scientific).
2. Inverted epifluorescence or confocal microscope with 60, 63, or 100× oil-immersion lens.
3. Lipofectamine 2000, Opti-MEM (Thermo Fisher).
4. Fiji (ImageJ distribution available at fiji.sc).
5. Computer running Windows, Mac OS, or Linux.

3 Methods

3.1 Labeling and Visualization of Mitochondria

All methods in this chapter involve primary, dissociated hippocampal or cortical cultures from embryonic rats or neonatal mice, but should be adaptable to other cell or tissue preparations. High-resolution images are key to accurate mitochondrial shape measurements. To permit the imaging of cultures using high-magnification (60–100×), low working-distance oil immersion lenses, neurons are cultured on a 4-chambered cover glass (e.g., Nunc Lab-Tek) coated with poly-L-lysine. We follow routine methods for preparing primary neuronal cultures that are described in detail elsewhere [12].

Mitochondria are readily visualized by fluorescence microscopy after labeling with mitochondria-targeted cationic dyes (e.g., MitoTracker, Rhodamine 123), with antibodies to mitochondrial antigens (e.g., cytochrome oxidase, Tom20) and fluorescent secondary antibodies, or by expression of mitochondria-targeted fluorescent proteins (FPs), such as GFP or mRFP. Mito-FPs and many cationic dyes can be visualized both in live cells and after fixation of the cultures with paraformaldehyde. Imaging fixed cultures is usually more convenient and may be the only option if autofluorescence of the culture medium overlaps with the emission of the fluorescent dye or protein.

We typically cotransfect neuronal cultures with a protein or shRNA of interest and a mito-FP in order to examine the impact of manipulating a specific fission/fusion enzyme, and its regulators, on mitochondrial shape. Compared to cationic dyes and immunofluorescence staining, mito-FP expression provides higher signal-to-noise ratios and eliminates interference by mitochondria of adjacent, non-transfected cells. We have had excellent success labeling the mitochondrial matrix with dsRed, EYFP, and turboBFP fused to the N-terminus of cytochrome-oxidase subunit 8 (COX8, addgene # 58425). The outer mitochondrial membrane can be labeled with FP fusions of the N-termini of Mas70p (yeast TOM70) or AKAP1. We routinely achieve 5% transfection efficiency in younger cultures (2–5 days in vitro, DIV) using Lipofectamine 2000 (Invitrogen) in the following protocol. More mature cultures are most efficiently transfected using magnetic beads (Magnetofection, Ozbiosciences) according to the manufacturer's instructions.

Protocol: chemical transfection and preparation of primary neuronal cultures for imaging.

This protocol describes transfection of the four compartments of a 4-chambered cover glass with the same plasmid DNA (encoding for example mitochondria-targeted GFP). For convenience, plasmid stocks are prepared at 100 ng/μl in sterile buffer (10 mM Tris pH 8.0).

1. Mix 23 μl plasmid stock with 112 μl Opti-MEM, and in a separate tube mix 1.6 μl Lipofectamine 2000 with 133.4 μl Opti-MEM.
2. Incubate separately at room temperature for 5 min.
3. Meanwhile, carefully draw off culture medium from cover glass, add back 300 μl to each compartment, and save the rest of the conditioned medium at 4 °C.
4. Combine DNA and Lipofectamine dilutions and incubate at room temperature for 10–15 min.
5. Add 60 μl DNA–Lipofectamine mix to each cover glass chamber.
6. Incubate for 2–3 h. in tissue culture incubator (37 °C, 5% CO_2).
7. To the remainder of the conditioned medium from step 3, add fresh medium to 2.2 ml and warm to 37 °C.
8. Aspirate off transfection mix and replace with 0.5 ml conditioned/fresh medium per chamber.
9. After 2–5 days, proceed to live imaging or fix as follows:
10. Add 0.5 ml 8% paraformaldehyde in phosphate-buffered saline (PBS) to each chamber (final concentration 4%). If cultures are to be imaged without prior immunofluorescence staining, nuclei can be counterstained during this step by adding Hoechst 33342 (2 μg/ml) to the fixative.
11. Incubate for 15 min at 37 °C.
12. Wash 5–6 times with PBS; add 0.5 ml PBS/chamber after last wash.
13. Proceed with imaging after optional immunofluorescence labeling.

Options for immunofluorescence include FP-directed antibody/fluorescent secondary antibody combinations to boost intrinsic fluorescence, or MAP2B, β3-tubulin, or NeuN to positively identify neurons. Nuclei are conveniently stained during incubation with fluorescent secondary antibodies; DAPI, Hoechst, or TO-PRO3 is used depending on channel availability. Wide-field epifluorescence or confocal images are captured at maximum resolution (no binning) and using 60–100× oil immersion lenses. While we typically capture single images of the neuron soma and proximal dendrites, it is also possible to capture the entire neuronal arbor at high resolution by merging overlapping images using ImageJ's "stitching" function (*Plugins* ▶ *Stitching*). Generally, between 30 and 60 images per experimental condition are sufficient to detect moderate effects.

It is near impossible to obtain reliable morphometric data from marginal images, so neuronal culturing, mitochondria labeling, and image acquisition should be optimized for best results. Optimizations include but are not limited to:

1. Assuring complete dissociation of the tissue and plating cells at a density appropriate for a particular mitochondria labeling technique, e.g., mitochondria labeled with cationic dyes or antibodies are best imaged in sparsely seeded cultures.
2. Minimizing temperature fluctuations, fluid shear, and other neuronal stressors that could artifactually impact mitochondrial morphology.
3. Adjusting dye and antibody concentration, incubation times, and wash and fixation conditions, and minimizing the time between labeling and imaging.
4. Adjusting imaging settings (light intensity, exposure time, focus, pinhole diameter, detector gain and offset) to minimize noise and to use the full range of pixel intensities without saturating the image.

3.2 Image Preprocessing

Measuring mitochondrial shape with ImageJ requires transforming grayscale or RGB images into binary (black and white) images to define objects and background. This digitization process, referred to as segmentation or thresholding, invariably entails loss of information. Filtering or preprocessing can reduce this information loss to improve morphometric data obtained from even high quality images. Improvements may be evident as a decrease in variability within an experimental group, an increase in effect size between groups, or both. ImageJ/Fiji implements a variety of image filters that are useful to this end, including noise reduction, background subtraction, and contrast and feature enhancement filters. Particularly useful prior to mitochondrial shape analyses are:

(a) *Process*▶ *Subtract Background.* This filter removes smooth and continuous, including unevenly illuminated backgrounds using a "rolling ball" algorithm [13]. The "Rolling Ball Radius" is an important parameter as it partitions the image into foreground and background. For 8-bit or RGB images, it should be larger than the largest object in the image. For a 1392 x 1040 pixel image of mitochondria imaged with a 100× objective, a radius of 50 works well.

(b) *Process* ▶ *Noise* ▶ *Despeckle.* As the name implies, this filter removes granular noise from the image. It replaces each pixel with the median value of its 3 × 3 neighborhood and is particularly helpful when analyzing confocal images acquired with high detector gain.

(c) *Process*▶ *Enhance Local Contrast (CLAHE).* Mitochondria are often unevenly labeled, so that weakly labeled parts of mitochondria may be lost after segmentation. Contrast-limited histogram equalization, or CLAHE, mitigates this problem by dividing an image into small squares, or tiles, each of which is contrast-enhanced separately. Bilinear interpolation then eliminates edge artifacts between tiles. The parameter "blocksize" specifies the size of the tiles in pixels, while "maximum slope" limits the contrast stretch during intensity transfer ("1" retains the original image). A combination of blocksize = 9 and slope = 4 results in an aggressively enhanced image that loses little information after segmentation.

(d) *Tubeness* and *Vesselness* (*Plugins* ▶ *Analyze* ▶ *Tubeness* and *Plugins* ▶ *Process* ▶ *Frangi Vesselness*). These two Hessian matrix-based filters are object-aware and were originally developed to enhance images of blood vessels [14, 15]. However, they also detect other tubular structures such as mitochondria or neurites. At the time of this writing, the Frangi Vesselness implementation is still experimental, so we will limit our discussion to the Tubeness filter. The "Sigma" parameter tunes the filter to detect thinner or thicker mitochondria. Processing an uncalibrated image, a value of 3 (pixels) yields acceptable results.

(e) *Process* ▶ *FFT* ▶ *Bandpass Filter.* This filter performs a fast Fourier transform (FFT) followed by Gaussian filtering of the power spectrum. As a bandpass filter, it can remove both small and large objects, filtering noise and uneven background in one step. Unique to this filter, the "Suppress stripes" option removes horizontal or vertical lines by masking frequency components near the coordinate axes of the power spectrum.

Figure 1 illustrates the effects of CLAHE and Tubeness filters on an epifluorescence image of a hippocampal pyramidal neuron expressing mitochondria-targeted GFP. Exposure was adjusted for

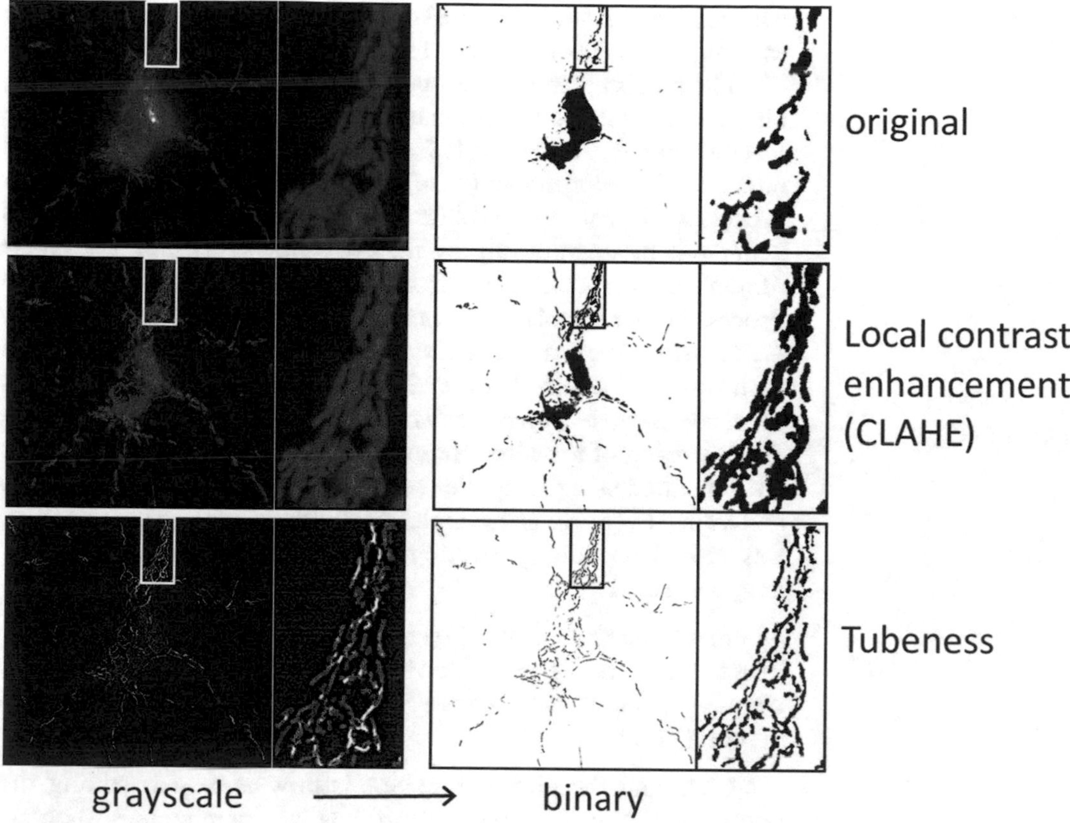

Fig. 1 Filters to enhance binary images of mitochondria. Because of large intensity variation in the original grayscale image (shown here in *pseudocolor*), local contrast enhancement or the tubeness filter is necessary to prevent digitization from eliminating dendritic mitochondria

optimal imaging of dendritic mitochondria, which leaves those in the cell body slightly overexposed. The left column shows 8-bit grayscale images pseudo-colored using the "Fire" color lookup table (*Image* ▶ *Lookup Tables* ▶ *Fire*). The right column shows the corresponding binary images obtained by thresholding according to ImageJ's default method (*Process* ▶ *Binary* ▶ *Make Binary*). Thresholding the original image eliminates most mitochondria in dendrites and renders mitochondria in the cell body as an indistinct mass.

Processing the original image by CLAHE with a block size of 9 and maximum slope of 4 dramatically improves the binary image such that it now visually approximates the original, unprocessed image. Local contrast enhancement retains peripheral mitochondria and at least partially resolves overlapping mitochondria in the soma. The Tubeness filter accentuates isolated mitochondria even more and strongly suppresses the mass of overlapping mitochondria in the soma. In this example, both CLAHE and Tubeness filters were preceded by rolling-ball background subtraction (radius 50) with optional smoothing enabled. This step suppresses artifacts resulting from enhancement of noise and background.

The next step after experimenting with and settling on a combination of filters is to automate image preprocessing. There are several ways to do this with ImageJ. For instance, calls to these filters can be encapsulated in a "wrapper" script (or macro) that processes all images in a folder. A template for such a wrapper is generated by opening a new macro window (*Plugins* ▶ *New* ▶ *Macro*) and then selecting from the macro editor window the Process Folder template (*Templates* ▶ *IJ1 Macro* ▶ *Process Folder*). A second option is to save the filter commands as a text file and then invoke *Process* ▶ *Multiple Image Processor.* A similar option is *Process* ▶ *Batch* ▶ *Macro,* which opens a text window that allows direct pasting of the filter commands. The actual filter commands are generated using ImageJ's command recorder (*Plugins* ▶ *Macro* ▶ *Record*). By way of an example, the following command sequence was recorded while generating the CLAHE-enhanced grayscale image in Fig. 1:

```
  run("Subtract Background...", "rolling=50");
run("Enhance Local Contrast (CLAHE)", "block-
size=9 maximum=4 mask=*None*");
  run("Fire");
```

Changing the filter parameters is now as easy as editing the commands. A more explicit way to change parameters is to define them as variables:

```
  rollingBall = 50; // radius.
  blockSize = 9; // 3x3 tile size.
  slope = 4; // contrast stretch limit.
  run("Subtract Background...", "rolling=" +
rollingBall);
run("Enhance Local Contrast (CLAHE)", "block-
size=" + blockSize + " maximum=" + slope + "
mask=*None*");
run("Fire"); // pseudocolor the image
```

Voila, you have written your first ImageJ macro! The two slashes start an in-line comment; multi-line comments are bracketed with /* and */. This may be familiar, because ImageJ's macro language is based on JavaScript, which in turn borrows its syntax from the classic C programming language.

3.3 Morphometry

Now that our images have been cleaned up, we can proceed to measuring the size of mitochondria. A simple macro for this task is listed in Appendix as "morphometry." At the core of the macro is ImageJ's *Analyze* ▶ *Analyze Particles*... command, which takes as input a binary image and compiles a table of measurements containing one row per particle (Results window). The remainder of the macro prepares the image for analysis by cropping to the selected region-of-interest (ROI), blanking the area outside of the selection, and thresholding. Overexposed areas (often the cell body in epifluorescence images) need to be excluded from the ROI. After the call to *Analyze Particles*, the macro extracts values from the Results window to compute composite measurements and averages for the ROI. The macro encourages experimentation with different preprocessing filters and thresholding methods (see also *Image* ▶ *Adjust* ▶ *Auto Threshold*) by including automated analysis in batch mode (more on this below). Figure 2 summarizes the shape descriptor output of the morphometry macro.

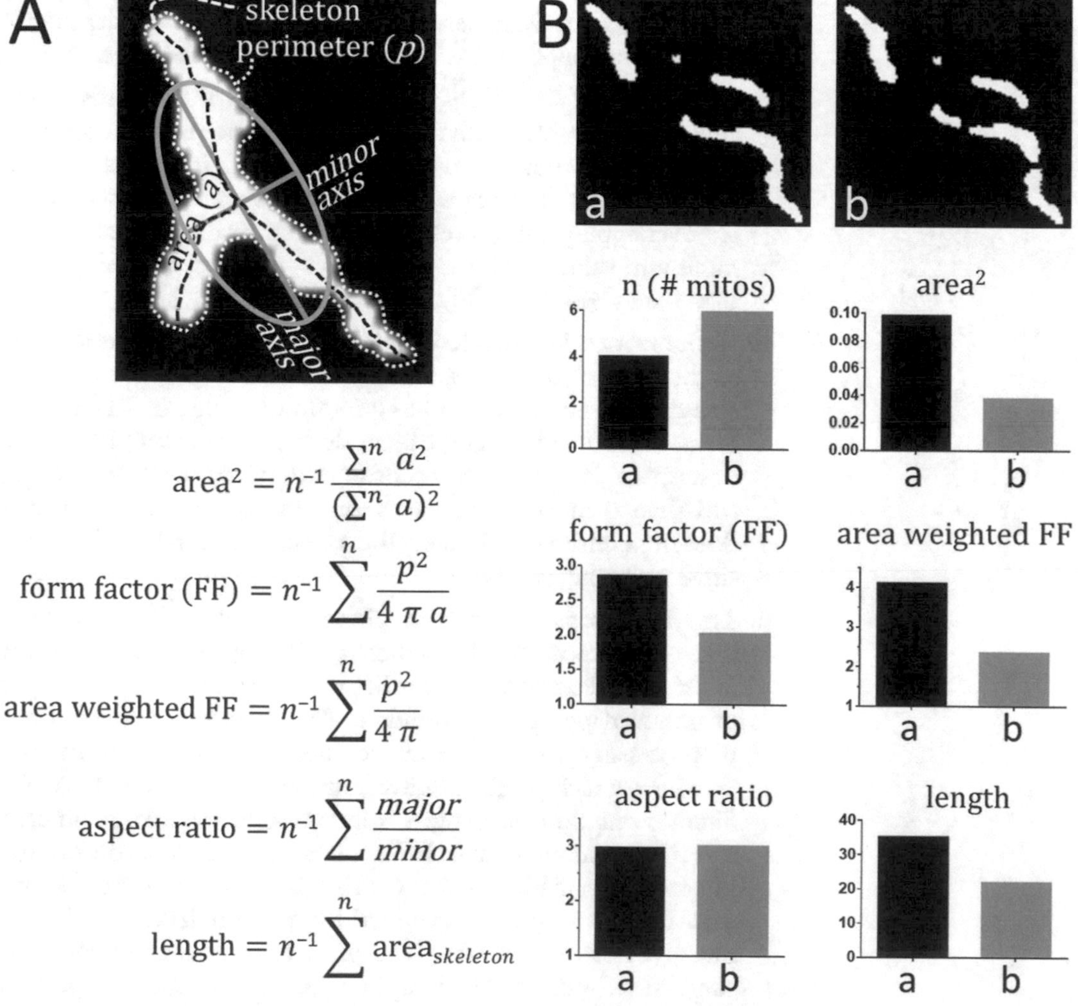

Fig. 2 Mitochondrial shape descriptors. (**a**) Definition of shape metrics generated by the morphometry macro in Appendix. Average metrics are calculated from area, perimeter, major and minor axis of an elliptical fit of binary particles, and area after skeletonizing the binary particles. (**b**) metrics obtained from binary images before (**a**) and after (**b**) two simulated fission events

These five descriptors were culled from a larger list of descriptors as those that most reliably reflect changes in the mitochondrial fission/fusion equilibrium:

1. *Area2* is computed by summing the squared areas of each particle and dividing by the square of the summed areas. This is essentially a measure of the average size of mitochondria weighted towards larger mitochondria.
2. *Form factor* also takes into account the mitochondrion's perimeter. As the inverse of circularity, it describes a particle's shape complexity. Its minimum value is 1 for a perfect circle. Form factor is a frequently reported mitochondrial shape metric that is particularly reliable for well-separated mitochondria. As noise and out-of-focus areas can artifactually increase an object's perimeter, form factor is very sensitive to image quality and is generally not applicable without pre-filtering the image.
3. *Area-weighted form factor* is a variant of form factor that averages across the ROI with a bias towards larger mitochondria or mitochondrial networks. Similar to area2, it provides more credible results in cases where highly elongated mitochondria are overlapping and therefore cannot be resolved. It also has a minimum value of 1, but produces larger, more variable averages than *form factor* (Fig. 2b).
4. *Aspect ratio* is independent of area and perimeter. It is defined as the ratio of the major and short axis of an ellipse fit to an object, and therefore also has a minimum value of 1 (Fig. 2a). Since bending an oblong object decreases its aspect ratio, this descriptor is primarily recommended for well separated, rod-shaped mitochondria in axons and dendrites. Because it does not consider perimeter, the aspect ratio is relatively insensitive to image quality.
5. *Length,* reported in units of pixels or microns (for calibrated images), the most intuitive of the shape descriptors. Unfortunately, it is also the least forgiving in that it requires high quality images of nonoverlapping mitochondria. This descriptor benefits from preprocessing with shape-aware filters such as *Tubeness* and *Vesselness,* which produce mitochondria of uniform width while filtering out clumps. Length is independent of area, perimeter, and major and minor axis. In fact, it is not even derived from the binary representation that gives rise to the other shape descriptors. Instead, length is computed by "skeletonizing" the binary image (*Process*▶ *Binary*▶ *Skeletonize*), reducing mitochondria to single-pixel-wide shapes. Length is then simply the average area (number of pixels) of these skeletons.

The morphometry macro also outputs the number of particles in the ROI (*n*), which may be useful in documenting fission and fusion events in a time lapse series. How do you interpret the macro's results and which descriptors do you report? We conclude that mitochondrial morphology has changed if (a) at least three of the five metrics show significant differences and if the other two at least trend in the same direction, (b) at a minimum trends are apparent when analyzing raw, unfiltered images, and (c) changes are readily apparent to a blinded observer. As mentioned before, visual inspection combined with some kind of categorical scoring is an important benchmark, especially when setting up mitochondrial morphometry for the first time. When presenting or publishing results, showing two independent shape measures adds credibility to your interpretations. Area^2, form factor, and area-weighted form factor have parameters in common, but aspect ratio and length are derived independently. As shown for aspect ratio and form factor in Fig. 3a, representing two independent metrics as an XY plot is particularly convincing if the data points fall on a straight line.

4 Results

Figure 3 illustrates the impact that filtering images prior to shape analysis has on the quality of the data. For this experiment, primary rat hippocampal cultures were transfected with wild-type or dominant-negative mitochondrial fission factor (Mff), the principal receptor for Drp1 at the outer-mitochondrial membrane. Mitochondrial RFP was co-transfected to label mitochondria. Images of fixed neurons (60/condition) were acquired using the 100× lens of an epifluorescence microscope. Neurons were outlined, avoiding the overlapping and overexposed mitochondria in the cell body (Fig. 3a).

Analysis of raw images shows small trends towards increased fission upon overexpression and decreased fission upon dominant-negative inhibition of Mff in comparison to neurons expressing outer mitochondrial GFP. Local contrast enhancement (CLAHE) alone or CLAHE followed by the Tubeness filter increases the effect size as well as the significance of manipulating Drp1 recruitment to mitochondria. Tubeness by itself results in only dominant-negative inhibition of Mff reaching statistical significance.

The choice of thresholding algorithm also has an impact on data quality. The morphometry macro calls the default algorithm (*Process*▶ *Binary*▶ *Make Binary*). However, there are 15 alternative methods available in the Fiji distribution under (*Image*▶ *Adjust*▶ *Auto Threshold*). Of those, the "Li" and "Otsu" algorithms have been found to be superior to the default under some conditions.

This example highlights the importance of choosing an optimal combination of preprocessing and thresholding methods. To simplify this process, the morphometry macro can be run in batch mode ("Batch Morphometry [F8]"), applying a set of previously

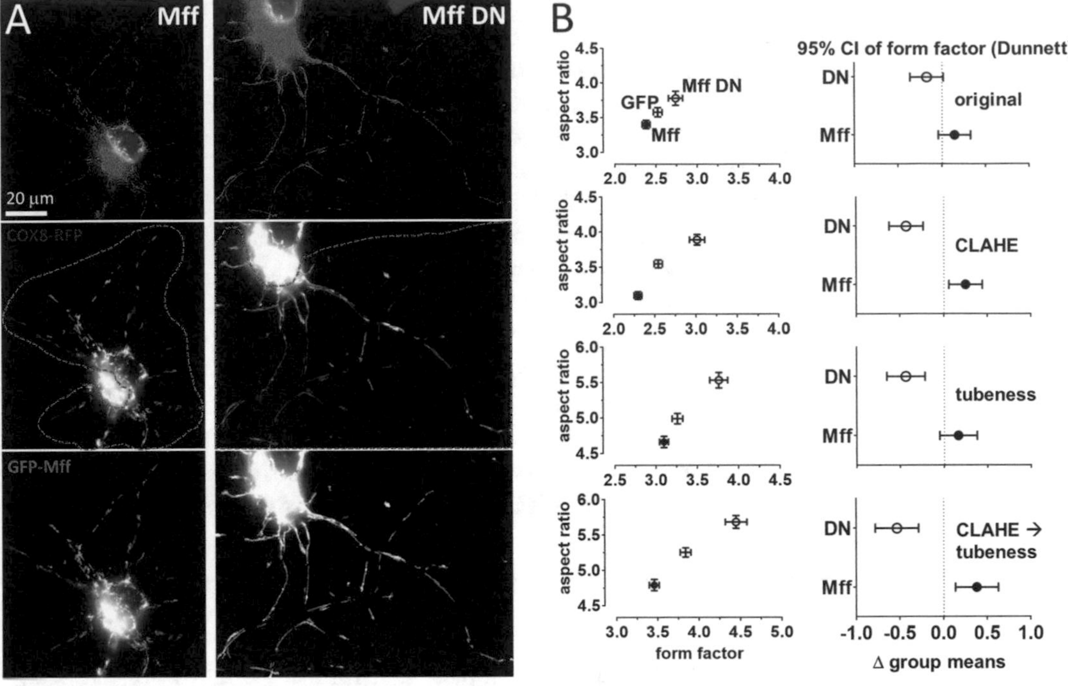

Fig. 3 Impact of Mff on mitochondrial shape. (**a**) Hippocampal neurons cotransfected with GFP-Mff (wild-type or dominant-negative, DN, mutant) and COX8-RFP were imaged by epifluorescence microscopy. ROIs excluding overexposed mitochondria in the soma (stippled *yellow line*) were analyzed for average mitochondrial shape. (**b**) form factor and aspect ratio differences improve significantly after image enhancement by CLAHE with or without tubeness filter (means ± SEM of 60 neurons per condition; form factor, Mff vs. GFP: $p < 0.01$, Mff DN vs. GFP: $p < 0.0001$ by one-way ANOVA with Dunnett's multiple comparisons test)

saved ROIs to a new set of images (or to the same images but using a different threshold method). With this feature, cells have to be outlined only once as long as the content of the ROI manager is saved after all images in a folder have been analyzed (*Window* ▶ *ROI Manager* ▶ *More* ▶ *Save*).

5 Conclusions

Mitochondrial dynamics is now well recognized as an important determinant of mitochondrial function, with disturbances of the mitochondrial fission/fusion equilibrium resulting first in neurological disorders. There is no reason that mitochondrial morphology should be investigated less rigorously and less quantitatively than respiration, ROS production, or any other aspect of mitochondrial function covered in this volume. ImageJ, an open-source environment supported by a large community of scientists, provides a platform for transparent image processing and analysis. ImageJ's command recorder and macro language enables scientists without programming experience to automate complex workflows. We hope that this chapter convinces the reader that fluorescence microscopy is compatible with exact science. Further, we hope to provide inspiration to automate other types of image analysis such as densitometry and colocalization.

Acknowledgments

This work is currently supported by NIH grants NS056244 and NS087908 to S.S. We thank past and present members of the laboratory for providing critical feedback for development of the methods described in this chapter.

6 Appendix—Morphometry Macro

```
var ch = 0; // channel to be analyzed for RGB
images
/*

* Measure mitochondrial morphology in the cur-
rent selection

* Ctrl+Shift+O closes current and opens next
image

*/

macro "Morphometry [F7]" {

   title = getTitle();

   morphometry(title, false); // not batch mode

}

/*

* Batch-apply a set of "named" ROIs to analyze
images with that file name

*/

macro "Batch Morphometry [F8]" {

   dir = getDirectory("Select an image direc-
   tory");

   while (roiManager("Count") == 0)

      waitForUser("Please open named ROIs into
      ROI manager");

   prevName = imgName = "";
```

```
      n = roiManager("Count");
      for (i = 0; i < n; ++i) { // loop
      through the ROI Manager table
         prevName = imgName;
         imgName = call("ij.plugin.frame.
         RoiManager.getName", i);
         if (isOpen(imgName)) { // named image
         is open
         selectWindow(imgName);
         } else { // done with current image,
         close and open next
           if (isOpen(prevName)) {
             selectWindow(prevName);
             close();
           }
           open(dir + imgName);
      }
      roiManager("Select", i);
      morphometry(imgName, true); // batch mode
      }
}
function morphometry(title, batchMode) {
   while (ch < 1 || ch > 3) { /* RGB chan-
   nel not yet selected, initialize; reinstall
   macro to change channel */
      ch = getNumber("Analyze RGB chan-
      nel(1-3):", 1);
      run("Set Measurements...", "decimal=5
      area perimeter fit");
      print("image\t n\t area2\t area-weight-
      ed ff\t form factor\t aspect ratio\t
      length"); /* header for results table */
   }
   if (bitDepth == 24) // RGB image
      run("Make Composite");
   if (isOpen("Binary")) {
```

```
    selectWindow("Binary");
    close();
} // close previous working image
if (isOpen("Skeleton")) {
    selectWindow("Skeleton");
    close();
} // close previous working image
selectWindow(title);
if (selectionType() == -1) // no selection
    run("Select All");
if (!batchMode) {
    roiManager("Add"); // save selection to
    ROI Manager for batch processing
    last = roiManager("Count") - 1;
    roiManager("Select", last);
    roiManager("Rename", title);
    /* roiManager("Save", File.directory +
    "named_ROIs.zip"); */ /* un-comment to
    save ROIs automatically */
}
// copy selection to new window and clear out-
side
setSlice(ch); // ignored if grayscale
run("Duplicate...", "title=Binary");
run("Make Inverse");
if (selectionType != -1) { // outside of ROI
is selected
    run("Duplicate...", " "); // make a mask
    of the background
    run("Convert to Mask");
    run("Create Selection");
    run("Make Inverse");
    roiManager("Add");
    close();
    n = roiManager("Count");
```

```
        roiManager("Select", n - 1);
        getRawStatistics(_area, backG); // mean is
        background
        setColor(backG);
        run("Restore Selection"); // fill outside
        of selection with background
        fill();
        run("Gaussian Blur...", "radius=64"); //
        smooth abrupt background transition
        roiManager("Delete"); /* delete masking
        selection (ROI manager has cell selec-
        tions) */
    }
    run("Select None");
    // subtract background and threshold
    run("Subtract Background...", "rolling=50");
    /* non-destructive filter even if already ap-
    plied */
    run("Make Binary");
    // also try other threshold methods includ-
    ed with Fiji, e.g.: run("Auto Threshold",
    "method=Li white");
    // create Results table of metrics, one line/
    particle
    run("Analyze Particles...", "size=9-Infini-
    ty circularity=0.00-1.00 show=Masks pixel
    clear");
    awff = ff = ar = sum_a = a2 = len = 0;
    for (i = 0; i < nResults; i++) { // for every
    particle in table
        a = getResult("Area", i);
        p = getResult("Perim.", i);
        ar += getResult("Major", i) /
        getResult("Minor", i); /* aspect ratio =
        length / width */
        sum_a += a;
        a2 += a * a; // area2 = a2 / (sum_a *
        sum_a)
```

```
        awff += b = (p * p) / (4 *
        3.14159265358979); // awff = ff * (a /
        sum_area)
        ff += b / a; // ff = p^2 / (4 * pi * a)
}
nParticles = nResults;
// skeletonize to get length
selectWindow("Mask of Binary"); /* created by
Analyze Particles .., excludes noise (< 9 pix-
els) */
rename("Skeleton");
run("Skeletonize");
run("Analyze Particles...", "size=0-Infinity
show=Nothing pixel clear");
for (i = 0; i < nResults; i++)
        len += getResult("Area", i);
// average and output
a2 /= sum_a * sum_a;
awff /= sum_a;
ff /= nParticles;
ar /= nParticles;
len /= nResults;
print(title + "\t " + nParticles + "\t " + a2
+ "\t " + awff + "\t " + ff + "\t " + ar + "\t
" + len);
selectWindow(title);
```

References

1. Popov V, Medvedev NI, Davies HA, Stewart MG (2005) Mitochondria form a filamentous reticular network in hippocampal dendrites but are present as discrete bodies in axons: a three-dimensional ultrastructural study. J Comp Neurol 492(1):50–65. doi:10.1002/cne.20682
2. De Stefani D, Rizzuto R, Pozzan T (2016) Enjoy the trip: calcium in mitochondria back and forth. Annu Rev Biochem. doi:10.1146/annurev-biochem-060614-034216
3. Bertholet AM, Delerue T, Millet AM, Moulis MF, David C, Daloyau M, Arnaune-Pelloquin L, Davezac N, Mils V, Miquel MC, Rojo M, Belenguer P (2016) Mitochondrial fusion/fission dynamics in neurodegeneration and neuronal plasticity. Neurobiol Dis 90:3–19. doi:10.1016/j.nbd.2015.10.011
4. Alexander C, Votruba M, Pesch UE, Thiselton DL, Mayer S, Moore A, Rodriguez M, Kellner U, Leo-Kottler B, Auburger G, Bhattacharya SS, Wissinger B (2000) OPA1, encoding a dynamin-related GTPase, is mutated in autosomal

dominant optic atrophy linked to chromosome 3q28. Nat Genet 26(2):211–215

5. Zuchner S, Mersiyanova IV, Muglia M, Bissar-Tadmouri N, Rochelle J, Dadali EL, Zappia M, Nelis E, Patitucci A, Senderek J, Parman Y, Evgrafov O, Jonghe PD, Takahashi Y, Tsuji S, Pericak-Vance MA, Quattrone A, Battaloglu E, Polyakov AV, Timmerman V, Schroder JM, Vance JM (2004) Mutations in the mitochondrial GTPase mitofusin 2 cause Charcot-Marie-Tooth neuropathy type 2A. Nat Genet 36(5):449–451
6. Waterham HR, Koster J, van Roermund CW, Mooyer PA, Wanders RJ, Leonard JV (2007) A lethal defect of mitochondrial and peroxisomal fission. N Engl J Med 356(17):1736–1741
7. Sheffer R, Douiev L, Edvardson S, Shaag A, Tamimi K, Soiferman D, Meiner V, Saada A (2016) Postnatal microcephaly and pain insensitivity due to a de novo heterozygous DNM1L mutation causing impaired mitochondrial fission and function. Am J Med Genet A. doi:10.1002/ajmg.a.37624
8. Koch J, Feichtinger RG, Freisinger P, Pies M, Schrodl F, Iuso A, Sperl W, Mayr JA, Prokisch H, Haack TB (2016) Disturbed mitochondrial and peroxisomal dynamics due to loss of MFF causes Leigh-like encephalopathy, optic atrophy and peripheral neuropathy. J Med Genet 53(4):270–278. doi:10.1136/jmedgenet-2015-103500
9. Shamseldin HE, Alshammari M, Al-Sheddi T, Salih MA, Alkhalidi H, Kentab A, Repetto GM, Hashem M, Alkuraya FS (2012) Genomic analysis of mitochondrial diseases in a consanguineous population reveals novel candidate disease genes. J Med Genet 49(4):234–241. doi:10.1136/jmedgenet-2012-100836
10. Fahrner JA, Liu R, Perry MS, Klein J, Chan DC (2016) A novel de novo dominant negative mutation in DNM1L impairs mitochondrial fission and presents as childhood epileptic encephalopathy. Am J Med Genet A. doi:10.1002/ajmg.a.37721
11. Schneider CA, Rasband WS, Eliceiri KW (2012) NIH Image to ImageJ: 25 years of image analysis. Nat Methods 9(7):671–675
12. Lim IA, Merrill MA, Chen Y, Hell JW (2003) Disruption of the NMDA receptor-PSD-95 interaction in hippocampal neurons with no obvious physiological short-term effect. Neuropharmacology 45(6):738–754
13. Sternberger SR (1983) Biomedical image processing. IEEE Comput 18:22–34
14. Frangi AF, Niessen WJ, Vincken KL, Viergever MA (1998) Multiscale vessel enhancement filtering. In: Wells WM, Colchester A, Delp SL (eds) Medical image computing and computer-assisted intervention, Lecture notes in computer sciences, vol 1496. Springer, Berlin, pp 130–137
15. Sato Y, Nakajima S, Shiraga N, Atsumi H, Yoshida S, Koller T, Gerig G, Kikinis R (1998) Three-dimensional multi-scale line filter for segmentation and visualization of curvilinear structures in medical images. Med Image Anal 2(2):143–168

Chapter 3

Live Imaging Mitochondrial Transport in Neurons

Meredith M. Course, Chung-Han Hsieh, Pei-I Tsai, Jennifer A. Codding-Bui, Atossa Shaltouki, and Xinnan Wang

Abstract

Mitochondria are among a cell's most vital organelles. They not only produce the majority of the cell's ATP but also play a key role in Ca^{2+} buffering and apoptotic signaling. While proper allocation of mitochondria is critical to all cells, it is particularly important for the highly polarized neurons. Because mitochondria are mainly synthesized in the soma, they must be transported long distances to be distributed to the far-flung reaches of the neuron—up to 1 m in the case of some human motor neurons. Furthermore, damaged mitochondria can be detrimental to neuronal health, causing oxidative stress and even cell death, therefore the retrograde transport of damaged mitochondria back to the soma for proper disposal, as well as the anterograde transport of fresh mitochondria from the soma to repair damage, are equally critical. Intriguingly, errors in mitochondrial transport have been increasingly implicated in neurological disorders. Here, we describe how to investigate mitochondrial transport in three complementary neuronal systems: cultured induced pluripotent stem cell-derived neurons, cultured rat hippocampal and cortical neurons, and *Drosophila* larval neurons in vivo. These models allow us to uncover the molecular and cellular mechanisms underlying transport issues that may occur under physiological or pathological conditions.

Key words Neurodegenerative disease, Mitochondrial transport, Microtubules, Axonal transport, Embryonic neurons, iPSCs, *Drosophila*, Larva, Live imaging

1 Introduction

A neuron's complex function is mirrored by its complex form: its processes can be long and intricately branched. This complicated form makes it challenging for neurons to transport the materials required for structure and function maintenance, both in and out of cell bodies and to far-flung neurites. Mitochondria are among the most crucial organelles requiring proper transport, as they perform aerobic respiration for ATP production, provide a reservoir for Ca^{2+} ions, synthesize heme compounds and some steroids, produce reactive oxygen species (ROS), and signal apoptosis. Neurons contain many areas of sub-specialization, like the presynapse, postsynapse, soma, axon, and dendrites, so the optimal density of mitochondria differs between and within cells as demands of the

Stefan Strack and Yuriy M. Usachev (eds.), *Techniques to Investigate Mitochondrial Function in Neurons*, Neuromethods, vol. 123, DOI 10.1007/978-1-4939-6890-9_3, © Springer Science+Business Media LLC 2017

neuron shift to different subcellular domains [1]. These changing requirements for mitochondria are reflected in the fact that at any given time in a healthy neuron, 30–40% of mitochondria are in motion [2–4]. In addition to the services they provide to the cell, when mitochondria are damaged or dysfunctional, they can produce an excess of ROS, which can in turn destroy cells. Damaged mitochondria must either be replenished by newly synthesized mitochondria transported from the soma, or the entire mitochondrion must be transported back to the soma for degradation, or cleared inside neurites in situ [5–7]. Precise regulation of mitochondrial transport in neurons is therefore critical to neuronal health.

Most mitochondrial movement occurs on microtubules. Movement towards the "plus" end of the microtubule—in the case of an axon, the axon terminal—is called anterograde motion, and occurs using the kinesin motors. Movement towards the "minus" end of the microtubule—in this case, the cell body—is called retrograde motion, and occurs using the dynein motor [1, 3, 8, 9]. Some mitochondria may need to go long distances without pausing, while others may need to go short distances with frequent pausing. For example, Ca^{2+} can cause mitochondria to detach from microtubules, thus allowing mitochondria to stop in areas of high Ca^{2+}-buffering demand or low ATP supply [10–12]. In this case, the mitochondrial ability to arrest motion is just as important as its ability to move.

Given the vital importance of mitochondria to cellular health and the complexity of controlling their proper localization, it is unsurprising to learn that impaired mitochondrial motility can seriously endanger a cell. Disruptions in neuronal mitochondrial transport and distribution have been closely correlated with many psychiatric and neurodegenerative disorders, like Charcot–Marie–Tooth disease [13], amyotrophic lateral sclerosis (ALS) [14], Alzheimer's disease [15–18], Huntington's disease [19–21], hereditary spastic paraplegia [22, 23], Parkinson's disease [5, 24], and schizophrenia [25–27]. Studying the fundamental principles that control mitochondrial motility in model systems may help elucidate the causes of neurological pathologies. Perhaps one of the clearest examples of this is found in the case of Parkinson's disease, where two familial Parkinson's disease-causing genes, *PINK1* (*PTEN-induced putative kinase 1*) and *Parkin*, have been found to function in the same pathway to arrest mitochondrial movement [5, 24]. This finding suggests that some Parkinsonian neurodegeneration may result from failure to arrest mitochondrial transport.

Using static imaging to study mitochondrial motility is insufficient for understanding how mitochondria distribution is achieved, and by what mechanisms their aberrant distribution may be pathogenic; therefore, we must live image mitochondrial transport in neurons. Huge strides have been made in live imaging axonal mitochondria in the past few years. One advantage of live imaging axons is that they have long, parallel arrays of microtubules, which makes following long distance movement of mitochondria straightforward. Also, because axons have largely uniform polarity, with "plus" ends pointing towards the axon terminals and "minus" ends towards the cell bodies, it is easy to distinguish kinesin- versus dynein-mediated movement. This homogeny is not found in dendrites, which in mammals exhibit mixed polarity [28, 29]. Difficulties with live imaging mitochondria in axons involve finding adequate mitochondrial markers, keeping neurons alive during imaging, preventing the neuronal network from obscuring individual mitochondrial movement, and getting light to penetrate in vivo tissues.

Here, we introduce methods for live imaging mitochondrial transport in axons, and address these concerns. We describe three complementary systems that have been used successfully in our laboratory: cultured induced pluripotent stem cell (iPSC)-derived neurons, cultured rat hippocampal and cortical neurons, and *Drosophila* larval neurons which are an in vivo model. Together, these systems allow for clear observation of mitochondrial movement, with the potential for easy genetic or chemical manipulation.

2 Methods

2.1 Live Imaging Mitochondrial Transport in iPSC-Derived Neuronal Axons

Since their introduction in 2007, iPSCs have revolutionized our ability to explore mechanistic pathways in physiologically relevant models. To generate iPSCs, terminally differentiated somatic cells from a subject (usually skin fibroblasts) are reprogrammed to a pluripotent cell state. There are several ways to reprogram the cells; the original and most common method employs cocktails of embryonic transcription factors, as discovered simultaneously in the laboratories of Dr. James Thomson and Dr. Shinya Yamanaka [30, 31]. iPSCs have emerged as an extraordinary human disease model, as they allow us to observe and experiment in human cells, while obviating ethical qualms with embryonic stem cells, and giving more flexibility for cell type differentiation than adult stem cells. Derivation of iPSCs from patients with diseases to the cell types most vulnerable in their diseases makes findings highly translatable. Here, we briefly discuss how to derive iPSCs to dopaminergic neurons, the cell type which degenerates in Parkinson's disease, and how to image and analyze mitochondrial transport in iPSC-derived neuronal axons.

iPSC-derived neurons exhibit characteristic morphologies and after 28 days of neuronal differentiation, axons can be easily identified by their unique structure, which is long and thin with uniform diameters and a lack of branching [29, 32]. Cultured neurons can be transiently transfected with constructs encoding a mitochondrial targeting sequence fused with a fluorescent protein, such as the red fluorescent mito-DsRed, to label live mitochondria. To simultaneously visualize neuronal morphology, presynaptic or postsynaptic specializations, or other organelles, cultured neurons can be co-transfected with an additional marker fused with green or blue fluorescent protein [5, 9, 10, 33]. In this way, neurons are sparsely transfected, allowing the process of a single transfected neuron to be examined and the orientation to be determined. Dense transfection, like with viral infection, can obscure individual axons and the origin of the cell body. Mitochondrial-specific dyes can also be used to aid in visualization, but just as with dense transfection, dyes mask observation of individual mitochondrial movement since they stain all mitochondria in the dish.

Additionally, agonists, antagonists, toxins, and ionophores can be easily applied to the cultures to alter intracellular signaling or ion concentrations, trigger cell death, or mimic pathological effects. Genetic manipulation by knocking down a gene with RNAi or by over-expressing a wild type or mutant gene can also be achieved by co-transfection. Thus, the neuronal culture system is a powerful platform for revealing mechanistic principles underlying mitochondrial transport.

2.1.1 Methods for iPSC-Derived Axons

Coverslip Preparation for Neuronal Culture

Cleaning: Glass coverslips (12 mm diameter, #1.5) are used for neuronal culture. To help with cell adhesion, acid washing with 70% nitric acid is used to pre-clean and micro-etch the surface of the coverslips (caution: wear gloves and be extremely careful). The coverslips are placed in a 500 ml beaker and 50 ml of 70% nitric acid is added. After incubating in acid for 2 h, with occasional mild agitation by shaking, acid is carefully decanted into a waste bottle. Excess acid can be washed with running tap water, and then rinsed extensively with double distilled water five times. After washes, excess water is absorbed with paper towels, and coverslips are transferred into a beaker containing 70% ethanol to incubate for 30 min. Coverslips can be dried between sheets of filter paper (1001-070, Whatman) and kept in a clean glass container. Once the coverslips are dry, the container can be wrapped with aluminum foil and autoclaved.

Coating: Autoclaved coverslips are UV-sterilized in a culture laminar flow hood for at least 30 minutes, and then transferred to a sterile petri dish. Filter-sterilized poly-ornithine (1 mg/50 ml, Sigma) is added to cover the surface of coverslips. After 2 h of incubation at room temperature (22 °C), the solution is aspirated and coverslips are washed three times with phosphate buffered saline (PBS: 137 mM NaCl, 2.7 mM KCl, 10 mM Na_2HPO_4, and 1.8 mM KH_2PO_4, pH = 7.4). After they are coated with poly-ornithine, coverslips are transferred into wells of a 24-well plate (Corning Life Sciences). Filter-sterilized laminin (170 μg/50 ml, Invitrogen) is then added to cover the surface of the coverslips, and they are incubated for at least 2 h at room temperature. Lastly, laminin is aspirated for removal, and coverslips are rinsed with PBS just before cell seeding.

Dopaminergic Neuron Derivation from iPSCs

iPSC lines can be converted from fibroblasts using standard procedures [31, 34–36]. Some control lines and lines from patients with neurodegenerative diseases are also available at cost in public repositories such as the Coriell Institute and the New York Stem Cell Foundation. Midbrain dopaminergic neurons are generated using a modified version of a previously described protocol [37]. For embryoid body (EB) formation, iPSC colonies are detached and cultured in suspension on non-adherent petri dishes using StemPro defined medium, without FGF2. At day 8, EBs are switched to neural induction medium containing Neurobasal medium (21103-049, Life Technologies) with GlutaMAX, Nonessential amino acids (NEAA), B27, and FGF2 (20 ng/ml) for an additional 2 days before being plated on Matrigel-coated dishes to form rosettes. The freshly formed rosettes are manually isolated and dissociated into single cells, and allowed to expand on Matrigel-coated dishes in order to generate a homogenous population of neural stem cells (NSCs) in the same neural induction medium for additional 2–5 days. For dopaminergic differentiation of NSCs, cells are cultured in Neurobasal medium supplemented with NEAA, L-glutamine (2 mM), B27, Shh (200 ng/ml), and FGF8 (100 ng/ml) for 10 days. At day 11, Shh and FGF8 are removed and replaced with brain-derived neurotrophic factor (BDNF, 20 ng/ml, 248-BD-025, R&D Systems), glial cell-derived neurotrophic factor (GDNF, 20 ng/ml, 450-10, Peprotech), transforming growth factor β-3 (TGFβ3, 1 M, AF-100-36E, Peprotech), ascorbic acid (200 μM, A5960, Sigma-Aldrich), and Dibutyryl-cAMP (1 mM, D0627, Sigma-Aldrich) for an additional 3 weeks. Alternatively, a dual-SMAD inhibition protocol is used to differentiate and maintain dopaminergic neurons [38–40]. Using this method, the neurons are maintained in Neurobasal medium supplemented with N-2 (17502-048, Life Technologies), BDNF (20 ng/ml), ascorbic acid (200 μM), GDNF (20 ng/ml), TGFβ3 (1 ng/ml), and dibutyryl-cAMP (500 μM).

NSCs can be confirmed by the expression of the neural progenitor markers Pax6, Sox1, or Nestin. Successful differentiation of dopaminergic neurons from NSCs can be confirmed by the expression of dopaminergic progenitor markers such as FoxA2 or Lmx1a/b after 14-day neuronal differentiation from NSCs, and mature dopaminergic neurons can be confirmed by the expression of β-tubulin isotype III and the rate-limiting enzyme in dopamine synthesis, tyrosine hydroxylase (TH), after 30-day neuronal differentiation from NSCs. The recommended antibody staining protocol is: Nestin (PRB-570C, Biolegend) at 1:500; Sox1 (AB5768, Chemicon) at 1:200; Lmx1a and b (139,736 and 139,726, Abcam) at 1:1500; FoxA2 (AF2400, R&D System) at 1:500; β-Tubulin isotype III (clone SDL.3D10, T8660, Sigma) 1:1000; and TH (P40101, Pel-Freeze) at 1:500.

Transient Transfection

Here we describe how to transfect with a single construct. Co-transfection with multiple constructs is also often used. A neuron that has absorbed one construct is likely to have taken up additional constructs [5, 10, 33]. For each well of a 24-well plate, 0.5 μg of DNA or 5 μl of Lipofectamine 2000 (Invitrogen) is added in serum-free Opti-MEM (Gibco) to a final volume of 50 μl in two separate tubes, and then gently mixed and incubated for 20 min at room temperature. During the incubation time, the culture medium of neurons for transfection is replaced with 200 μl pre-warmed Opti-MEM. After incubation, the 100 μl of DNA-Lipofectamine complex is added onto neurons with Opti-MEM. Then the neurons are returned into a 37 °C incubator with humidified atmosphere containing 5% CO_2. After transfection for 6 h, the DNA–Lipofectamine complex in Opti-MEM is replaced with the fresh culture medium. Transfected neurons are imaged 2–4 days later.

Image Acquisition

The transfected neurons on glass coverslips are placed in a 35 mm dish with pre-warmed CO_2-independent Hibernate E Low Fluorescence medium (HE-Lf, BrainBits) on a heated stage kept at 37 °C. Images are acquired by a Leica TCS SPE confocal microscope using a 63×/N.A.0.9 water-immersion objective and a Leica Application Suite (Version 2.6), with standard filter sets and excitation laser lines. The laser power is set minimally (1–10%) to reduce laser damage, and pinholes are opened maximally to image at a greater depth of the axon. Each individual image is 512 × 512 pixels. The objective and zoom factor results in a pixel dimension of approximately 0.23 μm. Time-lapse movies are obtained with 2–5 s intervals. Axons longer than 50 μm are selected for analysis.

Image Analysis

The confocal images of the time-lapse file collected by the Leica software are imported as an image stack into ImageJ (Version 1.48) using the LOCI plug-in, a Bio-Formats importer. To analyze the movement characteristics of individual mitochondria, kymographs are generated. First, the "straighten" plug-in or the "rectangular selection" function of ImageJ is used to choose and crop the axonal region to be analyzed from the image stack. Then, the "reslice" function is used to compress each cropped image to one line. Lastly, the "z-projection" function is used to compile all time-lapse image converted lines into one single picture, with time running from top to bottom. Thus in each kymograph, the *x*-axis represents mitochondrial displacement, and the *y*-axis represents time. Vertical lines indicate stationary mitochondria and diagonal lines are motile mitochondria moving either anterograde or retrograde.

With the aid of macros written in additional software such as LabVIEW and MATLAB [5, 10, 33], one can determine multiple features of movement, such as: (1) the instantaneous velocity of each mitochondrion, (2) the average velocity of those mitochondria that are in motion, (3) the percent of time each mitochondrion is in motion, (4) stop frequency, and (5) turn back frequency. The mitochondrial length and intensity can also be determined using ImageJ.

2.1.2 Expected Results for IPSC-Derived Axons

Neuronal mitochondria are labeled with the red fluorescent mito-dsRed by transient transfection. Figure 1 shows an example of an iPSC-derived neuronal axon from a healthy human subject. Movement of mito-dsRed was recorded and converted to a kymograph, which can be further used to analyze mitochondrial movement parameters such as velocity, percentage of time mitochondria in motion, stop frequency, and turn back frequency. Here, about 50% of the mitochondria are in motion within the first 100 s. This method can assist researchers in analyzing mitochondrial transport

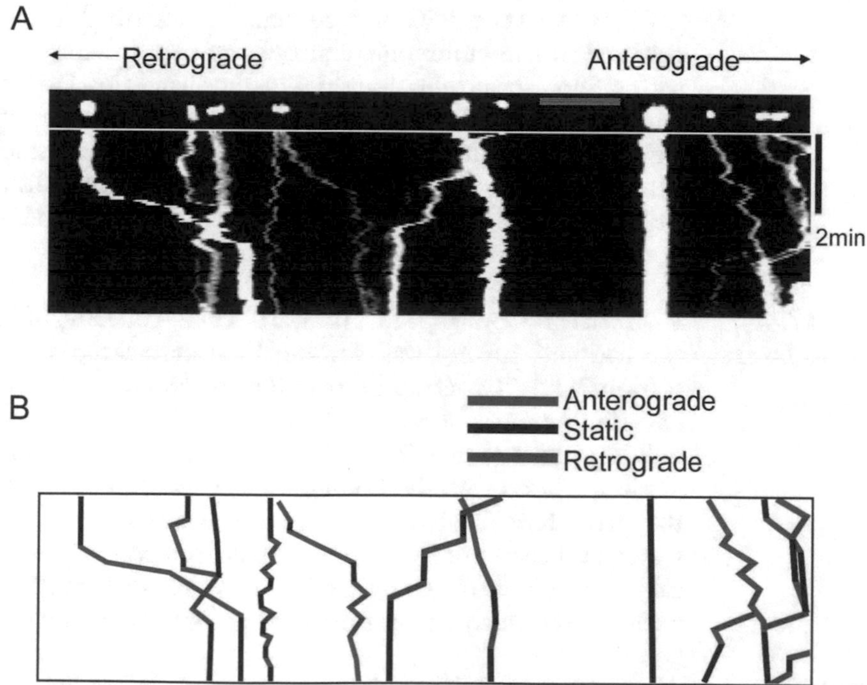

Fig. 1 Mitochondrial movement in an iPSC-derived neuronal axon. This representative iPSC-derived neuronal axon from a healthy subject was transfected with mito-dsRed. (**A**) The first frame of the time-lapse image series is shown above of a kymograph generated from the movie. The *x*-axis corresponds to mitochondrial position and the *y*-axis corresponds to time (progressing from *top* to *bottom*). Vertical white lines represent stationary mitochondria and diagonal lines are moving mitochondria. Scale bar = 10 μm. (**B**) A hand-drawn depiction of the kymograph shown in (**A**). Black lines label mitochondria when they are static, and blue or red lines label mitochondria when moving anterograde or retrograde, respectively

in human neurons derived from controls or patients with various neurological disorders, or in neurons with targeted genetic or chemical manipulation. It is a powerful tool for revealing regulatory mechanisms underlying mitochondrial motility, and for shedding light on pathological pathways altering mitochondrial transport in neurons.

2.2 Live Imaging Mitochondrial Transport in Rat Embryonic Neuronal Axons

Cultured rodent embryonic neurons are a well-established system and have been widely exploited for studying mitochondrial transport [2–4, 10, 41–44]. For example, both the roles of Ca^{2+} and syntaphilin in arresting mitochondrial motility have been elucidated using rat or mouse embryonic hippocampal neurons [10, 43, 44]. These neurons exhibit the key in vivo features of forming complex neuronal networks and relaying synaptic transmission [45]. After 3–4 days in vitro, axons differentiate from dendrites and exhibit unique morphological characteristics for live identification. Axons can reach a few millimeters, allowing for observation of long-range movements. The unipolarity of the microtubules in the axons makes it easy to orient and analyze mitochondrial transport [29].

Similar to the iPSC-derived neurons described in section 2.1, cultured rodent embryonic neurons can be transiently transfected with a fluorescent mitochondrial marker like mito-DsRed to label mitochondria, and with additional fluorescent markers for visualizing other structures. Both individual mitochondrial tracking and genetic or chemical manipulation can be easily accomplished. Here we describe the methods for live imaging mitochondrial transport in rat embryonic hippocampal and cortical neurons.

2.2.1 Methods for Rat Embryonic Neuronal Axons

Coverslip Preparation for Neuronal Culture

RD German Coverslips (12 mm, #1, 1943-10012A, Bellco Glass) are acid and ethanol washed, and then autoclaved as detailed in section 2.1.1 "Coverslip Preparation for Neuronal Culture". The day before use, coverslips are coated with filter-sterilized poly-ornithine (1 mg/50 ml) and laminin (170 μg/50 ml), and left in the hood with the UV light left on overnight. On the day of the procedure, coated coverslips are washed twice with sterile water, and then once with pre-warmed Neurobasal medium (also called Plain Neurobasal or PNB). The third wash of PNB is left in wells to prevent coverslips from drying out until the cells are plated.

Dissection and Culture of Rat Embryonic Neurons

All surgical tools should be autoclaved before use. A timed-pregnant Sprague Dawley Rat (day 18-19, Charles River) is anesthetized with CO_2 (first in a chamber, and then with a nose piece), and the belly is sprayed with 70% ethanol. Once unresponsive to toe pinch, an incision is made at the midline to expose the abdominal cavity. The uterine horn is removed by holding the uterus with forceps below the oviduct, cutting along the mesometrium, and pulling. The embryos are placed in a petri dish containing chilled

PBS, and the anesthetized rat is then euthanized by decapitation or cervical dislocation. The embryos are separated by cutting between implantation sites and removing the muscle layer and the decidua tissue with forceps. The embryos are then decapitated with scissors, and the heads placed in chilled PBS. The brains are isolated by carefully inserting Vannas spring scissors (15000-00, Fine Science Tools) into the spinal column, making a sagittal cut from the spinal cord to the nose, peeling back the skin and the skull with forceps, and gently detaching the brain with an Iris spatula (Fine Science Tools). Each brain is placed in Hank's Balanced Salt Solution and dissected under a dissecting microscope (at 10× or 15×) by removing the cerebellum and the olfactory bulbs, separating the hemispheres, stripping away the meninges, and removing the striatum with Dumont #5/45 Forceps (Fine Science Tools). The hippocampus and cortex are gently separated and are incubated in separate conical tubes containing chilled Hibernate E with Ca^{2+} (HE, BrainBits), penicillin (100 U/ml), and streptomycin (100 mg/ml). Once the dissection is complete, hippocampal or cortical tissue is carefully collected with a Pasteur pipette, transferred into pre-warmed Hibernate E without Ca^{2+} (HE-Ca, Brainbits), with 2 mg/ml of papain (PAP, Brainbits), and incubated at 37 °C for 10 min. The digested tissue is transferred via a Pasteur pipette into a new tube with 1 ml Hibernate E with Ca^{2+} and further dissociated by triturating 20–30 times with an autoclaved fire-polished Pasteur pipette. In order to separate the cells from the undigested tissue, the tissue is allowed to settle for 2 min, and then the supernatant without debris is transferred into a clean 15 ml conical tube and centrifuged at 1100 rpm for 1 min. The pellet is resuspended in 1 ml of filter-sterilized Neurobasal medium supplemented with penicillin–streptomycin–glutamine (100×) and B27 (50×). Cells are counted and seeded onto the prepared coverslips described in section 2.1.1 "Coverslip Preparation for Neuronal Culture" at a density of 100,000–180,000 cells/well of a 24-well plate, and kept in an incubator at 37 °C with 5% CO_2 in a humidified environment for up to 21 days. The medium is completely replaced 1–18 h after plating, and partially replaced every 2–3 days thereafter with pre-warmed Neurobasal medium containing B27 and antibiotics. This modified culturing protocol was derived from two sources [33, 46].

Transient Transfection

Mito-dsRed DNA (0.5 μg) is added to PNB (up to 50 μl) and mixed gently. In a separate tube, Lipofectamine 2000 (2.5 μl) is added to PNB (47.5 μl) and mixed. The tubes are combined and left in the hood for a minimum of 25 min. During this time, the conditioned media from each well of cultured neurons designated for transfection are collected and left in a water bath or an incubator at 37 °C. The wells are immediately washed twice with autoclaved water and once with 500 μl of pre-warmed PNB. The last wash with 500 μl PNB is left in the well. After the mixing period, 100 μl of DNA–Lipofectamine is added to each designated well with PNB, and the neurons are returned to the incubator. After 2 h, the DNA–Lipofectamine mixture is removed and the wells are washed three times with PNB. Then, 500 μl of the conditioned medium is added back to each well. The transfected neurons are imaged 2–4 days later, following the method outlined in section 2.1.1 "Image Acquisition", and can be analyzed following the method outlined in section 2.1.1 "Image Analysis".

2.2.2 Expected Results for Rat Embryonic Neuronal Axons

Figure 2 shows an axon from a rat cortical neuron transfected with mito-dsRed. Mitochondrial movement labeled by mito-dsRed was recorded and transformed into a kymograph. While the majority of mitochondria are stationary, about 35% are moving during the interval shown. This model is especially advantageous for investigating the influences of genetic manipulation, neurotrophic factors, toxins, and stress conditions on mitochondrial morphology and axonal transport under normal and pathological conditions.

2.3 Live Imaging Mitochondria in Vivo: Drosophila Larval Neurons

Drosophila provides a great in vivo model for live imaging mitochondrial motility in axons. In addition to being cheap and easy to house—even in a non-*Drosophila* lab—access to genetic mutants is extensive. Libraries of null alleles and RNAi lines, like the Bloomington Drosophila Stock Center (BDSC) and Vienna Drosophila Resource Center (VDRC) are all searchable on Flybase (http://www.flybase.com) and lines can be easily obtained. If a gene of interest does not have an extant null allele and thus a knockout is required, it is much simpler to generate null alleles in *Drosophila* as compared to in mammals. Among many methods for generating targeted gene deletion, one of the most traditional is imprecise excision, whereby a *P*-element insertion in or close to the site of interest is excised by a transposon [47, 48]. New systems like the CRISPR-Cas9 system may make this process even quicker [49, 50]. Null mutations of essential genes, which cause lethality when present throughout the full body, can be turned on in just a

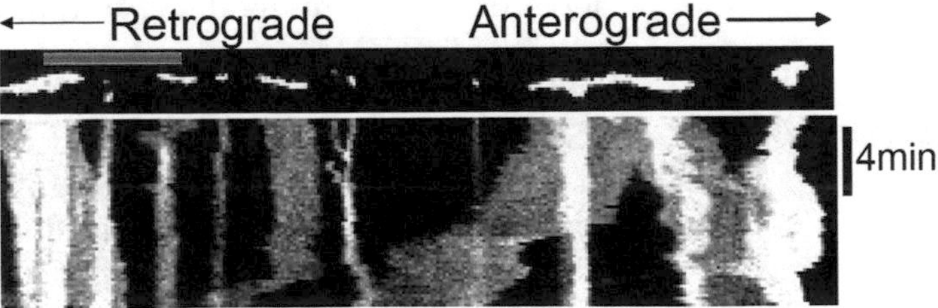

Fig. 2 Mitochondrial movement in a rat cortical neuronal axon. This representative cortical neuronal axon is transfected with mito-dsRed at 8 days in vitro. The kymograph is generated as in Fig. 1A. Scale bar = 10 μm

subset of cells or tissues. Several systems, like the MARCM [51, 52], Q [53], InSITE [54], and FLP/FRT [55] enable such tissue-specific manipulation. Over-expression of a gene can be easily achieved by using the UAS/GAL4 system, for selective temporal and/or spatial expression of a gene, or selective knockdown when coupled with RNAi [56]. It is even easy to combine mutations and transgenes in one animal, allowing for identification of epistatic versus additive phenotypes in double mutants, as well as allowing for rescue analysis with transgenes.

The fly has always been a powerful model for studying genetics, development, and behavior, harkening back to their use in uncovering the chromosomal theory of inheritance by T.H. Morgan. It has now also become an established model for studies of nervous system function and human neurodegenerative diseases. Due to the high genomic similarity between *Drosophila* and humans [57], many of the identified familial genes underlying disorders like Huntington's disease [58], Alzheimer's disease [59–63], Parkinson's disease [64, 65], and ALS [66] have been knocked out or their pathogenic forms have been expressed in *Drosophila*. For example, when the familial Parkinson's disease-causing gene *PINK1* is knocked out in *Drosophila*, the flies exhibit the locomotor defects and dopaminergic neurodegeneration also found in human patients. Significantly, *PINK1* null flies display defects in mitochondrial morphology and function and susceptibility to oxidative stress [67–72]. These studies in flies provide in vivo evidence of the cellular functions of PINK1, and point to mitochondria as critical in Parkinson's disease pathogenesis.

The nervous system of *Drosophila* larvae is highly organized [73]. The cell bodies of their central nervous system form the cortex of their brain lobes and ventral nerve cord (VNC; akin to our spinal cord). Two parallel zones of neuropil follow the midline of the VNC, and this is where the axonal and dendritic projections of cell bodies lie. Because these projections comingle, they are difficult to image; instead, the best structures to image are the long axons issuing out of the VNC and into the body wall. These axons have their cell bodies located in the VNC, and conjoin sensory neuronal axons which extend from the body wall back into the VNC, forming axonal bundles. For this reason, it is necessary to selectively label mitochondria in a subset of neurons, whose polarity can be determined unambiguously. Researchers have expressed a mitochondrial import signal fused with GFP (termed mito-GFP) selectively in motor neurons driven by the motor neuronal D42-GAL4 [74]. Motor neurons project axons from the VNC to the neuromuscular junctions (NMJs) in the body wall muscles; however, the high density of motor neuronal axons in axonal bundles makes it difficult to trace individual mitochondrial movement without photo-bleaching [33, 74]. Here, we provide a method for expressing mito-GFP sparsely using crustacean cardioactive peptide (CCAP)-GAL4, which selectively labels mitochondria in almost only one neuropeptidergic axon per axonal bundle, originating from the VNC and terminating at the NMJs of muscle 12 [75, 76]. In this way, mitochondrial fluorescent intensity can be greatly reduced in axonal bundles, and tracing and analyzing individual mitochondrial motility over a long distance can be achieved. We also introduce a method to live stain mitochondria with the mitochondrial membrane potential-dependent dye TMRM (tetramethylrhodamine). These are useful tools for determining both mitochondrial motility and function.

2.3.1 Methods for Drosophila Larval Neurons

Fly Stocks and Culture

Wild type, *CCAP-GAL4* [75], and *UAS-mito-GFP* [74] flies are used. Flies are cultured at 25 °C on molasses/yeast medium (6.7% molasses, 1.1% yeast, 5.5% cornmeal, 0.5% agar, and 0.13% *p*-hydroxy-benzoic acid methyl ester).

Dissection of Live Larvae

Late third instar wandering larvae are dissected in freshly prepared Schneider's medium (pH 7.4, S9895, Sigma) with 5 mM EGTA. Larval dissection is as previously described [77] in a chamber glass slide coated with a thin layer of Sylgard (Dow Corning Corporation, MI). The larva is opened along the dorsal midline with Vannas spring scissors (15000-00, Fine Science Tools), then the body wall muscles are pinned on the slide and the main guts and salivary glands are carefully removed. The larva is then washed gently three times using fresh Schneider's medium with 5 mM EGTA, and kept in the same medium for imaging. For TMRM staining, 20 nM TMRM (Molecular Probes) is added to Schneider's medium with 5 mM EGTA and mixed evenly, and then applied to the dissected larva for 20 min in the dark. After TMRM staining, the larva is washed three times again with fresh Schneider's medium with 5 mM EGTA and kept in the same medium with 5 nM TMRM for imaging.

Imaging Acquisition and Analysis

The freshly dissected larva is maintained at 22 °C and imaged within an hour of dissection. Time-lapse images of 100 s are taken every 2 s from each axon on a Leica DFC365 FX CCD camera for SPE II system with a Leica N2.1 filter LP 590 nm or an I3 filter LP 515 nm (JH Technologies). Kymograph conversion and motility analysis are similar to those described in section 2.1.1 "Image Analysis."

2.3.2 Expected Results for Drosophila Larval Neurons

Two different methods can be applied for monitoring mitochondrial transport in fly larval axons. Mitochondria can either be fluorescently labeled by mito-GFP in CCAP neurons, or stained with TMRM, a mitochondrial membrane potential-dependent dye. Individual mitochondrial movement can be traced in axons of *CCAP-GAL4, UAS-mito-GFP* larvae (Fig. 3), because CCAP

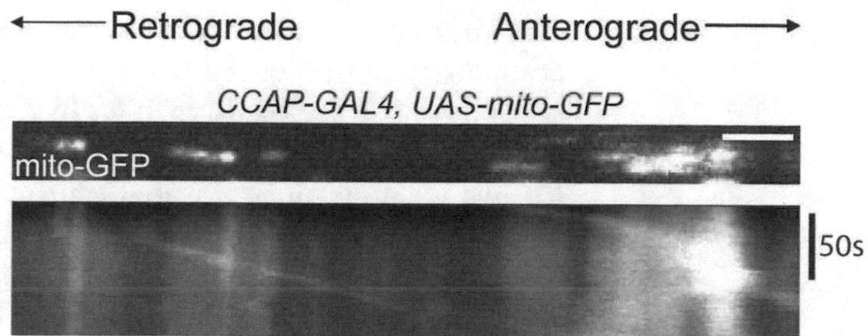

Fig. 3 Mitochondrial movement in a fly larval axon. Mito-GFP is expressed in a CCAP-expressing neuropeptidergic axon of a third instar larva. The representative segmental nerve passes hemisegment A3, with the VNC to the left. The kymograph is generated as in Fig. 1A. Scale bar = 10 μm

neurons project almost only one axon per axonal bundle [75, 76]. TMRM labels all mitochondria in axons with a polarized membrane potential, and thus long-range mitochondrial movement is hard to follow. This caveat can be circumvented by imaging the distal segment of the axon close to the NMJs, where fewer mitochondria are present. Combined with the power of fly genetics, these methods allow analysis of mitochondrial motility in a semi-intact in vivo system, to reveal the regulatory roles of genes and cellular pathways. In addition, TMRM intensity can be quantified and compared among different genetic backgrounds to determine the mitochondrial membrane potential as an indicator of mitochondrial function.

3 Conclusion

Increasing evidence suggests that mitochondrial transport defects may underlie many neurological disorders [5, 13–27], and therefore it is vital to understand the fundamental mechanisms controlling mitochondrial movement. Here, we have described how to live-image and analyze mitochondrial motility in axons in both cultured and in vivo systems.

It is difficult to explore physiological or pathological regulatory mechanisms in intact human or rodent brains due to the complexity and heterogeneity of the central nervous system and multiple inputs to individual regions. Cultured neuronal systems provide alternative models that can be used to study regulatory mechanisms without effects being masked by complicated circuitry. The advancements of modern microscopy and neuronal culturing systems have greatly facilitated our ability to image axonal transport of mitochondria in live neurons. The two cultured neuronal systems we have introduced here are iPSC-derived neurons and rat embryonic neurons. The rat neurons are easier to access, while the iPSC-derived neurons lend translational relevance to findings.

These cultured neuronal systems may allow us to bypass the complexity of the central nervous system, but we still need to dissect molecular mechanisms in an in vivo system to relate our findings to living organisms. *Drosophila* larvae are the optimal system for live imaging mitochondrial transport in vivo. They are easy to obtain and care for, easy to dissect, easy to image, and allow for easy genetic manipulation of molecular pathways.

These systems are especially powerful when used together, and will significantly forward our understanding of the mechanisms by which mitochondrial motility is regulated, and how its disruption can contribute to psychiatric and neurodegenerative diseases.

Acknowledgments

This work was supported by the Alfred P. Sloan Foundation (X.W.), the Klingenstein Foundation (X.W.), the California Institute of Regenerative Medicine (X.W.), the National Institutes of Health (X.W. RO1NS089583; RO0 NS067066), the Graduate Research Fellowship Program of the National Science Foundation (M.M.C.), the Postdoctoral Research Abroad Program of the National Science Council, Taiwan (P.T.), and the National Institutes of Health Ruth L. Kirschstein National Research Service Award (J.A.C. F32NS089155).

References

1. Hollenbeck PJ, Saxton WM (2005) The axonal transport of mitochondria. J Cell Sci 118(Pt 23):5411–5419. doi:10.1242/jcs.02745
2. Chen S, Owens GC, Crossin KL, Edelman DB (2007) Serotonin stimulates mitochondrial transport in hippocampal neurons. Mol Cell Neurosci 36(4):472–483. doi:10.1016/j.mcn.2007.08.004
3. Overly CC, Rieff HI, Hollenbeck PJ (1996) Organelle motility and metabolism in axons vs dendrites of cultured hippocampal neurons. J Cell Sci 109:971–980
4. Waters J, Smith SJ (2003) Mitochondria and release at hippocampal synapses. Pflugers Arch 447(3):363–370. doi:10.1007/s00424-003-1182-0
5. Wang X, Winter D, Ashrafi G, Schlehe J, Wong YL, Selkoe D, Rice S, Steen J, LaVoie MJ, Schwarz TL (2011) PINK1 and Parkin target Miro for phosphorylation and degradation to arrest mitochondrial motility. Cell 147(4):893–906. doi:10.1016/j.cell.2011.10.018
6. Ashrafi G, Schlehe JS, LaVoie MJ, Schwarz TL (2014) Mitophagy of damaged mitochondria occurs locally in distal neuronal axons and requires PINK1 and Parkin. J Cell Biol 206(5):655–670. doi:10.1083/jcb.201401070
7. Hoye AT, Davoren JE, Wipf P, Fink MP (2008) Targeting mitochondria. Acc Chem Res 41(1):87–97
8. Schnapp BJ, Reese TS (1989) Dynein is the motor for retrograde axonal transport of organelles. Proc Natl Acad Sci U S A 86(5): 1548–1552
9. Chang DTW, Honick AS, Reynolds IJ (2006) Mitochondrial trafficking to synapses in cultured primary cortical neurons. J Neurosci 26(26):7035–7045. doi:10.1523/JNEUROSCI.1012-06.2006
10. Wang X, Schwarz TL (2009) The mechanism of Ca2+-dependent regulation of kinesin-mediated mitochondrial motility. Cell 136(1):163–174. doi:10.1016/j.cell.2008.11.046
11. Saotome M, Safiulina D, Szabadkai G, Das S, Fransson A, Aspenstrom P, Rizzuto R, Hajnoczky G (2008) Bidirectional Ca2+-dependent control of mitochondrial dynamics by the Miro GTPase. Proc Natl Acad Sci U S A 105(52):20728–20733. doi:10.1073/pnas.0808953105

12. Macaskill AF, Rinholm JE, Twelvetrees AE, Arancibia-Carcamo IL, Muir J, Fransson A, Aspenstrom P, Attwell D, Kittler JT (2009) Miro1 is a calcium sensor for glutamate receptor-dependent localization of mitochondria at synapses. Neuron 61(4):541–555. doi:10.1016/j.neuron.2009.01.030
13. Baloh RH, Schmidt RE, Pestronk A, Milbrandt J (2007) Altered axonal mitochondrial transport in the pathogenesis of Charcot-Marie-Tooth disease from mitofusin 2 mutations. J Neurosci 27(2):422–430. doi:10.1523/JNEUROSCI.4798-06.2007
14. De Vos KJ, Chapman AL, Tennant ME, Manser C, Tudor EL, Lau KF, Brownlees J, Ackerley S, Shaw PJ, McLoughlin DM, Shaw CE, Leigh PN, Miller CC, Grierson AJ (2007) Familial amyotrophic lateral sclerosis-linked SOD1 mutants perturb fast axonal transport to reduce axonal mitochondria content. Hum Mol Genet 16(22):2720–2728. doi:10.1093/hmg/ddm226
15. Pigino G, Morfini G, Pelsman A, Mattson MP, Brady ST, Busciglio J (2003) Alzheimer' s presenilin 1 mutations impair kinesin-based axonal transport. J Neurosci 23(11):4499–4508
16. Rui Y, Tiwari P, Xie Z, Zheng JQ (2006) Acute impairment of mitochondrial trafficking by beta-amyloid peptides in hippocampal neurons. J Neurosci 26(41):10480–10487. doi:10.1523/JNEUROSCI.3231-06.2006
17. Thies E, Mandelkow E-M (2007) Missorting of tau in neurons causes degeneration of synapses that can be rescued by the kinase MARK2/Par-1. J Neurosci 27(11):2896–2907. doi:10.1523/JNEUROSCI.4674-06.2007
18. Calkins MJ, Reddy PH (2011) Amyloid beta impairs mitochondrial anterograde transport and degenerates synapses in Alzheimer's disease neurons. Biochim Biophys Acta 1812(4):507–513. doi:10.1016/j.bbadis.2011.01.007
19. Trushina E, Dyer RB, Ii JDB, Ure D, Eide L, Tran DD, Vrieze BT, Legendre-guillemin V, McPherson PS, Mandavilli BS, Houten BV, Zeitlin S, McNiven M, Aebersold R, Hayden M, Parisi JE, Seeberg E, Dragatsis I, Doyle K, Bender A, Chacko C, McMurray CT (2004) Mutant huntingtin impairs axonal trafficking in mammalian neurons in vivo and in vitro. Mol Cell Biol 24(18):8195–8209. doi:10.1128/MCB.24.18.8195
20. Chang DTW, Rintoul GL, Pandipati S, Reynolds IJ (2006) Mutant huntingtin aggregates impair mitochondrial movement and trafficking in cortical neurons. Neurobiol Dis 22:388–400. doi:10.1016/j.envpol.2005.09.025
21. Shirendeb U, Reddy AP, Manczak M, Calkins MJ, Mao P, Tagle DA, Reddy PH (2011) Abnormal mitochondrial dynamics, mitochondrial loss and mutant huntingtin oligomers in Huntington's disease: implications for selective neuronal damage. Hum Mol Genet 20(7):1438–1455. doi:10.1093/hmg/ddr024
22. Ferreirinha F, Quattrini A, Pirozzi M, Valsecchi V, Dina G, Broccoli V, Auricchio A, Piemonte F, Tozzi G, Gaeta L, Casari G, Ballabio A, Rugarli EI (2004) Axonal degeneration in paraplegin- deficient mice is associated with abnormal mitochondria and impairment of axonal transport. J Clin Invest 113(2):231–242. doi:10.1172/JCI200420138
23. Kasher PR, De Vos KJ, Wharton SB, Manser C, Bennett EJ, Bingley M, Wood JD, Milner R, McDermott CJ, Miller CCJ, Shaw PJ, Grierson AJ (2009) Direct evidence for axonal transport defects in a novel mouse model of mutant spastin-induced hereditary spastic paraplegia (HSP) and human HSP patients. J Neurochem 110(1):34–44. doi:10.1111/j.1471-4159.2009.06104.x
24. Liu S, Sawada T, Lee S, Yu W, Silverio G, Alapatt P, Millan I, Shen A, Saxton W, Kanao T, Takahashi R, Hattori N, Imai Y, Lu B (2012) Parkinson's disease-associated kinase PINK1 regulates Miro protein level and axonal transport of mitochondria. PLoS Genet 8(3):e1002537. doi:10.1371/journal.pgen.1002537
25. Atkin TA, MacAskill AF, Brandon NJ, Kittler JT (2011) Disrupted in Schizophrenia-1 regulates intracellular trafficking of mitochondria in neurons. Mol Psychiatry 16(2):122–124 . doi:10.1038/mp.2010.110121
26. Atkin TA, Brandon NJ, Kittler JT (2012) Disrupted in Schizophrenia 1 forms pathological aggresomes that disrupt its function in intracellular transport. Hum Mol Genet 21(9):2017–2028. doi:10.1093/hmg/dds018
27. Ogawa F, Malavasi ELV, Crummie DK, Eykelenboom JE, Soares DC, Mackie S, Porteous DJ, Millar JK (2014) DISC1 complexes with TRAK1 and Miro1 to modulate anterograde axonal mitochondrial trafficking. Hum Mol Genet 23(4):906–919
28. Baas PW, Deitch JS, Black MM, Banker GA (1988) Polarity orientation of microtubules in hippocampal neurons: uniformity in the axon and nonuniformity in the dendrite. Proc Natl Acad Sci U S A 85(21):8335–8339
29. Baas PW (1989) Changes in microtubule polarity orientation during the development of hippocampal neurons in culture. J Cell Biol 109(6):3085–3094. doi:10.1083/jcb.109.6.3085
30. Yu J, Vodyanik MA, Smuga-Otto K, Antosiewicz-Bourget J, Frane JL, Tian S, Nie J, Jonsdottir GA, Ruotti V, Stewart R, Slukvin

II, Thomson JA (2007) Induced pluripotent stem cell lines derived from human somatic cells. Science 318(5858):1917–1920. doi:10.1126/science.1151526

31. Takahashi K, Tanabe K, Ohnuki M, Narita M, Ichisaka T, Tomoda K, Yamanaka S (2007) Induction of pluripotent stem cells from adult human fibroblasts by defined factors. Cell 131(5):861–872. doi:10.1016/j.cell.2007.11.019
32. Cord BJ, Li J, Works M, McConnell SK, Palmer T, Hynes MA (2010) Characterization of axon guidance cue sensitivity of human embryonic stem cell-derived dopaminergic neurons. Mol Cell Neurosci 45(4):324–334. doi:10.1016/j.mcn.2010.07.004
33. Wang X, Schwarz TL (2009) Imaging axonal transport of mitochondria. Methods Enzymol 457(09):319–333. doi:10.1016/S0076-6879(09)05018-6
34. Park IH, Zhao R, West JA, Yabuuchi A, Huo H, Ince TA, Lerou PH, Lensch MW, Daley GQ (2008) Reprogramming of human somatic cells to pluripotency with defined factors. Nature 451(7175):141–146. doi:10.1038/nature06534
35. Okita K, Nakagawa M, Hyenjong H, Ichisaka T, Yamanaka S (2008) Generation of mouse induced pluripotent stem cells without viral vectors. Science 322(5903):949–953. doi:10.1126/science.1164270
36. Stadtfeld M, Nagaya M, Utikal J, Weir G, Hochedlinger K (2008) Induced pluripotent stem cells generated without viral integration. Science 322(5903):945–949. doi:10.1126/science.1162494
37. Swistowski A, Peng J, Liu Q, Mali P, Rao MS, Cheng L, Zeng X (2010) Efficient generation of functional dopaminergic neurons from human induced pluripotent stem cells under defined conditions. Stem Cells 28(10):1893–1904. doi:10.1002/stem.499
38. Chambers SM, Fasano CA, Papapetrou EP, Tomishima M, Sadelain M, Studer L (2009) Highly efficient neural conversion of human ES and iPS cells by dual inhibition of SMAD signaling. Nat Biotechnol 27(3):275–280. doi:10.1038/nbt.1529
39. Fasano CA, Chambers SM, Lee G, Tomishima MJ, Studer L (2010) Efficient derivation of functional floor plate tissue from human embryonic stem cells. Cell Stem Cell 6(4):336–347. doi:10.1016/j.stem.2010.03.001
40. Nguyen HN, Byers B, Cord B, Shcheglovitov A, Byrne J, Gujar P, Kee K, Schule B, Dolmetsch RE, Langston W, Palmer TD, Pera RR (2011) LRRK2 mutant iPSC-derived DA neurons demonstrate increased susceptibility to oxidative stress. Cell Stem Cell 8(3):267–280. doi:10.1016/j.stem.2011.01.013
41. Chada SR (2003) Mitochondrial movement and positioning in axons: the role of growth factor signaling. J Exp Biol 206(12):1985–1992. doi:10.1242/jeb.00263
42. Chada SR, Hollenbeck PJ, Lafayette W (2004) Nerve growth factor signaling regulates motility and docking of axonal mitochondria. Curr Biol 14:1272–1276. doi:10.1016/j
43. Kang JS, Tian JH, Pan PY, Zald P, Li C, Deng C, Sheng ZH (2008) Docking of axonal mitochondria by syntaphilin controls their mobility and affects short-term facilitation. Cell 132(1):137–148. doi:10.1016/j.cell.2007.11.024
44. Chen Y, Sheng ZH (2013) Kinesin-1-syntaphilin coupling mediates activity-dependent regulation of axonal mitochondrial transport. J Cell Biol 202(2):351–364. doi:10.1083/jcb.201302040
45. Kaech S, Banker G (2006) Culturing hippocampal neurons. Nat Protoc 1(5):2406–2415. doi:10.1038/nprot.2006.356
46. Kim KM, Vicenty J, Palmore GT (2013) The potential of apolipoprotein E4 to act as a substrate for primary cultures of hippocampal neurons. Biomaterials 34(11):2694–2700. doi:10.1016/j.biomaterials.2013.01.012
47. Voelker RA, Greenleaf AL, Gyurkovics H, Wisely GB, Huang SM, Searles LL (1984) Frequent imprecise excision among reversions of a P element-caused lethal mutation in Drosophila. Genetics 107(2):279–294
48. Daniels SB, McCarron M, Love C, Chovnick A (1985) Dysgenesis-induced instability of rosy locus transformation in *Drosophila melanogaster*: analysis of excision events and the selective recovery of control element deletions. Genetics 109(1):95–117
49. Cong L, Ran FA, Cox D, Lin S, Barretto R, Habib N, Hsu PD, Wu X, Jiang W, Marraffini LA, Zhang F (2013) Multiplex genome engineering using CRISPR/Cas systems. Science 339(6121):819–823. doi:10.1126/science.1231143
50. Mali P, Esvelt KM, Church GM (2013) Cas9 as a versatile tool for engineering biology. Nat Methods 10(10):957–963. doi:10.1038/nmeth.2649
51. Lee T, Luo L (1999) Mosaic analysis with a repressible cell marker for studies of gene function in neuronal morphogenesis. Neuron 22(3):451–461
52. Zong H, Espinosa JS, Su HH, Muzumdar MD, Luo L (2005) Mosaic analysis with dou-

ble markers in mice. Cell 121(3):479–492. doi:10.1016/j.cell.2005.02.012

53. Potter CJ, Tasic B, Russler EV, Liang L, Luo L (2010) The Q system: a repressible binary system for transgene expression, lineage tracing, and mosaic analysis. Cell 141(3):536–548. doi:10.1016/j.cell.2010.02.025
54. Gohl DM, Silies MA, Gao XJ, Bhalerao S, Luongo FJ, Lin CC, Potter CJ, Clandinin TR (2011) A versatile in vivo system for directed dissection of gene expression patterns. Nat Methods 8(3):231–237
55. Golic KG, Lindquist S (1989) The FLP recombinase of yeast catalyzes site-specific recombination in the Drosophila genome. Cell 59(3):499–509
56. Brand AH, Perrimon N (1993) Targeted gene expression as a means of altering cell fates and generating dominant phenotypes. Development 415:401–415
57. Adams MD, Celniker SE, Holt RA, Evans CA, Gocayne JD, Amanatides PG, Scherer SE, Li PW, Hoskins RA, Galle RF, George RA, Lewis SE, Richards S, Ashburner M, Henderson SN, Sutton GG, Wortman JR, Yandell MD, Zhang Q, Chen LX, Brandon RC, Rogers YH, Blazej RG, Champe M, Pfeiffer BD, Wan KH, Doyle C, Baxter EG, Helt G, Nelson CR, Gabor GL, Abril JF, Agbayani A, An HJ, Andrews-Pfannkoch C, Baldwin D, Ballew RM, Basu A, Baxendale J, Bayraktaroglu L, Beasley EM, Beeson KY, Benos PV, Berman BP, Bhandari D, Bolshakov S, Borkova D, Botchan MR, Bouck J, Brokstein P, Brottier P, Burtis KC, Busam DA, Butler H, Cadieu E, Center A, Chandra I, Cherry JM, Cawley S, Dahlke C, Davenport LB, Davies P, de Pablos B, Delcher A, Deng Z, Mays AD, Dew I, Dietz SM, Dodson K, Doup LE, Downes M, Dugan-Rocha S, Dunkov BC, Dunn P, Durbin KJ, Evangelista CC, Ferraz C, Ferriera S, Fleischmann W, Fosler C, Gabrielian AE, Garg NS, Gelbart WM, Glasser K, Glodek A, Gong F, Gorrell JH, Gu Z, Guan P, Harris M, Harris NL, Harvey D, Heiman TJ, Hernandez JR, Houck J, Hostin D, Houston KA, Howland TJ, Wei MH, Ibegwam C, Jalali M, Kalush F, Karpen GH, Ke Z, Kennison JA, Ketchum KA, Kimmel BE, Kodira CD, Kraft C, Kravitz S, Kulp D, Lai Z, Lasko P, Lei Y, Levitsky AA, Li J, Li Z, Liang Y, Lin X, Liu X, Mattei B, McIntosh TC, McLeod MP, McPherson D, Merkulov G, Milshina NV, Mobarry C, Morris J, Moshrefi A, Mount SM, Moy M, Murphy B, Murphy L, Muzny DM, Nelson DL, Nelson DR, Nelson KA, Nixon K, Nusskern DR, Pacleb JM, Palazzolo M, Pittman GS, Pan S, Pollard J, Puri V, Reese MG, Reinert K, Remington K, Saunders RD, Scheeler F, Shen H, Shue BC, Siden-Kiamos I, Simpson M, Skupski MP, Smith T, Spier E, Spradling AC, Stapleton M, Strong R, Sun E, Svirskas R, Tector C, Turner R, Venter E, Wang AH, Wang X, Wang ZY, Wassarman DA, Weinstock GM, Weissenbach J, Williams SM, Woodage T, Worley KC, Wu D, Yang S, Yao QA, Ye J, Yeh RF, Zaveri JS, Zhan M, Zhang G, Zhao Q, Zheng L, Zheng XH, Zhong FN, Zhong W, Zhou X, Zhu S, Zhu X, Smith HO, Gibbs RA, Myers EW, Rubin GM, Venter JC (2000) The genome sequence of *Drosophila melanogaster*. Science 287(5461):2185–2195
58. Green EW, Giorgini F (2012) Choosing and using Drosophila models to characterize modifiers of Huntington's disease. Biochem Soc Trans 40(4):739–745. doi:10.1042/BST20120072
59. Prussing K, Voigt A, Schulz JB (2013) *Drosophila melanogaster* as a model organism for Alzheimer's disease. Mol Neurodeg 8:35. doi:10.1186/1750-1326-8-35
60. Mhatre SD, Paddock BE, Saunders AJ, Marenda DR (2013) Invertebrate models of Alzheimer's disease. J Alzheimers Dis 33(1):3–16. doi:10.3233/jad-2012-121204
61. Bonner JM, Boulianne GL (2011) Drosophila as a model to study age-related neurodegenerative disorders: Alzheimer's disease. Exp Gerontol 46(5):335–339. doi:10.1016/j.exger.2010.08.004
62. Moloney A, Sattelle DB, Lomas DA, Crowther DC (2010) Alzheimer's disease: insights from *Drosophila melanogaster* models. Trends Biochem Sci 35(4):228–235. doi:10.1016/j.tibs.2009.11.004
63. Iijima-Ando K, Iijima K (2010) Transgenic Drosophila models of Alzheimer's disease and tauopathies. Brain Struct Funct 214(2-3):245–262. doi:10.1007/s00429-009-0234-4
64. Guo M (2012) Drosophila as a model to study mitochondrial dysfunction in Parkinson's disease. Cold Spring Harb Perspect Med 2(11). doi:10.1101/cshperspect.a009944
65. Whitworth AJ (2011) Drosophila models of Parkinson's disease. Adv Genet 73:1–50. doi:10.1016/B978-0-12-380860-8.00001-X
66. Bonini NM, Gitler AD (2011) Model organisms reveal insight into human neurodegenerative disease: ataxin-2 intermediate-length polyglutamine expansions are a risk factor for ALS. J Mol Neurosci 45(3):676–683. doi:10.1007/s12031-011-9548-9
67. Clark IE, Dodson MW, Jiang C, Cao JH, Huh JR, Seol JH, Yoo SJ, Ba H, Guo M (2006) Drosophila pink1 is required for mitochon-

drial function and interacts genetically with parkin. Nature 441(7097):1162–1166. doi:10.1038/nature04779

68. Park J, Lee SB, Lee S, Kim Y, Song S, Kim S, Bae E, Kim J, Shong M, Kim JM, Chung J (2006) Mitochondrial dysfunction in Drosophila PINK1 mutants is complemented by parkin. Nature 441(7097):1157–1161. doi:10.1038/nature04788
69. Yang Y, Gehrke S, Imai Y, Huang Z, Ouyang Y, Wang JW, Yang L, Beal MF, Vogel H, Lu B (2006) Mitochondrial pathology and muscle and dopaminergic neuron degeneration caused by inactivation of Drosophila Pink1 is rescued by Parkin. Proc Natl Acad Sci U S A 103(28):10793–10798. doi:10.1073/pnas.0602493103
70. Liu W, Acin-Perez R, Geghman KD, Manfredi G, Lu B, Li C (2011) Pink1 regulates the oxidative phosphorylation machinery via mitochondrial fission. Proc Natl Acad Sci U S A 108(31):12920–12924. doi:10.1073/pnas.1107332108
71. Vos M, Esposito G, Edirisinghe JN, Vilain S, Haddad DM, Slabbaert JR, Van Meensel S, Schaap O, De Strooper B, Meganathan R, Morais VA, Verstreken P (2012) Vitamin K-2 is a mitochondrial electron carrier that rescues pink1 deficiency. Science 336(6086):1306-1310. doi:10.1126/science.1218632
72. Vilain S, Esposito G, Haddad D, Schaap O, Dobreva MP, Vos M, Van Meensel S, Morais VA, De Strooper B, Verstreken P (2012) The yeast complex I equivalent NADH dehydrogenase rescues pink1 mutants. PLoS Genet 8(1):e1002456. doi:10.1371/journal.pgen.1002456
73. Gorczyca M, Budnik V (2006) Appendix: anatomy of the larval body wall muscles and NMJs in the third instar larval stage. In: Vivian B, Catalina RC (eds) International review of neurobiology, vol 75. Academic, New York, NY, pp 367–373. doi:http://dx.doi.org/10.1016/S0074-7742(06)75016-4
74. Pilling AD, Horiuchi D, Lively CM, Saxton WM (2006) Kinesin-1 and dynein are the primary motors for fast transport of mitochondria in Drosophila motor axons. Mol Biol Cell 17:2057–2068. doi:10.1091/mbc.E05
75. Park JH, Schroeder AJ, Helfrich-Forster C, Jackson FR, Ewer J (2003) Targeted ablation of CCAP neuropeptide-containing neurons of Drosophila causes specific defects in execution and circadian timing of ecdysis behavior. Development 130(12):2645–2656
76. Vomel M, Wegener C (2007) Neurotransmitter-induced changes in the intracellular calcium concentration suggest a differential central modulation of CCAP neuron subsets in Drosophila. Dev Neurobiol 67(6):792–808. doi:10.1002/dneu.20392
77. Budnik V, Gorczyca M, Prokop A (2006) Selected methods for the anatomical study of drosophila embryonic and larval neuromuscular junctions. In: Vivian B, Catalina RC (eds) International review of neurobiology, vol 75. Academic, New York, NY, pp 323–365. doi:http://dx.doi.org/10.1016/S0074-7742(06)75015-2

Chapter 4

Techniques to Investigate Bioenergetics of Mitochondria

William I. Sivitz

Abstract

In this chapter we review basic mitochondrial physiology as related to respiration, membrane potential, and ATP production and the relationship to ROS and calcium. Methods for measuring various functional parameters are described. Techniques differ according to whether one is assessing isolated mitochondria or intact cells. Although isolated mitochondria remain the only way to evaluate properties intrinsic to the organelles per se, mitochondrial function clearly is dependent on a multitude of cellular and whole body factors.

Key words Electron transport chain, Electron transport system, Respiration, Isolated mitochondria, ATP synthesis, Energy clamp, Uncoupling, Respiratory coupling, Mitochondrial membrane potential, Reactive oxygen species, Oxidative damage

1 Introduction

Although the phrase "mitochondrial function" is frequently used in the biomedical literature, it is important to remember that this terminology is broad and might refer to a wide range of issues. In this chapter, we consider several functional properties of mitochondria along with methods used for investigation. We include some long established techniques as well as more recent or novel methodology. The emphasis in this chapter is on assessing mitochondrial respiration (oxygen consumption), potential, ATP production, respiratory uncoupling, reactive oxygen species (ROS), and some aspects of calcium dynamics. The techniques to be described are adaptable to neurons and brain mitochondria as well as to a wide variety of cells and tissues. Mitochondrial biogenesis, autophagy (mitophagy), and dynamic changes such as fission, fusion, and distribution are mentioned but not discussed in detail. Such details may be found in other chapters within this book.

Stefan Strack and Yuriy M. Usachev (eds.), *Techniques to Investigate Mitochondrial Function in Neurons*, Neuromethods, vol. 123, DOI 10.1007/978-1-4939-6890-9_4, © Springer Science+Business Media LLC 2017

2 Electron Transport, Proton Pumping, and Respiration

2.1 Basic Physiology

Mitochondria generate energy as electrons are passed from donors at lower to acceptors at higher redox potential through various protein complexes. Along with this process, protons are pumped from the matrix outward generating a potential difference across the inner membrane. The resulting potential energy is used for ATP synthesis or dissipated as heat as protons leak back towards the matrix.

In this regard, it has been stated that mitochondria are "intrinsically on" meaning that the organelles are assembled in such manner that electron flow can only go from high to low potential. An analogy, although oversimplified, can be made with water flowing downward over a falls. In this regard, streams of water (electrons) are envisioned to enter the falls at more than one point varying in potential according to the height in which they enter. As water flows downward energy is consumed to turn paddle wheels that might be used to generate and store electrical charge (proton pumps generating a charge across the inner mitochondrial membrane). The waterfall, of course, is structured so that water, once entered, can only flow down, i.e., the falls is "intrinsically on."

Figure 1, depicts the process in schematic form in mitochondria. As indicated, electrons enter at one of four sites flowing downward with potential to generate potential energy ($\Delta\Psi$) as charge across the inner membrane. The term electron transport "chain" has been criticized as misleading since it implies a linear progression along a single pathway. Actually, electrons enter the electron transport *system* (ETS) or *branched* electron transport chain (ETC) at four separate sites which are convergent in that all eventuate in the reduction of coenzyme Q. Electrons donated by NADH enter at complexes I (NADH ubiquinone reductase) while succinate conversion to fumarate generates electrons at complex II (succinate dehydrogenase). Electrons derived from $FADH_2$ may also enter the convergent pathway through the electron transport flavoprotein (ETF). Electrons from glycerol 3-phosphate enter by way of a mitochondrial form of glycerol-3-phosphate dehydrogenase (GAPDH) located on the outer face of the inner membrane.

Electron flow from entry sites is directed through the intermediate ubiquinone to complex III (ubiquinol-cytochrome *c* reductase) [1] followed by electron transfer to another mobile intermediate, cytochrome c, which directs flow to complex IV (cytochrome *c* oxidase) where oxygen is finally consumed. ATP synthase (F_o•F_1-ATPase) or complex V consists of a joined membrane bound F_0-ATPase and rotatory F_1-ATPase. The complex is capable of "coupling" proton flow to conversion of ADP to ATP in an intricate manner that still remains incompletely understood [2].

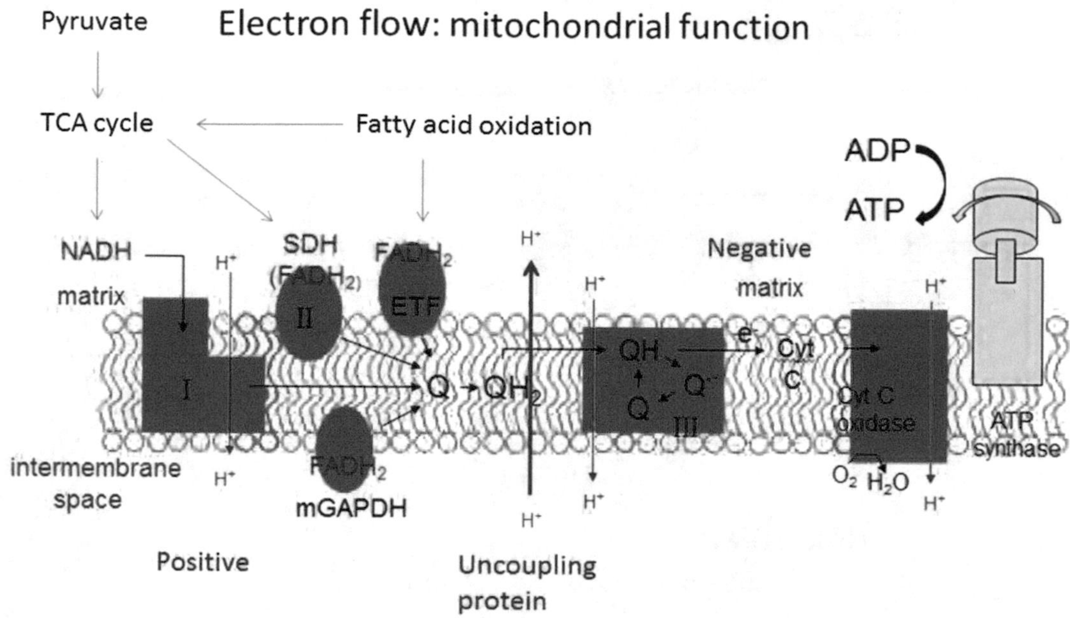

Fig. 1 Mitochondrial electron transport. The schematic illustration depicts the convergent nature of electron donation at one of four sites, complex I (NADH ubiquinone reductase), complex II (succinate dehydrogenase), the electron transfer flavoprotein (ETF), or a mitochondrial form of GAPDH. Reduced ubiquinone is processed through the Q-cycle in complex III where protons are pumped and electrons passed to mobile cytochrome c and then cytochrome oxidase. ATP formation by ATP synthase is coupled to mitochondrial potential generated by proton pumping at complexes I, III, and IV and offset by proton transfer in the opposite direction (proton leak) mediated in part by uncoupling proteins (UCPs). *Black arrows* depict electron transport

Substrates for the TCA cycle enter the mitochondrial matrix through pyruvate dehydrogenase, carrier proteins, or one of multiple shuttle mechanisms. Fatty acyl-CoAs enter through the carnitine palmitoyl transferase system (CPT-I and CPT-II) for β-oxidation. Metabolism of different substrates results in electron donation to specific complexes or sites. Oxidation of glutamate, malate, and pyruvate provides NADH for electron entry at complex I, while succinate donates electrons to complex II. β-oxidation of fatty acyl-CoAs generates electrons for entry at complex I or complex II by generation of acetyl-CoA. In addition, fatty acyl-CoA metabolism provides electrons through the ETF via $FADH_2$, which is a product of β-oxidation independent of the TCA cycle.

2.2 Measurement and Interpretation of Mitochondrial Respiration

Classically, mitochondrial oxygen consumption is determined by monitoring the rate of decrease in oxygen tension in the respiratory buffer by polarography using a Clarke type electrode (Fig. 2a). Using a small volume chamber, which can be less than 1 ml; it is easy to measure oxygen consumption by isolated mitochondria over time, a process that remains linear over several minutes even as the chamber oxygen content decreases. One or more energy substrates

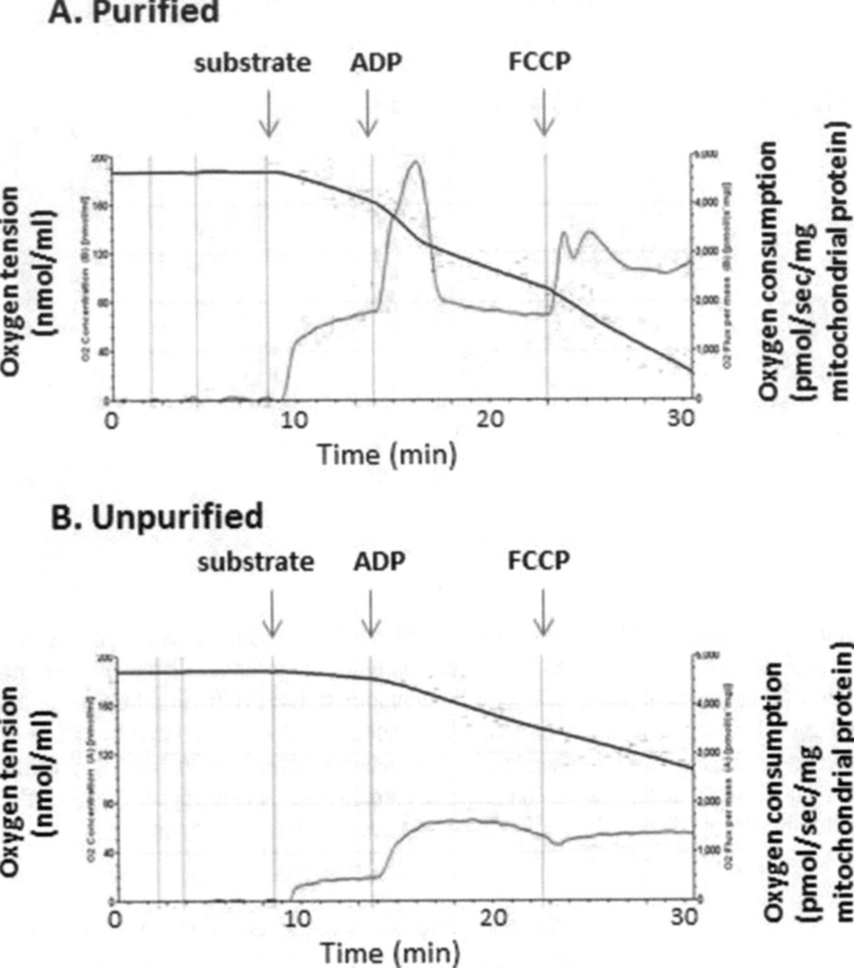

Fig. 2 Hind limb respiration by muscle mitochondria of a normal mouse. Mitochondria were isolated by differential centrifugation with or without gradient (Percoll) purification and incubated in an Oroboros O2K (Oxygraph) respirometer. Respiration buffer consisted of 120 mM KCl, 5 mM KH_2PO_4, 2 mM $MgCl_2$, 1 mM EGTA, 3 mM HEPES (pH 7.2) with 0.3% fatty acid-free BSA. (**a**) Oxygen tension (*blue line*) over time in purified mitochondria with sequential additions of respiratory substrates (5 mM glutamate, 1 mM malate, and 5 mM succinate), 0.3 mM ADP, and the chemical uncoupler, carbonyl cyanide *p*-[trifluoromethoxy]-phenyl-hydrazone (FCCP). *Red line* represents the derivative function or rate of oxygen consumption. State 4 respiration is induced by substrate addition while state 3 is evident after addition of ADP. The subsequent inflection represents the point of depletion of added ADP after which respiration returns to state 4. Respiration then reaches maximal after addition of FCCP. (**b**) Relatively poor responsiveness of unpurified mitochondria. The tracings shown represent the actual data as graphically output using the Oxygraph respirometer. However, the text along the right and left Y-axes is too small to read in the reproductions shown. Therefore, appropriate legends are added in larger print

are added to initiate respiration through action at specific entry points. Direct addition of NADH to isolated mitochondria would not be expected to initiate respiration unless the outer membrane is disrupted, which is not uncommon since preparation of mitochondria may not produce perfectly intact organelles.

Mitochondria are usually prepared through differential centrifugation using protocols that have been optimized for various tissue or cell systems. Additional purification is advised using gradient systems such as Percoll. In fact, we have noted much improved respiration in Percoll purified as opposed to unpurified mitochondria (compare Fig. 2a, b) as well as easier detection of calcium dynamics (*see* below). Mitochondrial integrity can be measured by the release of cytochrome C [3] or by assessing respiration on exogenous NADH or cytochrome c, which should not easily permeate mitochondria [4].

Depending on the choice of substrates, for example glutamate plus malate for complex I or succinate for complex II, one can study electron flow as oxygen consumption specifically through a single complex. This is possible since, in the presence of complex I or complex II substrates, flux through the entire TCA cycle is limited by shuttle systems which do not allow a fully operational (closed) cycle. When using succinate as a fuel for complex II, it is common practice to inhibit complex I (generally done by adding rotenone) in order to prevent reverse or backflow electron transport to complex I [5]. In studies using fatty acyl-CoAs as substrate, malate can be added to maintain the TCA cycle by replenishing oxaloacetate for reaction with acetyl-CoA at the citrate synthase step.

Beyond isolated mitochondria, one can also study permeabilized cells using the above methodology. Substances such as saponin or digitonin are commonly used to permeabilize the cells. In this case the cytoplasm is replaced by the respiratory buffer with presumably intact connections between the sarcoplasmic or endoplasmic reticulum.

Mitochondrial respiratory states were originally defined by Chance and Williams [6]. Respiration with substrate added in excess and during ADP conversion to ATP is referred to as state 3. In the absence of ADP or in the presence of oligomycin to block ATP synthase or after consumption of all added ADP, respiration is referred to as state 4. Chance and Williams further defined respiration with no ADP or substrate as state 1, with added ADP and before endogenous substrate exhaustion as state 2, and after exhaustion of oxygen (anaerobic respiration) as state 5. Although originally defined as above, the term "state 2 respiration" has also been used to imply respiration in the presence of substrate but without added ADP [7].

2.2.1 Measurement of State 3 and State 4 Respiration

After addition of energy substrate, the rate of oxygen consumption in the absence of added ADP or in the presence of oligomycin to block ATP synthesis defines state 4 respiration. The respiratory rate after adding ADP as substrate for ATP synthesis determines state 3 respiration (Fig. 2). The relative rates of oxygen used to generate ATP is represented by the ratio of state 3 to state 4 respiration referred to as the respiratory control ratio [8]. The efficiency of oxygen use to generate ATP can be estimated by the ratio of the added ADP concentration to the amount of oxygen consumed during state 3, in other words, the ADP to O ratio [8].

2.2.2 Measurement of Respiration Over Intermediate Respiratory States (ADP Recycling Methodology)

Most studies of isolated mitochondria or permeabilized cells as affected by physiologic or pathophysiologic conditions (including diabetes and obesity) have been carried out under state 4 or state 3 conditions as illustrated in Fig. 2 [8]. However, a problem with such studies is that mitochondria in vivo do not function at either state 3 or 4, but rather in between. Moreover, when ADP is added to measure state 3 respiration, the nucleotide is usually added in amounts that induce near maximal O_2 flux until nearly all is consumed. Thereafter, respiration quickly returns to state 4, leaving no indication of mitochondrial function under actual intermediate respiratory conditions.

However, it is possible to assess mitochondrial function over intermediate respiratory states by methods designed to regenerate ADP for phosphorylation. In effect, this clamps the ADP concentration at a level determined by the amount of ADP added. Such methods include the use of creatine plus creatine kinase [9–11], ATPase with excess ATP [9, 12–14], and glucose plus hexokinase (HK) [11, 15]. In recent work, we used 2-deoxyglucose (2DOG) plus hexokinase as illustrated in Fig. 3a (HK/2DOG energy clamp). ATP generated from ADP under these conditions drives the conversion of 2DOG to 2DOG phosphate (2DOGP) while regenerating ADP. The reaction occurs rapidly and irreversibly, thereby effectively clamping ADP concentrations. ADP conversion to ATP utilizes membrane charge and, at a steady ADP concentration, membrane potential ($\Delta\Psi$) is set at a steady level. Since $\Delta\Psi$ is determined by the relative generation and utilization of membrane charge, clamping ADP at constant respiration (determined by substrate addition) also clamps $\Delta\Psi$. Therefore, in the presence of given energy substrate conditions, $\Delta\Psi$ should fall with incremental additions of ADP but reach plateau levels after each addition. This is as illustrated in the example shown in Fig. 3b. In past work we have demonstrated this for mitochondria of rat [16] and mouse [17] skeletal muscle and mouse liver and heart [18].

Figure 4 depicts respiration (O_2 flux) by isolated hind limb and brain mitochondria incubated under the conditions indicated. As shown, respiration is incrementally increased as ADP is added in incremental amounts generating respiratory states intermittent between 4 and 3.

The hexokinase 2DOG recycling methodology has an additional advantage in that the generation of 2DOG phosphate (2DOGP), which can be determined with high sensitivity and specificity, serves as a measure of the rate of ATP production. This is detailed below in sections 5 and 9.

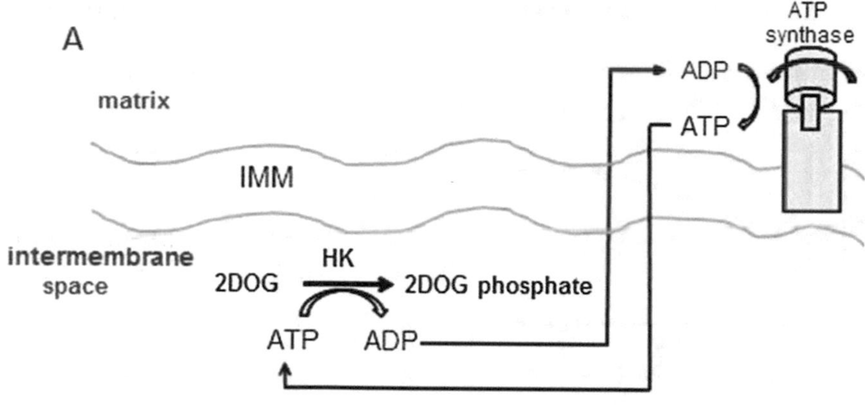

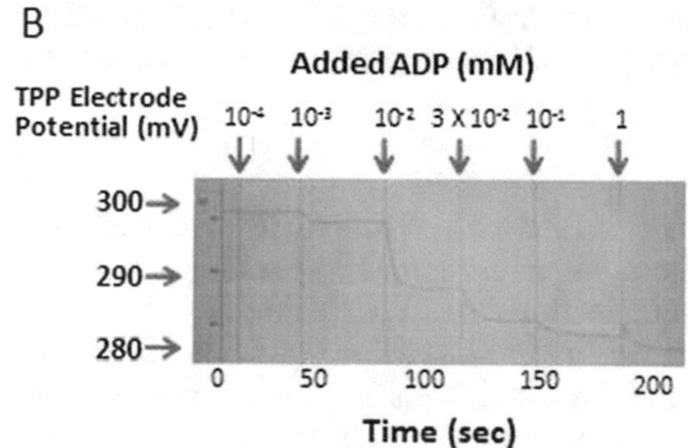

Fig. 3 The hexokinase (HK)/2-deoxyglucose (2DOG) energy clamp. (**a**) Saturating amounts of 2DOG and hexokinase recycle ATP back to ADP by irreversible formation of 2DOG phosphate which is not further metabolized. ADP is clamped at levels determined by the amount added. (**b**) Computer tracing (potential versus time) obtained on incubation of normal rat gastrocnemius mitochondria (0.25 mg/ml) fueled by combined substrates, glutamate (5 mM) + malate (1 mM) + succinate (5 mM). ADP was added in incremental amounts to generate the final total recycling nucleotide phosphate concentrations indicated (mM). After each addition, a plateau potential is reached consistent with recycling at a steady ADP concentration. Note the potential on the *y*-axis depicts negative electrode potential, not mitochondrial potential. The actual $\Delta\Psi$ follows a similar pattern after calculation using the Nernst equation based on the distribution of [TPP+] external and internal to mitochondria

2.3 Measurement of Mitochondrial Respiration in Intact Cells or Tissues

Studies of isolated mitochondria or permeabilized muscle fibers are carried out under conditions wherein the mitochondria are exposed to respiratory buffer rather than cytoplasm; the only way to directly determine mitochondrial respiration per se. However, this does not reproduce actual physiology or represent true in vivo mitochondrial function, since the organelles in vivo interact with the cytoplasm, endoplasmic and sarcoplasmic reticulum, and with other cellular components. Hence it is obviously important to understand mitochondrial function at the intact cell level. Oxygen consumption, as in isolated mitochondria, can be assessed in whole cells by measuring the drop in O_2 content in the medium.

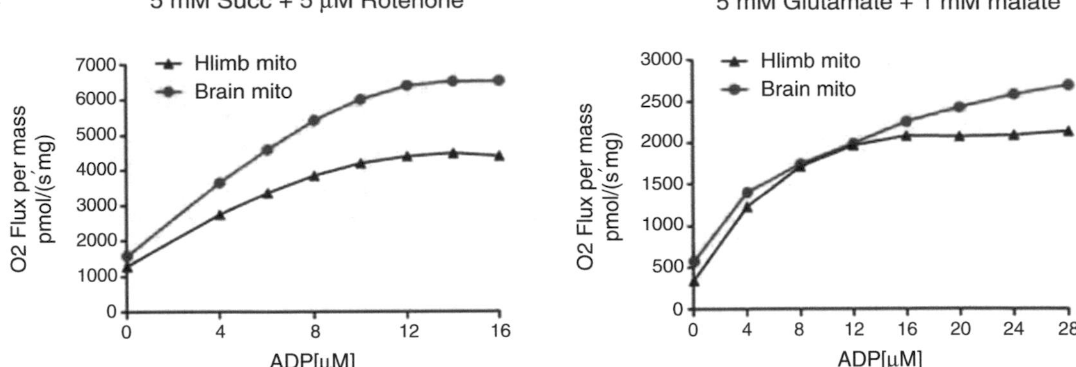

Fig. 4 Respiration by hind limb and brain mitochondria at different clamped concentrations of ADP representing a range of respiratory states from 4 to 3. Mitochondrial were incubated in respiratory buffer consisting of 99 mM KCl, 10 mM NaCl, 5 mM Na_2HPO_4, 2 mM $MgCl_2$, 6 mM KOH, 10 mM HEPES pH 7.2, 0.925 mM EGTA, 0.2% BSA and fueled by 5 mM glutamate + 1 mM malate (panel **a**) or by 5 mM succinate +5 μM rotenone (panel **b**)

However, this can be difficult since it depends on cell suspension, a problem when using cell types that are supported by tissue matrix connections in vivo or adherence to culture plates in vitro.

A better method is to perfuse cells using small oxygen electrodes proximal and distal to the cell preparation. This has been described even for cells perfused under microscopy [19] or as recently described by our laboratory [20] on bovine aortic endothelial cells grown on glass beads and perfused on columns. It is necessary to precisely calibrate the proximal and distal electrodes to each other which can be done using a shunt pathway around the cell preparation. By adding various inhibitors of mitochondrial respiration and/or ATP synthase it is possible to use this type of system to study mitochondrial function in intact cells [21].

Although good for mimicking physiologic conditions, the above perfusion technique is cumbersome and low in throughput. In recent years, newer respirometry techniques have emerged greatly increasing our ability to assess mitochondrial function within intact cells. A common technique typified by the Seahorse respirometer [22], measures oxygen consumption within an isolated microvolume within a larger volume of surrounding medium. The instrument does this by lowering a probe containing oxygen and pH sensors towards the bottom of wells (within 24- or 96-well plates) just above the cells in question; transiently, creating a small isolated volume in which the measurements are made (oxygen tension and pH). After measurement the probes are repositioned above the media allowing re-equilibration with the rest of the media volume, thereby avoiding the problem of loss of oxygen content over time.

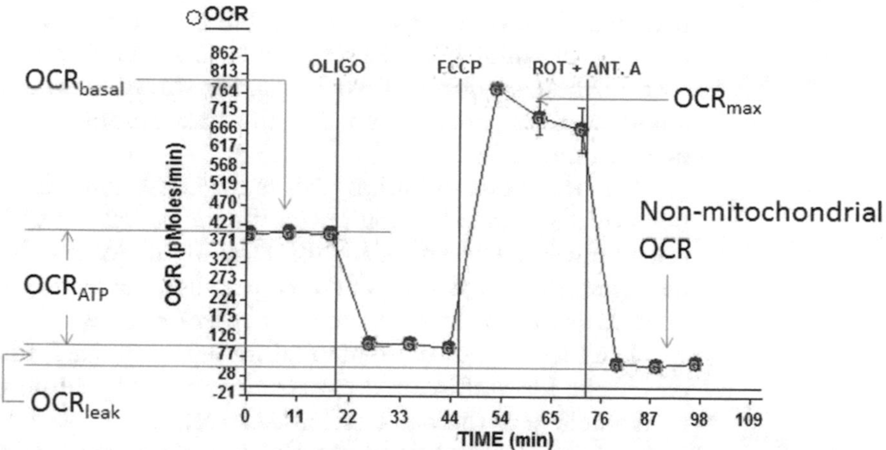

Fig. 5 Mitochondrial functional parameters assessed in bovine aortic endothelial (BAE) cells incubated using a 24-well Seahorse respirometer. Respiration medium consisted of medium M199 lacking sodium bicarbonate and pyruvate (Invitrogen). Oxygen consumption rates (OCR) were determined before and after sequential injections of 2 μM oligomycin (oligo) to inhibit ATP production (oligo), 2 μM FCCP to uncouple mitochondria generating maximal respiration, and 0.5 μM antimycin A (Ant A) plus 2 μM rotenone (Rot) to inhibit all mitochondrial respiration. As indicated, parameters of mitochondrial function including basal OCR, OCR directed at ATP production, maximal uncoupled OCR, non-mitochondrial OCR, and OCR directed at the proton leak can be calculated. Data represent mean ± SE, $n = 4$

The Seahorse respirometer measures both respiration as oxygen consumption rate (OCR) and the extracellular acidification rate (ECAR); the latter usually representing lactate accumulation and reflecting aerobic metabolism possibly associated with decreased mitochondrial oxidative metabolism. Additions of certain substances can be injected into the wells using channels positioned alongside the probes at desired preprogrammed times in order to specifically assess basal OCR, OCR directed at ATP production, maximal uncoupled OCR, non-mitochondrial OCR, and OCR attributable to the proton leak. Figure 5 depicts an experimental run illustrating the measurement of these parameters. In this example, bovine aortic endothelial (BAE) cells were seeded at a density of 5000–10,000 cells per well in 24-well plates designed for respirometer analyses. Cells reached confluency after 2 days and were subjected to respirometry 3 days after seeding. OCR and ECAR were determined in assay medium consisting of medium M199 (Invitrogen) lacking bicarbonate and pyruvate over 95 min with assessments at 8–10 min intervals. During respirometry, wells were sequentially injected at the times indicated in Fig. 5 with: oligomycin (2 μM) to block ATP synthase in order to assess respiration required for ATP turnover (OCR_{ATP}); carbonyl cyanide *p*-[trifluoromethoxy]-phenyl-hydrazone (FCCP, 2 μM), a proton ionophore, to induce chemical uncoupling and maximal respiration (OCR_{MAX}); and antimycin-A (0.5 μM) plus rotenone (2 μM) to completely inhibit electron transport and measure non-mitochondrial respiration. The FCCP concentration used in these studies was determined by titration with differing amounts of the uncoupler using the least amount required for maximal uncoupling.

For better quantification, values for OCR and ECAR should be normalized to DNA content of the individual wells [23] or to cell number if known. Also, for meaningful results, it is quite important that the density of adherent cells be uniform so that the samples quantified by the probes are representative.

The method has been adapted for assessing small tissue samples in the form of isolated pancreatic islets [24]. Although islets do not adhere to the well bottoms, a screen can be placed over the islets so that they remain in position for assessment by the probes when lowered into the wells.

3 Mitochondrial Membrane Potential

The electrical potential across the inner mitochondrial membrane (often referred to as ΔΨ) is generated by proton pumping at complexes I, III, and IV and offset by proton transfer in the opposite direction due to the proton leak or offset by utilization for ATP synthesis. Membrane potential can be quantified in different ways, some using estimated qualitative signals and other methods generating more precise quantitative results. We first discuss quantitative assessment followed by mention of qualitative, less precise, but useful methods based on uptake of certain fluorescent dyes.

3.1 Quantitative Assessment of ΔΨ

A common method is to incubate mitochondria in the presence of a measureable cationic molecule at a concentration too low to, itself, affect membrane potential. A common example is tetraphenylphosphonium (TPP) [25]. If the distribution of the cation inside and outside the mitochondrial matrix can be determined, ΔΨ can be calculated by the Nernst equation. One way to do this is to include a labeled form of the compound, for example, tetra[3H] phenylphosphonium bromide [5]. Mitochondria can then be separated by centrifugation and external and internal counts determined allowing calculation of the relative external and internal concentrations (C_e and C_i, respectively, in counts per unit volume). To determine C_i, it is also necessary to know the mitochondrial matrix volume and the extent to which TPP^+ is nonspecifically bound to mitochondrial protein as opposed to freely present within the matrix. These values can be determined using isotopic labelling [26]. However, for practical purposes on can use estimated values for these parameters (typically about 1 μl for the matrix volume and 33% binding). This is possible because the logarithmic nature of the Nernst equation (*see* below) minimizes the impact of these parameters on calculated $\Delta\Psi$. Moreover, if one is comparing $\Delta\Psi$ between two physiologic states and if these states do not affect matrix volume or cation binding, then any errors would cancel in the comparative values. Having calculated C_i, one can determine C_e as the external counts with the denominator being the respiratory volume (the internal volume of the mitochondria is small in comparison).

Once the values for C_e and C_i are calculated they can be entered into the Nernst equation ($\Delta\Psi = RT/zF \cdot \ln C_e/C_i$) to calculate $\Delta\Psi$. z = valence of cation, T and F are standard thermodynamic quantities (R designates the gas constant or 8.314 J/°K, T the temperature, and F represents Faraday's constant or 96,494 cal/mol). So, at a temperature of 310 °K (37 °C) at one atmosphere and for a univalent cation, $\Delta\Psi = 61.5 \log (C_e/C_I)$. Note that only the relative values for C_e and C_i are needed to enter the ratio, so if this is determined isotopically the concentrations can simply be expressed as counts per unit volume.

In practice, an easier way to determine the relative distribution of TPP internal and external to the mitochondrial matrix is to carry out incubations in the presence of an electrode sensitive to TPP. Such an electrode can be constructed as described [25] or commercially purchased. The electrode potential outside the mitochondria is then monitored during incubation and C_e for TPP determined from the recorded electrode potential based on prior calibration of the electrode as a function of TPP^+ added in the absence of mitochondria. The amount of TPP internal to the matrix is then determined from the known amount of TPP^+ added and external concentration. Note that it is still necessary to know or estimate the mitochondrial matrix volume and the extent to which TPP^+ is bound to mitochondrial protein.

3.2 Qualitative Estimation of $\Delta\Psi$

Although precise calculation of $\Delta\Psi$ is desirable, this is more cumbersome than needed for many studies. Moreover, the above TPP method is difficult, although not impossible [27] to apply to intact cells. Alternatively, $\Delta\Psi$ or changes in $\Delta\Psi$ over time can be estimated using a variety of fluorescent lipophilic cation dyes as reviewed [28]. Fluorescent probes are often used in a non-quenching mode meaning that low concentrations are used that do not aggregate or quench the signals. In other cases, using higher concentrations quenching may be desirable as a means to observe signal reduction due to accumulation within mitochondria which might occur as a result of certain manipulations.

Tetramethylrhodamine methyl ester (TMRM) and the ethyl ester (TMRE) are commonly used. Detection of fluorescent signals can be done by microscopy and signal quantification using a charge coupled device (CCD camera) to count intensity, by confocal microscopy, or using a plate reader. TMRM and TMRE accumulate over short time periods, so when used in intact cells, plasma membrane effects need to be considered. But these probes remain good choices especially for effects over longer time periods (several minutes to hours) or to assess chronically different conditions. These dyes can inhibit electron transport, less with TMRM, but at low nM concentrations this is generally not an issue.

Rhodamine 123 is another probe commonly used for estimating $\Delta\Psi$. Slow equilibration allows studies over minutes to assess mitochondrial uptake mitigating concern over plasma membrane accumulation [29]. Rhodamine 123 also appears useful for whole cell long-lasting quenching studies wherein it is possible to monitor a decrease in signal intensity as the dye is shifted to mitochondria.

Another approach to measuring $\Delta\Psi$ in intact cells is to use the dichromatic probe JC-1. The fluorescent signal from this dye shifts from green to red with increasing aggregation as occurs with higher potential and mitochondrial uptake. This appears useful for experiments wherein one desires a qualitative assessment of a sizeable effect of a given manipulation to alter $\Delta\Psi$. On the other hand, the dependence on aggregation may limit more subtle changes. Moreover, limitations include sensitivity to concentration due to aggregation and slow permeation and equilibration leading to possible discrepancies in fluorescence from different cell components or from localization of mitochondria [28].

3.3 Effect of pH

One often neglected consideration is that $\Delta\Psi$ as referred to above represents the difference in charge or electrical gradient across the inner membrane but does not represent the total force driving protons, i.e., what has been termed the proton-motive force (pmf). There is a pH difference which constitutes a portion of the pmf (in the ballpark of about 1/6th under typical physiologic conditions). The common probes discussed above that assess $\Delta\Psi$ do just that, measuring charge but neglecting the pH component. If desired, one can assess this by using pH sensitive fluorescent fusion reporter proteins, certain fluorophores, or probes that directly measure mitochondrial pH [30]. It is also possible to abolish the pH gradient using nigericin to convert the pH gradient to a potassium gradient, then measuring the pmf and subtracting $\Delta\Psi$ (charge difference) to determine the pH component [31].

4 Mitochondrial Respiratory Uncoupling

ATP synthesis depends on the utilization the proton gradient by ATP synthase [8] consistent with the widely accepted chemiosmotic theory stating that ATP synthesis depends on the proton-motive force (pmf) [32]. Respiratory coupling would then be best defined as the relationship between electron transport and the pmf. Uncoupling of oxidative phosphorylation would result from any factor (other than ATP synthesis), leading to disruption of the proton gradient through transfer of protons opposite to the direction of proton pumping (proton leak). This is as classically induced by uncoupling protein 1 in brown adipose tissue as the molecule discharges the inner membrane utilizing the energy for heat production. Beyond the catalytic activity of uncoupling proteins, the proton leak can result from nonspecific leaks or through chemical induction by compounds such as dinitrophenol or FCCP.

4.1 Assessment of Uncoupling Activity

Respiratory uncoupling has been assessed in various ways although most methods are not exact in that they do not reflect the true relationship between electron transport and pmf. For example, one can simply measure ATP production and respiration and consider uncoupling as any decrease in the ratio of ATP to oxygen consumption. This information can be useful but does not consider the possibility that less ATP production might be due to impaired ATP synthase activity or transfer of phosphate or ADP into the mitochondrial matrix; that is, processes that reduce ATP production independent of the pmf [8]. Respiratory uncoupling has also been inferred in even less specific fashion simply as a decrease in membrane potential or an increase in oxygen consumption presumably to compensate for less effective ATP production.

A more direct way to assess respiratory uncoupling is to directly measure the relationship between electron transport (driving force for proton pumping) and mitochondrial membrane potential. This is discussed below.

4.2 Simultaneous Determination of Respiration and $\Delta\Psi$ and Respiratory Uncoupling

By inserting a potential sensitive TPP electrode along with a Clarke type oxygen sensor into a small volume (usually 1 ml or less) respiratory chamber, is possible to simultaneously monitor respiration and membrane potential during incubation of isolated mitochondria. Respiration on specific substrates generates stoichiometric proton pumping allowing determination of the ratio of proton transfer to membrane potential, in other words, proton conductance. In fact, the kinetic relationship between proton transfer and $\Delta\Psi$ has been described as an optimal measure of the proton leak [8, 33]. This relationship can be conveniently assessed by incubating isolated mitochondria in the absence of added ADP (so that ATP is not synthesized) on the complex II substrate, succinate (plus rotenone to block reverse transport to complex I) in the presence of titrated amounts of malonate to inhibit succinate dehydrogenase. This creates a range of values for respiration and potential. Since proton transfer under these conditions follow a 6:1 stoichiometry with respiration, the kinetic relationship of proton transfer to $\Delta\Psi$ can be plotted. To do this correctly, nigericin should be added to convert the pH gradient to a potassium gradient so that all contribution to the pmf is measured. It is best to also add oligomycin to block any ATP synthesis that might

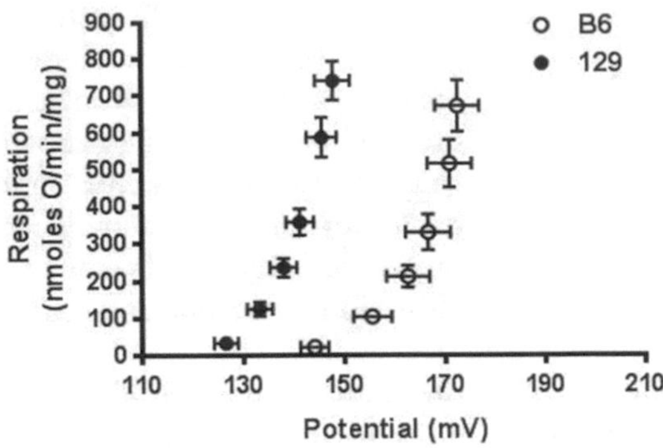

Fig. 6 Respiratory uncoupling demonstrated by simultaneous determination of oxygen consumption and membrane potential in mitochondria isolated from intrascapular brown adipose tissue of obesity prone, C57BL/6 (B6), and obesity resistant, 129S6/SvEvTac (129), mice. Mitochondria were fueled with succinate plus varying amounts of malonate to inhibit succinate dehydrogenase and, thereby, titrate membrane potential downward. Kinetic plots (proton leak kinetics) depict oxygen consumed as a function of membrane potential. Mitochondria were incubated in 120 mM KCl, 5 mM KH2P04, 2 mM $MgCl_2$, 1 mM EGTA, 3 mM HEPES, pH 7.2 with 0.3% fatty acid-free BSA, 2 μM oligomycin to inhibit ATP synthase, 5 μM rotenone to inhibit electron entry at complex-I, and 0.1 μM nigericin to abolish the pH gradient across the mitochondrial membrane. Under these conditions hydrogen transfer occurs with 6:1 stoichiometry to respiration, so data points represent hydrogen transfer as a function of membrane potential (i.e., proton conductance at each data point). Data show the shift upward and to the left for the 129 mouse mitochondria compared to mitochondria isolated from the B6 mice, indicating relatively greater respiratory uncoupling by the mitochondria of the obesity resistant strain. Data represent mean ± SE, $n = 6$ mice per group

result from contamination of the mitochondrial preparation with endogenous ADP. Generating these plots (H^+ transfer versus pmf) for mitochondria exposed to differing physiologic states allows one to assess the effect of these states on uncoupling activity. This is illustrated in Fig. 6, where we used this methodology to compare respiratory coupling by uncoupling protein 1 in brown adipose tissue between obesity prone and obesity resistant strains of mice [34].

5 Mitochondrial ATP Production

5.1 Conventional Methods for Quantification of ATP Production

There are several conventional methods to measure ATP production, each with advantages and limitations. Fluorescent and bioluminescent measurements available in commercial kit form are sensitive but not specific and prone to background interference,

For example, coupling ADP conversion to ATP with luciferase bioluminescence as carried out using commercially available kits is easily accomplished and sensitive but can be prone to variations in light emission [35] leading to loss of specificity. ^{31}P NMR is specific but not sensitive and requires long acquisition times unless large numbers of mitochondria are used. HPLC has been used where precise data is needed, but is cumbersome. ATP:O ratios can be obtained by incubating mitochondrial in a respiratory chamber and determining the amount of oxygen consumed while known amounts of ADP are converted to ATP. This is a simple procedure requiring only measurement of respiration. However, this does not measure ATP directly, and the ratio can be altered by any condition that affects uncoupling, ATP synthase, and respiration itself.

5.2 Novel Method for Assessing ATP Production by Isolated Mitochondria or Permeabilized Cells Using ADP Recycling by HK and 2DOG

We recently published [16] a novel nuclear magnetic resonance (NMR)-based assay that can be used to measure ATP production from isolated mitochondria or permeabilized cells with high efficiency in a manner that alleviates the above problems. The technique takes advantage of the ADP recycling methodology described in section 2.2.2. In this technique (Fig. 3), ADP is added at a chosen concentration and clamped using excess 2DOG and hexokinase to recycle the nucleotide as discussed in section I.B.2. We then use two-dimensional NMR spectroscopy to quantify ATP formation as the conversion of 2DOG to 2DOG-P (Fig. 7) [16].

This method has major advantages over prior assays. Based on signal to noise ratios and direct comparison [16], the 2DOGP NMR spectra are 41-fold more sensitive for ATP detection than phosphorous NMR. This enables assay in small volumes of mitochondria typically used for respiratory studies (not possible for ^{31}P NMR). The NMR signals are highly specific, unlike luciferase fluorescent assays. There is also good throughput since we can add mitochondria or permeabilized cells to multiple wells of a 96-well plate, incubate under the desired conditions, spin off the mitochondria, and save the samples in "straws" (NMR assay tubes) which are batched and sent to our NMR facility where the samples can be run overnight with automated sample changing. We can also simultaneously (with assay of ATP production) carry out fluorescent detection of other phenomena, e.g., ROS, in the same wells of the 96-well plates. These plates fit our plate reader and can be shaken during the assay between reading cycles which typically involve only 13 s of a 60 s cycle. Another powerful aspect discussed in section 2.2.2, not inherent in any other methods, is that since mitochondrial ADP is clamped, $\Delta\Psi$ is also clamped (at least as affected by changes in ADP content that would otherwise occur as ADP is consumed). Hence, one can determine the rate of ATP production at known respiration and potential and at selected ADP concentrations. Specific methodology, as we previously published [36], for implementing the HK/2DOG energy clamp and determination of the ATP production rates is included in Sect. 9.

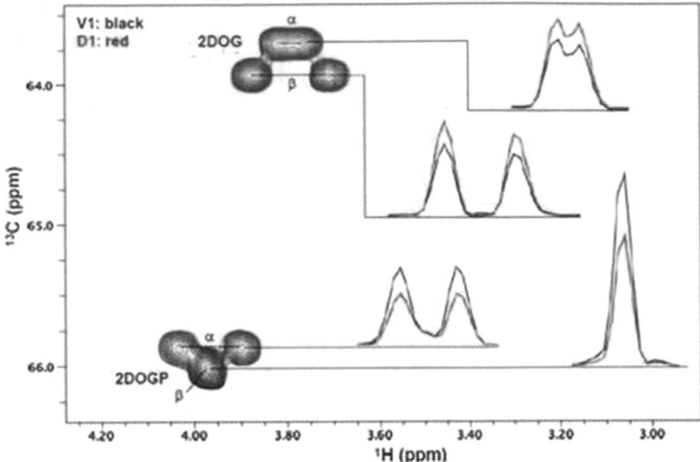

Fig. 7 ATP production by mitochondria of a control and diabetic rat quantified as 2-deoxyglucose phosphate (2DOGP) using the hexokinase/2-deoxyglucose energy clamp. Gastrocnemius mitochondria were isolated from a rat made diabetic with the β-cell toxin, streptozotocin, and from a control vehicle-treated rat. The data shown represent two dimensional NMR (^{13}C and ^{1}H chemical shifts) spectroscopic signals depicting 2DOG and 2DOGP from mitochondria of the diabetic (*red*) and control (*black*) mice. *Red* and *black* colored topographical spectra were overlaid and ^{1}H signals integrated at the selected ^{13}C chemical shift values. Thus, we can use this method with very high sensitivity and specificity. The reduction in conversion of 2DOG to 2DOGP by the diabetic mitochondria is clearly evident

6 ROS Production

6.1 ROS Production by Isolated Mitochondria

Mitochondria generate ROS mostly in the form of superoxide from multiple sites releasing the radical either to the matrix or externally to the inner membrane space and outside the organelles [37]. Complex I release to the matrix and complex III release to both the matrix and external space have received much attention. However, other sites include mitochondrial GAPDH, the electron transferring flavoprotein, pyruvate dehydrogenase, α-ketoglutarate dehydrogenase, and the iron–sulfur centers in the aconitase protein where conversion of superoxide to the hydroxyl radical results in inactivation of the enzyme. ROS production is maximal during state 4 respiration, wherein radical formation is enhanced as electron flow leads to high potential unmitigated by ATP generation [38]. This condition is easily reproduced in isolated mitochondria simply by not adding ADP or by using oligomycin to block ATP synthase. However, it is important to remember that mitochondria in vivo are rarely, if ever, in this unmitigated state.

When assessing mitochondrial superoxide production it is important to consider that matrix superoxide is rapidly converted to H_2O_2 by manganese superoxide dismutase (MnSOD). Matrix H_2O_2 is then released outside as it penetrates mitochondrial membranes. The most common way to assess mitochondrial superoxide in isolated mitochondria is to use a fluorescent technique to indirectly measure the radical as H_2O_2 after conversion. Some recommend adding exogenous MnSOD to the incubation medium to completely convert superoxide to H_2O_2 and to include the portion released externally. For isolated mitochondria, we and others have used 10-acety l-3,7-dihydroxyphenoxazine (DHPA or Amplex Red) which appears fairly specific for this radical when released from the isolated organelles. Using a plate reader or other instrument, one can then follow the formation of H_2O_2 over time under selected incubation conditions. To check for specificity, it is advisable to measure H_2O_2 production with and without added catalase. It is also important to test any substances added to the assay medium to be sure they do not themselves induce fluorescence of the probe.

A highly specific but somewhat cumbersome way to assess oxygen radical formation by isolated mitochondria is through EPR spectroscopy. This can be done by detecting specific signals resulting from free radical interactions with added compounds as spin traps [5, 39]. We have used, the spin trap , 5,5-dimethyl-L-pyrroline-*N*-oxide (DMPO) to detect superoxide generating a specific signal representing either this compound or the hydroxy radical [5]. These two possibilities can be separated by adding MnSOD, which should abolish the signal generated by superoxide. Past work [5] in our laboratory has shown that the DMPO signal represents superoxide released externally from mitochondria, as opposed to matrix superoxide which is converted to H_2O_2.

In theory, it is possible to assess complex III superoxide released to the cy*t*oplasmic side of isolated mitochondria simply by measuring H_2O_2 production (for example, as DHPA fluorescence) in the presence and absence of added SOD. SOD should increase fluorescence to the extent that it would then include the contribution of externally released superoxide to the H_2O_2 pool. Superoxide production has been effectively assessed in this way in studies of the topology of muscle, heart, and liver mitochondria [40], although that required mathematical correction for fluorescent interference.

6.2 Mitochondrial ROS Production in Intact Cells

Several studies have measured intact cell total ROS production as H_2O_2 using fluorescent probes such as carboxy dichlorodihydrofluorescein with more or less attention to radical specificity. However, most intact cell studies do not separate mitochondrial from cytoplasmic ROS. Some specificity for intact cell mitochondrial superoxide, as opposed to cytoplasmic, can be detected using mitochondrial targeted hydroethidine or "MitoSOX." MitoSOX is a hydroethidine derivative conjugated to the cation triphenylphosphonium resulting in potential dependent accumulation of the probe in the mitochondrial matrix. The accumulation in the matrix is very large since cationic triphenylphosphonium conjugated molecules accumulate many fold [41]. The difference in fluorescence between untargeted DHE and MitoSOX may provide a semiquantitative index of relative cytoplasmic and mitochondrial superoxide. One concern is that MitoSOX could undergo oxidation in the cytoplasm, which is difficult to ascertain. Since DHE and MitoSOX do not measure H_2O_2, treatment with a SOD mimetic should decrease fluorescence and may serve as a means of validation that superoxide is being measured. Another, important consideration with respect to mitochondrial-targeted DHE is that the probe is dependent on mitochondrial membrane potential to enter the organelles. Resolution of this requires that potential be monitored and an appropriate correction be applied. Although difficult, this has been accomplished using tetramethylrhodamine methyl ester (TMRM) to measure fluorescence in cerebellar granule neurons [19]. Since DHE has been criticized as nonspecific, some advocate analysis of the oxidation products by high pressure liquid chromatography (HPLC) to document specificity for superoxide as opposed to H_2O_2 [42]. In using rhodamine derivatives like TMRM, attention also has to be paid to the capacity for these compounds themselves to be a source of ROS [43].

6.3 Oxidative Damage

The term oxidative stress has been applied to both ongoing ROS production and oxidative damage. However, clearly these are different phenomena and need to be separately assessed. Distinct markers can be used to assess oxidative damage to proteins, lipids, and DNA. Further, chronic radical production is compensated by a variety of enzymatic and other mechanisms which can be assessed as evidence of oxidative stress.

A common way to detect oxidative damage to DNA is to measure 8-hydroxy-2′deoxyguanosine (8-OHdG), a compound formed by oxidation of deoxyguanosine in blood or urine, which can be analyzed by various analytical methods including HPLC, gas chromatography–mass spectrometry, and ELISA [44]. Unstable lipid peroxides derived from polyunsaturated fatty acids break down to several compounds amenable to biochemical assay. For example, isoprostanes are often measured as the marker, 8-isoprostane, formed by peroxidation of arachidonic acid [45]. Isoprostanes have adverse vascular effects including mitogenesis and altered vascular reactivity [45]. Other markers of lipid peroxidation are alkanals which can be measured as 4-hydroxy-2-nonenal (4-HNE), malondialdehyde (MDA), and acrolein [44].

Oxidative damage can also modify amino acids and, therefore, change structure and function and lead to cross-linking or protein breakdown [44]. Peroxynitrite, which results from oxygen radical interaction with nitric oxide, can cause nitration of tyrosine resulting in a formation of the marker compound, nitrotyrosine [46].

Certain methods can be used to measure oxidative damage specifically within mitochondria. One method is to determine the activity of the aconitase enzyme, a protein which is highly sensitive to oxidative damage [47]. Mitochondrial protein can also be evaluated for 4-HNE protein adducts by immunoblotting antibody [7], although specificity can be questioned.

7 Mitochondrial Calcium Dynamics

Several studies of isolated mitochondria have been carried out in the presence of calcium binding agents such as EGTA to avoid calcium interactions. However, calcium is a critical factor for in vivo regulation of mitochondrial function [48]. When studied in isolated mitochondria, low (nanomolar) concentration of calcium increase respiration and ATP production [48]; whereas higher concentrations lead to opening of the mitochondrial permeability transition pore reducing membrane potential [49, 50]. In fact, when present in high concentrations calcium will lead to irreversible opening of the pore with loss of potential and autophagy of the organelles (mitophagy). Figure 8 depicts data generated in our laboratory showing the effect of calcium on ATP production by mouse hind limb skeletal muscle mitochondria. We noted that we were unable to observe calcium-induced increases in respiration and ATP production unless mitochondria were gradient-purified beyond differential centrifugation and calcium depleted before study. Moreover, in order to properly control the ionized calcium concentration to which isolated mitochondria are exposed, it is critical to buffer calcium with the appropriate ratio of added calcium salt and EGTA or other calcium binding compound [51, 52].

Of course, calcium concentrations in intact cells are regulated in far different fashion, so it is difficult to actually be sure that experiments as depicted in Fig. 8 represent the true physiologic effects of calcium. Within cells, calcium is sequestered within the endoplasmic (ER) or sarcoplasmic reticulum (SR) maintained at concentrations estimated at 250–600 μM as opposed to cytoplasmic concentrations roughly 100 nM [53]. However, mitochondria are intermittently exposed to large concentration of free calcium within confined contact points between the ER or SR and mitochondrial membranes known as mitochondrial associated membranes (MAM) [53]. Within the MAMs, calcium release occurs in

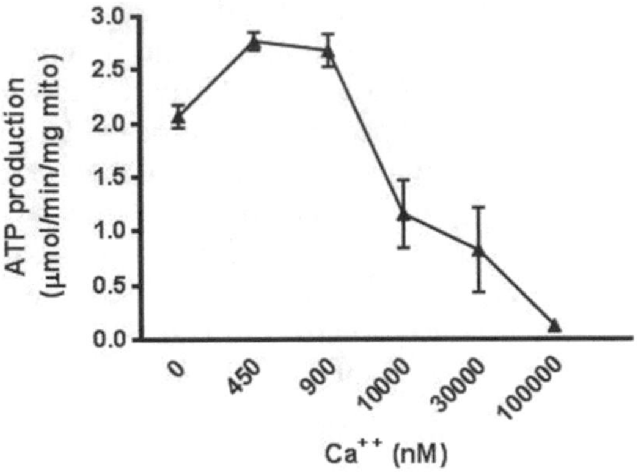

Fig. 8 Effect of calcium on ATP production by mitochondria isolated from mouse hind limb skeletal muscle. Mitochondria we incubated at 37 °C in respiratory buffer consisting of 99 mM KCl, 10 mM NaCl, 5 mM Na_2HPO_4, 2 mM $MgCl_2$, 6 mM KOH, 10 mM HEPES pH 7.2, 0.2% BSA plus ADP clamp conditions of 20uM ADP, 5 U/mL hexokinase, and 5 mM ^{13}C 2DOG. Mitochondria were incubated in wells of a 96-well plate with shaking and exposed to free calcium concentrations as indicated. The assay was performed using prior Ca^{2+}-depleted mitochondria. Free calcium concentrations were set by mixing proper portions of Ca^{2+} stock solutions (A: containing 1.0 mM EGTA only; B: containing 1.0 mM EGTA and 1.0 mM $CaCl_2$). Free Ca^{2+} was calculated using the web tool: http:/www.stanford.edu/~cpatton/CaMgATPEGTA-NIST.htm. Data represent mean ± SE, $n = 4$

discrete calcium "puffs" or "sparks" leading to intermittent localized high concentrations which are taken up by mitochondria through the recently described [54] calcium uniporter. This results in high levels of free calcium within the mitochondrial matrix that reach a peak and rapidly decline. Such calcium "puffs" or "sparks" can be observed using appropriate cellular indicators and microcopy with fast frame rates [55]. The decline follows as calcium is quickly extruded through sodium or proton dependent efflux channels, most prominently through the Na^+–Ca^{2+} exchanger [56].

8 Other Considerations

It is important to note that factors apart from the above impact mitochondrial function. Mitochondria are dynamic undergoing regulated fission and fusion and localize within certain compartments depending on cell type. Mitochondrial function is also highly dependent upon biogenesis and autophagy (mitophagy) which are regulated in accord with energetic needs. These aspects are beyond the current scope but may be found in other chapters within this book.

9 Specific Methodology [36]: Hexokinase/2DOG Energy Clamp and Determination of ATP Production Rates

9.1 Materials

9.1.1 Isolation of Mitochondria (See Note 1)

1. Isolation medium: 0.25 M sucrose, 5 mM HEPES (pH 7.2), 0.1 mM EDTA, 0.1% BSA (fatty acid-free).
2. Purification medium: 30% v/v Percoll®. Dilute 3 parts Percoll® with 7 parts of isolation medium. 2.4 ml of Percoll® + 5.6 ml of isolation medium = 8 ml, sufficient for two centrifuge tubes. Keep on ice.
3. Beckman XL-80 ultracentrifuge or similar instrument, precooled to 4 °C, SW60 swinging bucket rotor with caps and greased O-ring seals, polyallomer centrifuge tubes ~ 4.2 ml max capacity per tube.

9.1.2 Incubation

1. Respiration medium (of choice): One example is 105 mM KCl, 10 mM NaCl, 5 mM KH_2PO_4, 2 mM $MgCl_2$, 10 mM HEPES pH 7.2, 1 mM EGTA, 0.2% defatted BSA.
2. Microplate with 96-wells (for example, a Costar #3792 black round bottom plate).
3. Reagents: Hexokinase (HK), 2-deoxyglucose (2DOG).

9.1.3 Sample Preparation for NMR Spectroscopy

1. Sample dilution buffer: 120 mM KCl, 5 mM KH_2PO_4, 2 mM $MgCl_2$, pH 7.2.
2. Reagents: deuterium oxide (D_2O).
3. Standard 7 in. (length) × 5 mm (outer diameter) NMR tubes.

9.1.4 NMR Spectroscopy

1. NMR spectrometer equipped with a dual or triple resonance probe and capable of acquiring ^{1}H and $^{1}H/^{13}C$ HSQC NMR spectra.
2. NMR spectrometer equipped with an auto sample changer, thus capable of continuous data acquisition of multiple samples, e.g., 60 samples.

9.2 Methods

9.2.1 Isolation of Mitochondria from Tissue (See Note 2)

1. Carry out all procedures on ice or at 4 °C.
2. Rinse tissue in isolation medium.
3. Homogenize up to 1 g of tissue in 10–15 mL of isolation medium using a Potter-Elvehjem type tissue grinder in an ice bucket. Fibrous tissues should be minced with scissors to aid tissue disruption. The Teflon pestle is mounted on a drill set to approximately 300 rpm. Four to six passes are typically required. Optionally, a subsequent pass of the homogenate through a ground glass style homogenizer can increase the mitochondrial yield from fibrous tissues.
4. Centrifuge homogenate at 500 × *g* for 10 min (low-speed spin) (*see* **Note 3**).

5. Transfer supernatant to Sorvall type tube (Oakridge screw cap). Discard the pellet and centrifuge at 10,000 × *g* for 10 min (high-speed spin). Discard supernatant.
6. Wash mitochondrial pellet with isolation medium without BSA.
7. Resuspend the final pellet at ~50% v/v in isolation medium without BSA.
8. Add 3.8 ml of 30% Percoll® solution to each SW60 polyallomer tube on ice.
9. Resuspend mitochondrial crude prep pellets in 0.1 mL of isolation medium containing BSA.
10. Lay the mitochondria on top of the Percoll® solution and insert the tubes into the buckets.
11. Use a balance to precisely equalize the mass of the buckets + tubes + lids, adding isolation medium containing BSA to adjust the mass. Ensure that the contents of the tubes are within 3 mm of the top of the tubes.
12. Hang the buckets on the precooled SW60 rotor at 4 °C and spin for 30 min at 30,000 rpm (~90,000 × *g*).
13. The pure mitochondria band appears near the bottom of the tube, just above a clear and dense mass of Percoll®. Remove all contaminating fractions above the mitochondria band with a pipet.
14. Transfer the mitochondria band to a 1.5 mL centrifuge tube.
15. Add 1 mL of isolation medium without BSA. Spin in a microfuge at 8000 × *g* for 5 min at 4 °C. Remove the supernatant. Resuspend the pellet in 1 mL of isolation medium without BSA and spin in a microfuge at 8000 × *g* for 5 min at 4 °C.
16. Resuspend the final washed pellet in BSA-free isolation medium and keep on ice.

9.2.2 Assay Incubation

1. Warm a 96-well microplate to 37 °C for 10 min.
2. Preload all reagents (before adding mitochondria) to wells with 1.2× respiration medium (upon subsequent addition of mitochondria the medium will be 1×) in a total volume of 50 μL. For rat hind limb muscle mitochondria [16], we recommend the following final assay concentrations: 5 mM succinate + 5 mM glutamate + 1 mM malate (or other mitochondrial fuel selection and concentrations as desired), 5–10 U/mL hexokinase (HK) (*see* **Note 4**), 5 mM 2DOG (*see* **Note 4**), [6-^{13}C]2DOG for 2D NMR detection or unlabeled 2DOG for 1D NMR detection (*see* below), and ADP at desired concentration (up to 100 μM).
3. Wells containing no added substrate should be present on the plate to serve as background control.

4. Two to three wells should be included in the plate to serve as positive controls for ATP induced conversion of 2DOG to 2DOGP. These wells have the same final total volume of 60 μL as the others, but only contain 2DOG or [6-^{13}C]2DOG (depending on whether 1D or 2D NMR methods are used for measurement, respectively, *see* below), hexokinase, and ATP in 1× respiration medium. Use of twofold molar excess of ATP with respect to 2DOG is to ensure full conversion of 2DOG to 2DOGP.
5. To start the assay, add 10 μL of 6× concentrated mitochondria suspended in 1× respiration medium. The well has a total volume of 60 μL now. Final mitochondrial concentrations are typically 0.1–0.5 mg/ml. To carry out incubations, we place the multi-well plates in a FLUOstar Optima microplate reader with intermittent shaking. This enables simultaneous fluorescent assessment of other parameters if desired (*see* **Note 5**).
6. After incubation typically for 5–30 min, harvest the wells by transferring the well contents to 500 μL tubes containing 1 μL of 120 μM oligomycin. Then immediately centrifuge the tubes at 10,000 × *g* for 4 min at 4 °C.
7. Transfer the supernatants to new labeled tubes and hold them at −20 °C until NMR sample assembly.

9.2.3 Processing the Well Contents for NMR-Based ATP Assay

1. Add 0.39 mL of sample dilution buffer, 50 μL of deuterium oxide (D_2O), and 40 μL of assay well supernatant to a standard 7 in. (length) × 5 mm (outer diameter) NMR tube.
2. Deliver the prepared NMR samples (kept at 4 °C) to the NMR facility.

9.2.4 NMR Spectroscopy for Quantifying ATP Production (See Note 6)

1. In our studies, we use a Bruker Avance II 500 MHz NMR spectrometer equipped with a 5 mm TXI triple resonance non-cryoprobe operating at 37 °C. The spectrometer is also equipped with an automatic sample changer capable of holding a maximum of 60 samples. The amount of ATP produced by the mitochondria is quantified by measuring the amount of 2DOGP produced from 2DOG in the presence of hexokinase as described above.
2. For precise measurement of the amount of 2DOGP formed, duplicate or triplet control samples are prepared during the assay. The control samples have the same total volume of 60 μL as the other samples, but only contain 2DOG or [6-^{13}C]2DOG (depending on whether 1D or 2D NMR methods are used for measurement, respectively) (*see* **Note 7**), hexokinase, and 10 mM ATP in 1× respiration medium. These control samples are then subjected to the same protocol for NMR sample preparation. Due to the presence of excess ATP, these control samples

should show 2DOG fully converted to 2DOGP. Therefore, these control samples can serve as standards for quantification of 2DOGP formation within the mitochondrial samples and can also be used to check for assay reproducibility since duplicate or triplet control samples are used.

3. Load the control samples and mitochondrial samples onto the sample changer. We can run 60 samples continuously without interruption at this spectrometer.
4. Use a control sample or mitochondrial sample to lock, tune, and shim. Save the optimized shimming parameters which serve as the starting shimming setting for the subsequent automatic robot run. Also, use this sample to calibrate ^{1}H and ^{13}C channel pulse widths (^{13}C pulse calibration is needed only when [6-^{13}C]2DOG is used).
5. For Bruker Topspin software, start the automation program ICONNMR. Set up the automatic robot run by choosing 1D ^{1}H NMR experiment if unlabeled 2DOG is used or choosing 2D ^{13}C/^{1}H HSQC NMR experiment if [6-^{13}C]2DOG is used in the samples. Enter appropriate pulse program parameters such as pulse power level, pulse width, relaxation delay, number of scans, water presaturation parameters, and number of t1 increments. Also, set to tune ^{1}H channel and shim on every sample. It takes about 25 min to collect either a 1D ^{1}H spectrum or a 2D ^{13}C/^{1}H HSQC spectrum for each sample. Start the robot run.
6. Transfer the collected NMR data to another Linux computer where the acquired data are processed by using NMRPipe package [57] and analyzed using NMRView [58].
7. Using the assigned ^{1}H and ^{13}C NMR resonances of 2DOG and 2DOGP [16], measure the peak intensities in NMRView of 2DOGP in the processed control and mitochondrial samples.
8. Quantify the amount of 2DOGP present in the mitochondrial samples by comparing the peak intensity of 2DOGP of the mitochondrial sample with that of the control/standard samples. Since a known amount of 2DOG is added in the assay incubation, the amount of 2DOGP formed corresponds to the amount of reduction of 2DOG and thus can be expressed as percentage of conversion from 2DOG to 2DOGP. The amount of 2DOGP determined for the mitochondrial samples corresponds to the amount of ATP produced by the mitochondria. Representative data derived from the analysis by 2D NMR methods are shown in Fig. 7.

9.2.5 Calculation of ATP Production Rates

1. Determine the percent conversion of 2DOG to 2DOGP. From the known initial 2DOG concentration and percent conversion, determine the molar amount of 2DOGP formed which

equals to the molar amount of ATP generated. Calculate ATP production rates in the microplate assay wells based on the volume in microplate, the amount used in NMR, and incubation time.

10 Notes

1. For studies of permeabilized cells, the cytoplasm is replaced by the respiratory media enabling assessment of mitochondrial function independent of cytoplasmic events. The methods described herein are for isolated mitochondria. The methodology can also be applied to permeabilized cells grown in multi-well plates by adding reagents to the cells (after permeabilization) and following the procedures starting in section 9.2.2.
2. Crude mitochondria pellets obtained by differential centrifugation should be further purified using a self-generating Percoll® gradient. We use a published method [59] described for liver mitochondria (which we adapted for heart and skeletal muscle mitochondria) by using a centrifugal force of 90,000 × *g* to establish the gradient. Others recommend only 30,000 × *g*. In our initial attempts by using 30,000 × *g*, we did not get acceptable results as evidenced by loose, fluffy, and diffuse bands or mitochondria that stayed only at the top of the tube. When we increased the centrifugal force to 90,000 × *g*, we obtained excellent separation of mitochondria from contaminants. In this way the mitochondrial band (near the bottom of tube) is clearly separated from less-dense contaminants and broken mitochondria (upper and middle bands, respectively).
3. To increase the mitochondrial yield with fibrous tissues, save the first low-speed pellet for resuspension and regrinding using the ground glass type homogenizer. Spin the homogenate at low speed. Combine the supernatants from both low-speed spins prior to the high-speed spin.
4. Hexokinase and 2DOG should be present in excess. Trying different concentrations to determine the amount needed for maximum ATP production is advisable.
5. The procedure described is for ATP production. But it is possible to simultaneously measure other parameters such as H_2O_2 production as we have done in the past [16, 38]. For example, incubations can be carried out in a plate reader with shaking for simultaneous fluorescent detection of various parameters.
6. There are several advantages to our ATP assay. First, $\Delta\Psi$ is clamped, allowing assessment of ATP as a function of its direct driving force (i.e., $\Delta\Psi$). Second, the assay is sensitive enough to measure ATP production using small amounts of mitochondria.

We found [16] that the 1D ^{1}H NMR method is 34-fold more sensitive and the 2D $^{1}H/^{13}C$ HSQC NMR method is 41-fold more sensitive when compared to 1D ^{31}P NMR for ATP detection. The higher sensitivity of the 2D NMR method is due to the fact that the chemical shifts of the two H6 protons of the β-anomeric form of 2DOGP are degenerate, resulting in detection of one single C6/H6 HSQC cross peak with high intensity. Third, both the 1D and 2D NMR spectra are highly specific. Fourth, throughput is quite good since we add mitochondria to multiple wells of a 96-well plate, incubate, spin off the mitochondria, and then save the samples for NMR analysis. Fifth, a powerful aspect is that we can assess fluorescent signals (for example, derived from mitochondrial ROS) simultaneously with ATP quantification by NMR, since such probes are unlikely to interfere with mitochondrial ATP production or with NMR detection of 2DOGP for ATP quantification (*see* **Note 5**).

7. The 2D NMR method is preferred since it is more sensitive and has almost no background interference.

References

1. Scheffler I (1999) Mitochondrial electron transport and oxidative phosphorylation Mitochondria. Wiley-Liss, New York, NY, pp 141–245
2. Arechaga I, Ledesma A, Rial E (2001) The mitochondrial uncoupling protein UCP1: a gated pore. IUBMB Life 52(3–5):165–173
3. Wojtczak L, Zaluska H, Wroniszewska A, Wojtczak AB (1972) Assay for the intactness of the outer membrane in isolated mitochondria. Acta Biochim Pol 19(3):227–234
4. Kim C, Patel P, Gouvin LM, Brown ML, Khalil A, Henchey EM et al (2014) Comparative analysis of the mitochondrial physiology of pancreatic beta cells. Bioenergetics 3(1):110
5. O'Malley Y, Fink BD, Ross NC, Prisinzano TE, Sivitz WI (2006) Reactive oxygen and targeted antioxidant administration in endothelial cell mitochondria. J Biol Chem 281(52): 39766–39775
6. Chance B, Williams GR (1955) Respiratory enzymes in oxidative phosphorylation. III. The steady state. J Biol Chem 217(1):409–427
7. Boudina S, Sena S, Theobald H, Sheng X, Wright JJ, Hu XX et al (2007) Mitochondrial energetics in the heart in obesity-related diabetes: direct evidence for increased uncoupled respiration and activation of uncoupling proteins. Diabetes 56(10):2457–2466
8. Brand MD, Nicholls DG (2011) Assessing mitochondrial dysfunction in cells. Biochem J 437(3):297–312
9. Brawand F, Folly G, Walter P (1980) Relation between extra- and intramitochondrial ATP/ADP ratios in rat liver mitochondria. Biochim Biophys Acta 590(3):285–289
10. Walter P, Stucki JW (1970) Regulation of pyruvate carboxylase in rat liver mitochondria by adenine nucleotides and short chain fatty acids. Eur J Biochem 12(3):508–519
11. Wanders RJ, Groen AK, Van Roermund CW, Tager JM (1984) Factors determining the relative contribution of the adenine-nucleotide translocator and the ADP-regenerating system to the control of oxidative phosphorylation in isolated rat-liver mitochondria. Eur J Biochem 142(2):417–424
12. Davis EJ, Davis-van Thienen WI (1978) Control of mitochondrial metabolism by the ATP/ADP ratio. Biochem Biophys Res Commun 83(4):1260–1266
13. Davis EJ, Lumeng L, Bottoms D (1974) On the relationships between the stoichiometry of oxidative phosphorylation and the phosphorylation potential of rat liver mitochondria as functions of respiratory state. FEBS Lett 39(1):9–12
14. Davis EJ, Davis-Van Thienen WI (1984) Rate control of phosphorylation-coupled respiration

by rat liver mitochondria. Arch Biochem Biophys 233(2):573–581

15. Kuster U, Bohnensack R, Kunz W (1976) Control of oxidative phosphorylation by the extra-mitochondrial ATP/ADP ratio. Biochim Biophys Acta 440(2):391–402
16. Yu L, Fink BD, Herlein JA, Sivitz WI (2013) Mitochondrial function in diabetes: novel methodology and new insight. Diabetes 62(6): 1833–1842
17. Bai F, Fink BD, Yu L, Sivitz WI (2016) Voltage-dependent regulation of complex II energized mitochondrial oxygen flux. PLoS One 11(5): e0154982
18. Yu L, Fink BD, Herlein JA, Oltman CL, Lamping KG, Sivitz WI (2013) Dietary fat, fatty acid saturation and mitochondrial bioenergetics. J Bioenerg Biomembr 46(1):33–44
19. Jekabsons MB, Nicholls DG (2004) In situ respiration and bioenergetic status of mitochondria in primary cerebellar granule neuronal cultures exposed continuously to glutamate. J Biol Chem 279(31):32989–33000
20. Fink BD, O'Malley Y, Dake BL, Ross NC, Prisinzano TE, Sivitz WI (2009) Mitochondrial targeted coenzyme Q, superoxide, and fuel selectivity in endothelial cells. PLoS One 4(1): e4250
21. Nicholls DG, Johnson-Cadwell L, Vesce S, Jekabsons M, Yadava N, Johnson-Cadwell LI et al (2007) Bioenergetics of mitochondria in cultured neurons and their role in glutamate excitotoxicity 'Mild Uncoupling' does not decrease mitochondrial superoxide levels in cultured cerebellar granule neurons but decreases spare respiratory capacity and increases toxicity to glutamate and oxidative stress. J Neurosci Res 101(6):1619–1631
22. Ferrick DA, Neilson A, Beeson C (2008) Advances in measuring cellular bioenergetics using extracellular flux. Drug Discov Today 13(5–6):268–274
23. Fink BD, Herlein JA, Yorek MA, Fenner AM, Kerns RJ, Sivitz WI (2012) Bioenergetic effects of mitochondrial-targeted coenzyme Q analogs in endothelial cells. J Pharmacol Exp Ther 342(3):709–719
24. Wikstrom JD, Sereda SB, Stiles L, Elorza A, Allister EM, Neilson A et al (2012) A novel high-throughput assay for islet respiration reveals uncoupling of rodent and human islets. PLoS One 7(5):e33023
25. Kamo N, Muratsugu M, Hongoh R, Kobatake Y (1979) Membrane potential of mitochondria measured with an electrode sensitive to tetraphenyl phosphonium and relationship between proton electrochemical potential and phosphorylation potential in steady state. J Membr Biol 49(2):105–121
26. Fink BD, Hong YS, Mathahs MM, Scholz TD, Dillon JS, Sivitz WI (2002) UCP2-dependent proton leak in isolated mammalian mitochondria. J Biol Chem 277(6):3918–3925
27. Nobes CD, Brown GC, Olive PN, Brand MD (1990) Non-ohmic proton conductance of the mitochondrial inner membrane in hepatocytes. J Biol Chem 265(22):12903–12909
28. Perry SW, Norman JP, Barbieri J, Brown EB, Gelbard HA (2011) Mitochondrial membrane potential probes and the proton gradient: a practical usage guide. Biotechniques 50(2): 98–115
29. Nicholls DG, Ward MW (2000) Mitochondrial membrane potential and neuronal glutamate excitotoxicity: mortality and millivolts. Trends Neurosci 23(4):166–174
30. Takahashi A, Zhang Y, Centonze E, Herman B (2001) Measurement of mitochondrial pH in situ. BioTechniques 30(4):804–8, 10, 12 passim
31. Lambert AJ, Brand MD (2004) Superoxide production by NADH:ubiquinone oxidoreductase (complex I) depends on the pH gradient across the mitochondrial inner membrane. Biochem J 382(Pt 2):511–517
32. Mitchell P (1961) Coupling of phosphorylation to electron and hydrogen transfer by a chemi-osmotic type of mechanism. Nature 191:144–148
33. Brand MD, Chien LF, Ainscow EK, Rolfe DF, Porter RK (1994) The causes and functions of mitochondrial proton leak. Biochim Biophys Acta 1187(2):132–139
34. Fink BD, Herlein JA, Almind K, Cinti S, Kahn CR, Sivitz WI (2007) The mitochondrial proton leak in obesity-resistant and obesity-prone mice. Am J Physiol Regul Integr Comp Physiol 293:R1773–R1780
35. Manfredi G, Spinazzola A, Checcarelli N, Naini A (2001) Assay of mitochondrial ATP synthesis in animal cells. Methods Cell Biol 65:133–145
36. Yu L, Fink BD, Sivitz WI (2015) Simultaneous quantification of mitochondrial ATP and ROS production. Methods Mol Biol 1264:149–159
37. Brand MD (2010) The sites and topology of mitochondrial superoxide production. Exp Gerontol 45(7–8):466–472
38. Boss O, Hagen T, Lowell BB (2000) Uncoupling proteins 2 and 3: potential regulators of mitochondrial energy metabolism. Diabetes 49(2): 143–156
39. Fink BD, Reszka KJ, Herlein JA, Mathahs MM, Sivitz WI (2005) Respiratory uncoupling by UCP1 and UCP2 and superoxide generation in endothelial cell mitochondria. Am J Physiol Endocrinol Metab 288(1):E71–E79
40. St-Pierre J, Buckingham JA, Roebuck SJ, Brand MD (2002) Topology of superoxide

production from different sites in the mitochondrial electron transport chain. J Biol Chem 4277:44784–44790

41. Murphy MP (1997) Selective targeting of bioactive compounds to mitochondria. Trends Biotechnol 15(8):326–330
42. Laurindo FR, Fernandes DC, Santos CX (2008) Assessment of superoxide production and NADPH oxidase activity by HPLC analysis of dihydroethidium oxidation products. Methods Enzymol 441:237–260
43. Petrat F, Pindiur S, Kirsch M, de Groot H (2003) NAD(P)H, a primary target of 1O2 in mitochondria of intact cells. J Biol Chem 278(5):3298–3307
44. Piconi L, Quagliaro L, Ceriello A (2003) Oxidative stress in diabetes. Clin Chem Lab Med 41(9):1144–1149
45. Mezzetti A, Cipollone F, Cuccurullo F (2000) Oxidative stress and cardiovascular complications in diabetes: isoprostanes as new markers on an old paradigm. Cardiovasc Res 47(3):475–488
46. Mercuri F, Quagliaro L, Ceriello A (2000) Oxidative stress evaluation in diabetes. Diabetes Technol Ther 2(4):589–600
47. Gardner PR (1997) Superoxide-driven aconitase FE-S center cycling. Biosci Rep 17(1): 33–42
48. Glancy B, Willis WT, Chess DJ, Balaban RS (2013) Effect of calcium on the oxidative phosphorylation cascade in skeletal muscle mitochondria. Biochemistry 52(16):2793–2809
49. Siemen D, Ziemer M (2013) What is the nature of the mitochondrial permeability transition pore and what is it not? IUBMB Life 65(3): 255–262
50. Bernardi P. The mitochondrial permeability transition pore: a mystery solved? Front Physiol 2013;4:95
51. Bers DM (1982) A simple method for the accurate determination of free [Ca] in Ca-EGTA solutions. Am J Physiol 242(5):C404–C408
52. Dweck D, Reyes-Alfonso A Jr, Potter JD (2005) Expanding the range of free calcium regulation in biological solutions. Anal Biochem 347(2):303–315
53. Patergnani S, Suski JM, Agnoletto C, Bononi A, Bonora M, De Marchi E et al (2011) Calcium signaling around Mitochondria Associated Membranes (MAMs). Cell Commun Signal 9(1):19
54. De Stefani D, Raffaello A, Teardo E, Szabo I, Rizzuto R (2011) A forty-kilodalton protein of the inner membrane is the mitochondrial calcium uniporter. Nature 476(7360):336–340
55. Miyazaki K, Ross WN (2013) Ca2+ parks and puffs are generated and interact in rat hippocampal CA1 pyramidal neuron dendrites. J Neurosci 33(45):17777–17788
56. O-U J, Pan S, Sheu SS (2012) Perspectives on: SGP symposium on mitochondrial physiology and medicine: molecular identities of mitochondrial Ca2+ influx mechanism: updated passwords for accessing mitochondrial Ca2+-linked health and disease. J Gen Physiol 139(6):435–443
57. Delaglio F, Grzesiek S, Vuister GW, Zhu G, Pfeifer J, Bax A (1995) NMRPipe: a multidimensional spectral processing system based on UNIX pipes. J Biomol NMR 6(3):277–293
58. Johnson BA, Blevins RA (1994) NMR view: a computer program for the visualization and analysis of NMR data. J Biomol NMR 4(5): 603–614
59. Hovius R, Lambrechts H, Nicolay K, de Kruijff B (1990) Improved methods to isolate and subfractionate rat liver mitochondria. Lipid composition of the inner and outer membrane. Biochim Biophys Acta 1021(2):217–226

Chapter 5

Respirometry in Neurons

Liang Zhang and Eugenia Trushina

Abstract

Altered cellular energetics associated with mitochondrial dysfunction is implicated in the etiology of multiple neurodegenerative diseases emphasizing a strong need for reliable and reproducible methods of detection. The Extracellular Flux (XF) Analyzer from Seahorse Bioscience provides an outstanding opportunity for comprehensive assessment of mitochondrial function and cellular energetics by monitoring real-time aerobic respiration and glycolysis in intact, living cells. In this chapter, we provide a method to measure respiration in primary mouse cortical neurons using a Seahorse XF24 Analyzer.

Key words Neurons, Mitochondria, Bioenergetics, Respirometry, Seahorse Extracellular Flux Analyzer, Oxidative phosphorylation, Electron transport

1 Introduction

Cellular respiration is a process where biochemical energy from nutrients is converted into adenosine triphosphate (ATP) through a series of metabolic reactions. Oxidative phosphorylation (OXPHOS) takes place in the inner mitochondrial membrane, and is the most efficient metabolic pathway that produces ATP via aerobic respiration (Fig. 1) [1, 2]. During OXPHOS, electrons are transferred through the mitochondrial electron transport chain (ETC) via a series of redox reactions that are coupled to phosphorylation of ADP to ATP and oxygen consumption. The ETC (Fig. 1) includes four protein complexes composed of nicotinamide adenine dinucleotide (NADH) dehydrogenase (complex I), succinate dehydrogenase (complex II), cytochrome c oxidoreductase (complex III), and cytochrome c oxidase (complex IV) [3]. Electrons moving through ETC coupled with translocation of protons (H^+) from the matrix to the intermembrane space, establish an electrochemical gradient or proton-motive force (PMF) (Fig. 1). The energy generated from PMF is used to drive ATP synthesis in a phosphorylation reaction via ATP synthase. The PMF is regulated by the electrical potential difference established between

Stefan Strack and Yuriy M. Usachev (eds.), *Techniques to Investigate Mitochondrial Function in Neurons*, Neuromethods, vol. 123, DOI 10.1007/978-1-4939-6890-9_5, © Springer Science+Business Media LLC 2017

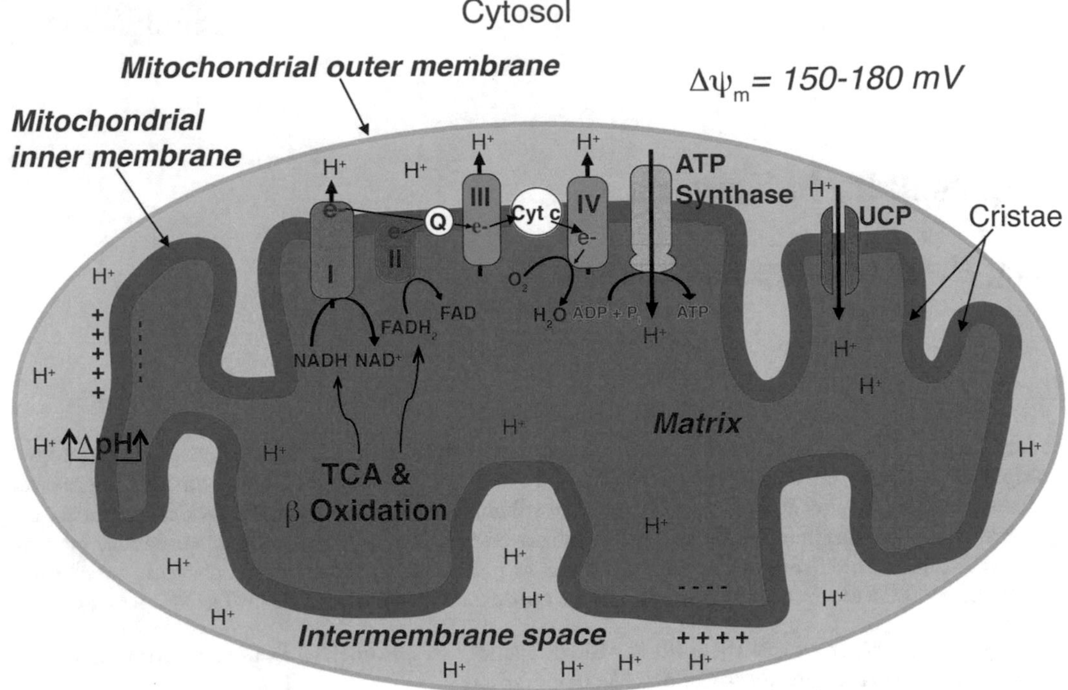

Fig. 1 Schematic representation of mitochondrion and OXPHOS machinery. Mitochondrion is a double-membrane organelle where the inner membrane forms cristae containing four ETC complexes and ATP synthase. During electron transport, the ETC complexes push protons from the matrix out to the intermembrane space creating a concentration gradient of protons (ΔpH) and an electrical potential across the membrane ($\Delta\psi_m$) creating a proton-motive force (PMF) that powers ATP synthase conversion of ADP to ATP. Uncoupling proteins (UCP) promote translocation of protons across the membrane depleting proton gradient

cytoplasm and the matrix known as mitochondrial membrane potential ($\Delta\psi_m$) and the proton gradient (ΔpH) across the inner mitochondrial membrane [2]. Under normal physiological conditions, PMF is dominated by $\Delta\psi_m$, which accounts for over 70% of total potential [4, 5]. Nevertheless, maintaining the PMF at high potential can lead to dielectric breakdown of the membrane and formation of reactive oxygen species (ROS) [6, 7]. Hence, the PMF buildup during OXPHOS is counterbalanced by ATP synthesis during which protons reenter the matrix diminishing the PMF. Under steady state conditions, the rate of electron transport equilibrates with proton translocation resulting in sufficient energy production and minimal generation of ROS [3]. Thus, the rate of oxygen consumption based on ATP synthesis is defined as coupled respiration. The effectiveness of this coupled reaction is known as coupling efficiency [8]. The OXPHOS system becomes uncoupled when protons enter the matrix without ATP synthesis (e.g., via proton leak), which stimulates electron transport and oxygen consumption [6]. During uncoupling, all energy is released as heat [2, 9]. The amount of oxygen consumed during aerobic respiration can be measured as oxygen consumption rate (OCR), often referred to as cellular respiratory rate, that could provide useful information on cell bioenergetics, metabolism, mitochondrial function in health and disease, and efficacy of therapeutic interventions.

Many drugs that target specific components of ETC could be used to dissect the machinery of OXPHOS and proton transport (Fig. 2). For example, rotenone is a commonly used inhibitor of complex I that interferes with the electron transfer between iron–sulfur clusters and ubiquinone [10]. Thenoyltrifluoroacetone (TTFA) is a potent inhibitor of complex II that binds at the quinone reduction site preventing ubiquinone binding [11]. Electron transfer from semi-quinone to complex III can be blocked with antibiotics antimycin A (AA) or myxothiazol. The AA binds to the Q_I sites of complex III located in the mitochondrial matrix [12] while myxothiazol binds at the sites located in the intermembrane space [13]. Chemicals such as cyanide or azide prevent transfer of electrons from complex IV to O_2 by reacting with the heme and inducing chemical hypoxia [14]. Oligomycin is often used to study function of ATP synthase since it inhibits the F_0 subunit that pumps protons back to the matrix [15]. Compounds that uncouple OXPHOS include 2,4-dinitrophenol (DNP) and trifluoromethoxy carbonylcyanide phenylhydrazone (FCCP) [16–18]. Consequently, application of individual or a combination of specific mitochondrial inhibitors and uncouplers with subsequent measurements of OCR

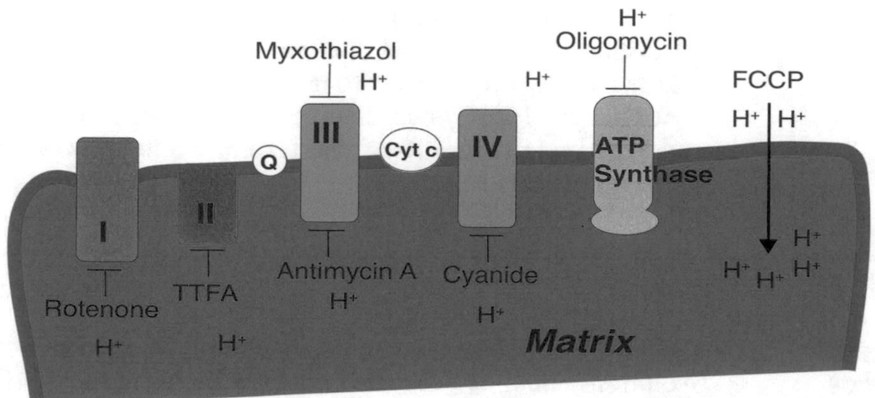

Fig. 2 Pharmacological inhibitors of OXPHOS and uncouplers. *TTFA* thenoyltrifluoroacetone, *FCCP* trifluoromethoxy carbonylcyanide phenylhydrazone

could provide key information on cellular energetics including rates of substrate utilization, activity of specific ETC complexes, rates of ATP synthesis and utilization, the extent of proton leak, coupling efficiency, spare respiratory capacity, ultimately providing a comprehensive assessment of mitochondrial function.

Traditionally, respiration was measured using isolated mitochondria. However, this procedure is tedious, time consuming, and does not account for the effect of cellular environment [19]. To overcome these obstacles, new tools have been developed to measure respiration in intact cells providing physiologically relevant data. One of the instruments to measure OCR is a sealed, stirred chamber with a built-in Clark-type oxygen electrode [20]. Despite its flexibility to measure mitochondrial function in both isolated mitochondria and intact cells, the Clark-type electrode is sensitive to electrical interference and requires frequent calibration [19, 21]. Another method to measure OCR noninvasively in intact cells includes the application of oxygen-dependent quenching phosphorescence dye. However, this technique was found to lead to an increase in intracellular ROS [19, 22]. Contrary to OCR, microphysiometer monitors cellular respiration by measuring an extracellular acidification rate (ECAR) that represents the concentration of lactic acid produced during glycolysis [23]. However, in many cases, changes in ECAR are too small to be reliably detected. Considering that cellular respiration constantly fluctuates in response to changes in the environment, a comprehensive assessment of cellular energetics should reflect contribution from both aerobic respiration (OCR) and glycolysis (ECAR). Extracellular Flux (XF) Analyzer invented by Seahorse Bioscience, which uses fluorescent oxygen and pH biosensors in a microplate assay format, allows simultaneously and noninvasively measure OCR and ECAR in intact, leaving cells [24]. Application of XF analyzer was instrumental in evaluating changes in cellular energetics associated with multiple neurodegenerative diseases including Alzheimer's, Huntington's, and Parkinson's diseases [25–27]. Previously, we described the application of a Seahorse XF24 Extracellular Flux Analyzer to measure cellular energetics in dorsal root ganglia (DRG) neurons [28]. Below, we provide a step-by-step method to measure cellular respiration in primary embryonic mouse cortical neurons.

2 Materials

2.1 Reagents and Solutions

All common reagents required for the preparation of buffers and solutions could be purchased from Sigma (St. Louis, MO) unless specified otherwise. All solutions are prepared using molecular biology grade water.

2.2 Solutions and Reagents for Culturing Neurons

1. *HEPES buffered saline (HBS)*. Dissolve 144 mM NaCl, 3 mM KCl, 10 mM HEPES, and 5.5 mM glucose in sterile deionized water; adjust pH to 7.3; filter-sterilize using a 0.22 μm filter; store up to 3 months at 4 °C.
2. *Papain*. Dilute papain (Worthington, cat. no. 3119) in warm (37 °C) HBS buffer to a final concentration of 2 mg/ml and store up to 6 months at 4 °C. On the day of neuronal plating, warm the solution to 37 °C prior to the use for tissue digestion.
3. *Borate buffer*. Dissolve 12.5 mM borax and 41 mM boric acid in sterile deionized water; adjust pH to 8.4 using 1 N NaOH. Filter-sterilize using a 0.22 μm filter, and store up to 3 months at 4 °C.
4. *Poly-*l *-ornithine (PO)*. Dissolve anhydrous PO (Sigma, cat no. P3655) to a stock concentration of 20 mg/ml using sterile deionized water. Aliquot and store at −20 °C. The day before neuronal plating, dilute an aliquot of PO to 0.5 mg/ml using borate buffer, and filter-sterilize in the sterile hood. Add 0.5 ml to each well of a Seahorse 24-well microplate. Close lids and leave plates with PO in the sterile tissue culture hood overnight. Next day, aspirate PO and rinse each well 2× with sterile deionized water. Air-dry for at least 2 h in the laminar flow hood prior to neuronal plating (**Note 1**).
5. *Heat-inactivated bovine calf serum (BCS)*. Thaw BCS serum (Fisher, cat no. SH3007203) in a 37 °C water bath. Swirl the bottle to ensure the solution is homogenous. Transfer serum to the 56 °C water bath for 30 min, swirl continuously. Chill the bottle in cold water for 5–10 min. Repeat twice. Under sterile conditions, aliquot serum into 50 ml tubes, store at −20 °C.
6. *Neuronal plating serum(+) media*. In a sterile tissue culture hood, combine 400 ml DMEM (Lonza, cat no. 12-709F), 50 ml heat-inactivated BCS, 50 ml F12/GlutaMAX (Invitrogen, cat no. 31765-035), and 1 ml pen/strep (Sigma, cat no. P0781). Mix well, aliquot into 50 ml tubes, and store at 4 °C.
7. *Neuronal feeding serum(−) media*. In a sterile tissue culture hood, combine 500 ml Neurobasal media (Invitrogen, cat no. 21103-049), 10 ml B27 supplement (Invitrogen, cat no. 17504-044), and 1 ml pen/strep (Sigma, cat no. P0781). Mix well, aliquot into 50 ml tubes, and store at 4 °C. Supplement media with 0.5 mM glutamine prior to feeding neurons. *See* **Notes 2** and **3**.
8. *Cytosine β-* d*-arabinofuranoside (AraC)*. Prepare 1 mM AraC *(Sigma, cat no. C1768)* solution in sterile deionized water. Filter-sterilize with a 0.22 μm syringe filter. Store up to 1 month at 4 °C.

2.3 Solutions and Reagents for Assaying Respirometry in Neurons

1. *Glucose.* Prepare 1 M glucose solution in deionized water; filter-sterilize with a 0.22 μm syringe filter; store as 1.5 ml aliquots at −20 °C.
2. *Sodium pyruvate.* Prepare 1 M of sodium pyruvate in sterile deionized water; filter-sterilize with a 0.22 μm syringe filter; store as 0.5 ml aliquots at −20 °C.
3. *XF assay media. This should be prepared fresh on the day of the assay.* Supplement XF base media (Seahorse Bioscience, part # 102353-100) with 25 mM glucose, 0.5 mM glutamine and 2 mM sodium pyruvate. Adjust pH to 7.4.
4. *Mitostress assay reagents (Sigma).* Prepare oligomycin, rotenone, and antimycin separately as 10 mM stock solutions in 100% DMSO. FCCP may be dissolved in either 100% DMSO or ethanol. Store as 50 μl aliquots up to a year at −20 °C.
5. *RIPA buffer.* Dissolve 0.15 M NaCl, 1 mM EDTA, 1 mM EGTA, 0.5% sodium deoxycholate, 0.1% SDS, 1% Triton X-100, and 50 mM Tris–HCl in deionized water. Adjust pH to 7.4; store up to 6 months at 4 °C. Supplement RIPA buffer with protease and phosphatase inhibitor cocktail prior to using for cell lysis.

2.4 Seahorse XF24 Analyzer Plates and Cartridges

1. *Microplates.* To assay neuronal respiration, you will need to purchase an assay kit from Seahorse Bioscience that includes microplates and disposable sensor cartridges, also known as a "FluxPak," and assay XF medium [24]. The 24-well microplates are specifically designed for the use with a Seahorse XF24 Analyzer and cannot be substituted with other plates. It is important to remember for proper cell plating that the surface area of each well of the 24-well microplate is equal to the area of a regular 96-well tissue culture dish. The Seahorse Bioscience provides three types of microplates that differ by volume and composition of plastic. Before conducting an experiment, it is important to identify what type of microplate will serve best to achieve desired experimental outcomes. The XF24 V7 microplate (Seahorse Bioscience, part # 100777-004) is made of polystyrene (PS) and will accommodate most cell types and applications [24]. However, PS is permeable to gases such as CO_2 and O_2 [29] that could lead to their accumulation and consequent release during measurements resulting in the underestimation of actual levels of OCR [30]. The XF24 V7 polyethylene terephthalate (PET) microplate (Seahorse Bioscience, part # 101037) does not absorb gases. While some reports suggest that polyethylene and polypropylene could be toxic to neurons [31], we did not detect any problems culturing neurons in any of these microplates [28]. Finally, the XF24 V28 microplate (Seahorse Bioscience, part # 100882-004) has larger volume of the mirochambers that could

be beneficial for experiments where high rates of respiration are anticipated. However, while these plates have increased dynamic range for OCR, they have reduced sensitivity for ECAR [24]. As a starting point, we recommend to utilize XF24 V7 PS microplates.

2. *Cartridges*. Along with the microplates, each FluxPak includes utility plates to hydrate sensor cartridges and XF calibrant solution. Each sensor cartridge contains one sensor probe per well with two embedded solid state fluorophores that measure levels of dissolved oxygen and protons in a microchamber created during measurements. Each sensor probe is equipped with four ports that allow a sequential injection of up to four different reagents within the assay. It is essential to load each port with equal volume of reagents to ensure that the air pressure required for injection is applied evenly across the plate during the assay. The fluorophores on the sensor probes need to be hydrated using utility plates and XF Calibrant Solution before running the assay.
3. *XF assay medium*. In order to achieve accurate OCR and ECAR measurements, it is important to use buffer-free XF assay medium. Buffering agents such as sodium bicarbonate could quench CO_2 produced by cells during the kinetic assay. Changes in pH over time could affect ECAR and proton production rate calculation. The XF assay medium may be purchased directly from the company or made using recipes provided on the company's website [24]. It is essential to choose a correct base medium and supplements such as glucose, sodium pyruvate or glutamine to obtain optimal respiratory response during the assay.

2.5 Microplate Coating Substrates

The type of coating substrate used for culturing primary neurons could significantly affect cellular morphology, differentiation, adhesion, and growth. To reliably and reproducibly measure OCR and ECAR, neurons have to be spread evenly across the bottom of the well forming a monolayer. While some of the neurons such as DRG may require a mixture of substrates [28], coating plates with poly-L-ornithine in our hands resulted in evenly distributed cortical and striatal neurons (Sect. 2. Section 2.2 item 4, Fig. 3a).

3 Methods

3.1 Culturing Cortical Neurons

First of all, make sure that your animal protocol is approved by the Institutional Animal Care and Use Committee. Next, we highly recommend that your laboratory establishes a regular schedule to culture neurons. This will ensure the consistency of cultures and reproducibility of bioenergetics data. For in-depth details on how

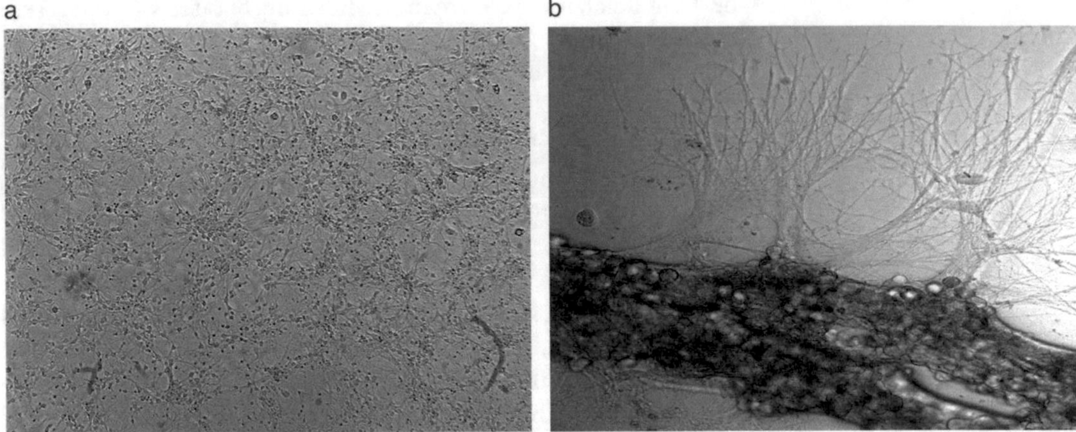

Fig. 3 Images of cortical neurons cultured in a 24-well microplate for 7 days. (**a**) Neurons are evenly distributed in the well covered with PO providing excellent conditions for bioenergetics measurements. (**b**) An example of neuronal clumps. This well should be excluded from the analysis

to isolate and culture various neuronal cells, we recommend an outstanding book by G. Banker [32]. In our laboratory, we routinely culture cortical neurons for bioenergetics measurements using the protocol described below [33].

Time-pregnant female mice on gestational day 17–18 are anesthetized with isoflurane, sacrificed by cervical dislocation, and fetuses are rapidly removed. Fetal brains are extracted and placed in sterile HBS buffer (Sect. 2.2, item 1). Cortices are dissected, minced and placed for digestion in papain solution (Sect. 2.2, item 2) for 20 min at 37 °C. Tissue is transferred to the conical tube with plating serum(+) media (Sect. 2.2, item 2), and triturated using Pasteur pipette. Cells are counted and plated in the PO-coated XF24 V7 microplate (Sect. 2.2, item 4) at 8–10 × 10^4 cell/well in 100 μl of plating serum(+) media. Four wells on the microplate (A1, B4, C3, D6) are always left empty to ensure a background correction during the assay (Fig. 4). Cells are transferred to a 5% CO_2 incubator and kept at 37 °C for an hour before the total volume in each well is brought to 500 μl (**Note 4**). Three days after plating, serum(+) media is replaced with serum(−) feeding media (Sect. 2.2, item 7) that support neurons but inhibit proliferation of non-neuronal cells. The control over proliferating cells could be further conducted by adding 1–2 μM of AraC (Sect. 2.2, item 8) every other day (**Notes 2**, **3**). Neurons should be fed every 3–4 days by replacing half of the media with fresh feeding serum(−) media. The ideal neuronal monolayer for bioenergetics measurements is presented in Fig. 3a.

3.2 Mitostress Assay

Mitochondrial stress (mitostress) assay is designed to assess key parameters of mitochondrial function by monitoring changes in OCR

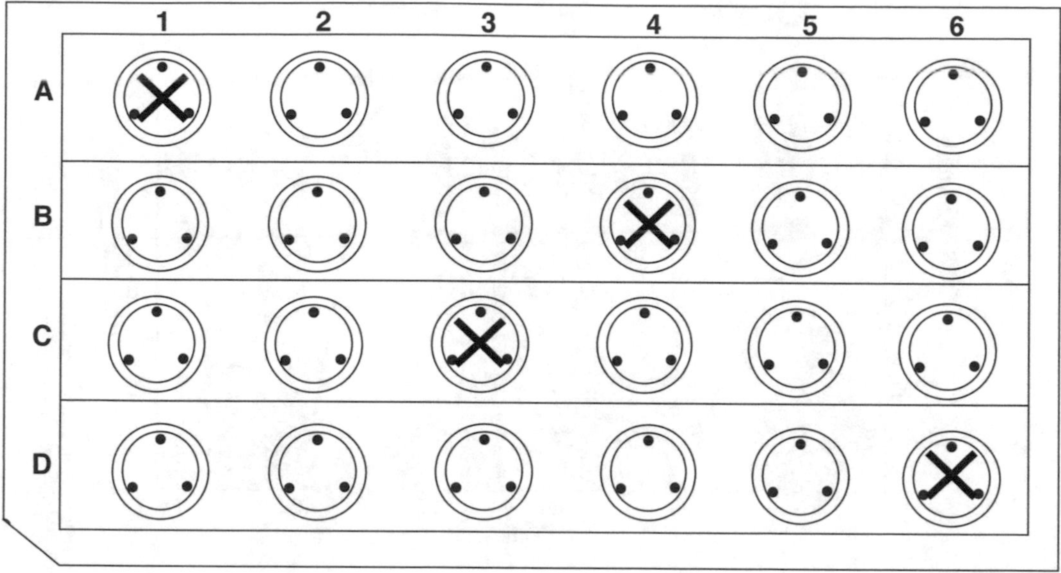

Fig. 4 An overview of the XF24 microplate. Each well of the XF24 microplate has a surface area of 0.32 cm^2 identical to that of a standard 96-well tissue culture dish. There are three molded stops on the bottom of each well for the sensor cartridge to rest upon during measurements. Four wells marked with a *cross* are always left empty for temperature control and background correction. The maximum volume of each well is 1000 μl

and ECAR in response to the addition of various ETC inhibitors and uncouplers (Fig. 5). Conventional mitostress assay includes the consecutive injections of oligomycin, FCCP, and a mixture of rotenone and antimycin A (AA) with simultaneous measurements of OCR and ECAR (Table 1). After the baseline OCR and ECAR have been established (Fig. 5i), the first injection of oligomycin, an ATP synthase inhibitor, leads to a rapid reduction in OCR due to the inhibition of electron transport (Fig. 5ii, circles). Oligomycin-induced ATP depletion in neurons leads to increased glycolysis in order to maintain energy supply. This is detected by an increase in ECAR (Fig. 5ii, squares). The next injection contains an uncoupler FCCP (carbonylcyanide-4-(trifluoromethoxy)-phenylhydrazone). FCCP promotes translocation of protons into the matrix leading to rapid increase in oxygen consumption without ATP synthesis. Consequently, both OCR and ECAR are increased (Fig. 5iii). The last injection includes a mixture of rotenone and AA. These reagents completely shut down electron transport and mitochondrial respiration resulting in a sharp reduction in OCR while ECAR remains at the high rate (Fig. 5iv). Interpretation of the data is provided in Sect. 4. This method could be modified to accommodate a particular experimental paradigm by electing various reagents and altering the order of injections.

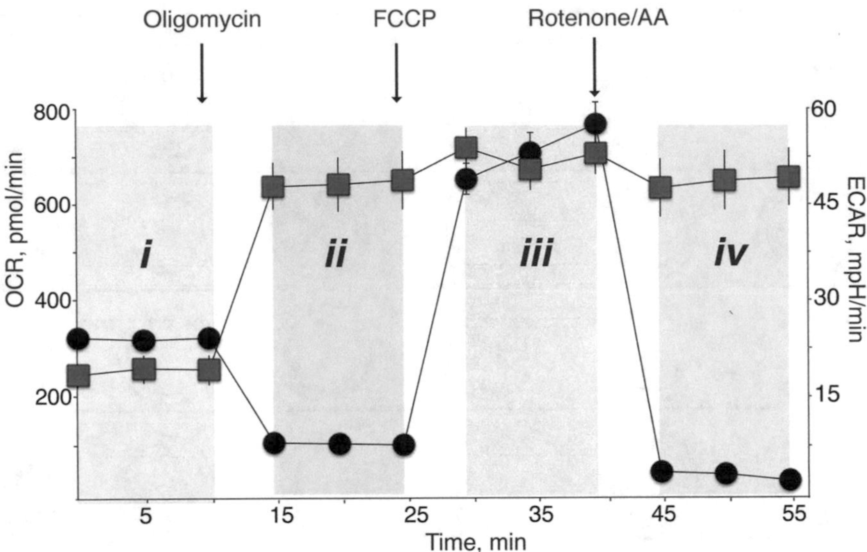

Fig. 5 Rates of oxygen consumption and glycolysis in primary neurons. The OCR (*circles*) and ECAR (*squares*) in primary cortical neurons measured within mitostress assay: (*i*) basal respiration, (*ii*) oligomycin-induced changes in respiration, (*iii*) respiratory response after an addition of FCCP; (*iv*) non-mitochondrial respiration measured after complete inhibition of electron transport by rotenone and AA

Table 1
The order of commands in mitostress assay

Command	No. of loops	Time (min)
Calibrate	–	–
Equilibrate	Auto	Auto
Mix, wait, measure	3	3, 2, 3
Inject A	–	–
Mix, wait, measure	3	3, 2, 3
Inject B	–	–
Mix, wait, measure	3	3, 2, 3
Inject C	–	–
Mix, wait, measure	3	3, 2, 3

3.3 Optimization of Experimental Conditions

The extent of cellular response in a mitostress assay could be affected by multiple parameters including culture conditions, cell density, concentration of the individual inhibitors and uncouplers, and composition of XF assay media. Therefore, it is necessary to optimize experimental conditions in preliminary experiments. First, we cannot emphasize enough how important it is to establish consistent and reproducible neuronal cultures to eliminate any possible effect of the variability in the age of the embryos on metabolic state of developing neurons. Another important parameter to consider is the ratio between neuronal and non-neuronal cells in the cultures. Since non-neuronal cells have different rate of metabolism and, contrary to neurons, are dividing cells, mixed cultures may produce various rates of OCR and ECAR compared to pure neuronal cultures. Using our protocol and ensuring that mice are time-pregnant, we are able to consistently culture primary neurons at E17 with the purity of 95% [33]. Next, it is important to establish optimum cell density to warrant that changes in OCR and ECAR are measured above the background noise and within the linear range of the detection. The relationship between different cell density and changes in OCR in conventional mitostress test is shown in Fig. 6. In this example, optimum cell density could be selected based on the greatest response to the addition of FCCP (Fig. 6, pink circles) shown as absolute rates (Fig. 6a) and as a percentage of a baseline (Fig. 6b). However, levels of OCR should not exceed 800 pmol/min, which is the detection limit of the XF24 Analyzer. We have achieved good results plating neurons at 8–10 × 10^4 cell/well. Next, it is necessary to optimize working concentrations of ETC inhibitors and uncouplers. Low concentrations could induce only partial response while high concentrations could crush the system compromising data collection. This is especially important with respect to FCCP. As shown in Fig. 7, levels of OCR dropped sharply after

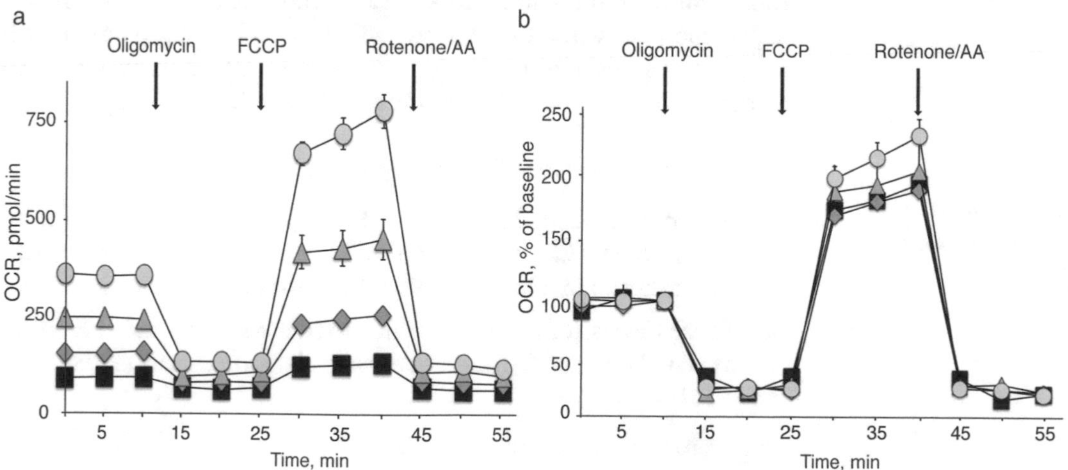

Fig. 6 Optimization of cell density. C2C12 cells (a mouse myoblast cell line) from the same cell culture flask seeded at a density of 10,000 (*black square*), 20,000 (*grey diamond*), 40,000 (*blue triangle*), or 80,000 (*pink circle*) cells per well and assessed using mitostress assay. (**a**) Absolute rates of OCR correlate with increased cell seeding density. (**b**) Results are expressed as a percentage of baseline confirming data generated in (**a**)

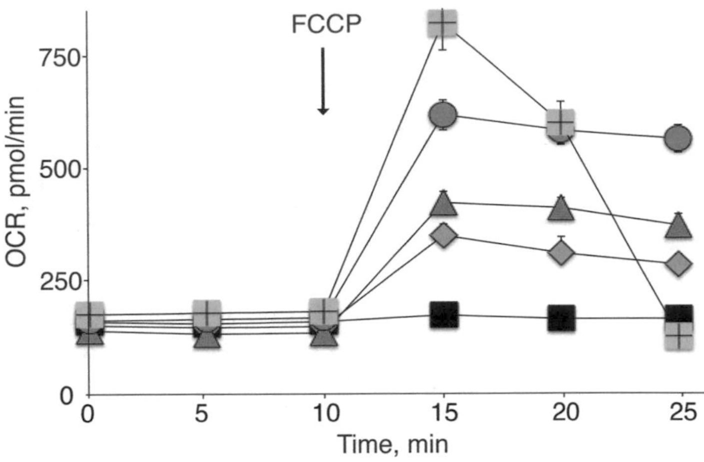

Fig. 7 Optimization of FCCP concentration. DRG neurons were plated at 80 K cell/well. Cells were assessed after the injection of uncoupler FCCP at the following concentrations: 0.1 μM (*black squares*), 0.5 μM (*gray diamonds*), 0.75 μM (*green triangles*), 1 μM (*blue circles*), and 1.5 μM (*purple squares*). Increased concentration of FCCP induced linear rise in maximal OCR. However, high concentration of FCCP (1.5 μM) caused rapid decrease of OCR during repetitive measurements

administration of 1.5 μM of FCCP to neurons (squares). In contrast, low doses of FCCP induced changes in OCR that were similar within three consecutive measurements and the maximal OCR response to FCCP increased linearly with concentration (Fig. 7). We normally use 1 μM FCCP, which produces an increase in OCR that is stable within three consecutive measurements. Finally, assay media utilized in the mitostress assay should be prepared fresh and supplemented with appropriate substrates. We supplement XF assay media with glucose, glutamine and sodium pyruvate to obtain optimal respiratory response (Section 2.3, item 3).

3.4 Protocol for Mitostress Assay in Neuronal Cultures (Fig. 8)

Culturing neurons in XF24 microplates (Day 0–Day 6)

1. Neurons are seeded at 8–10 × 10^4 cells/well in plating serum(+) media (Sect. 2.2, item 6) on PO-coated XF24 V7 microplate and cultured in a 5% CO_2, 95% humidified incubator at 37 °C.
2. Three days after plating, replace media with serum(−) feeding media. AraC may be added to control proliferation of glial cells (**Notes 2** and **3**).

Preparation for mitostress assay (Day 7)

3. The night before mitostress assay, feed neurons with fresh feeding media.
4. Set up the XF24 assay template using assay wizard according to the manufacturer's instructions. The protocol for mitostress assay utilized in our laboratory is presented in Table 1 (**Note 5**).

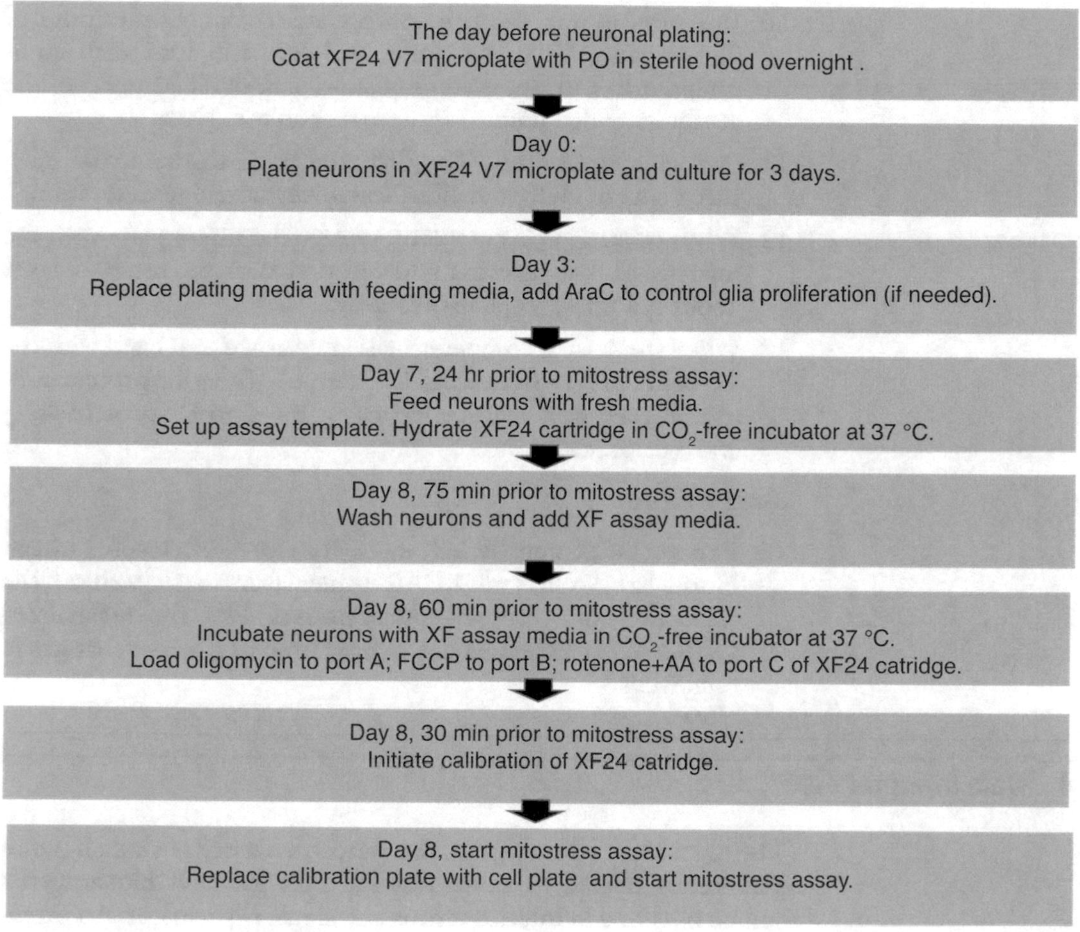

Fig. 8 Overview of the mitostress assay

5. Hydrate sensor cartridge by adding 1 ml of XF Calibrant Solution to each well of a utility plate; place the plate with the cartridge overnight in a *CO2-free* incubator maintained at 37 °C (**Note 6**).

Running mitostress assay (Day 8)

6. Prepare XF assay media as described in Sect. 2.3, item 3.
7. Gently wash neurons with 1 ml of warm, unbuffered XF assay media to remove traces of sodium bicarbonate (**Note 7**).
8. Add 675 μl of XF assay media (Table 2) to each well of the microplate that contains neurons (Fig. 4) and inspect each well under the light microscope. Mark the wells with significant clumping or detached cells (Fig. 3b).
9. Incubate neurons in a *CO2-free* incubator for 60 min at 37 °C to equilibrate assay medium temperature and pH (**Note 8**).

10. In the meantime, prepare appropriate stock concentrations of oligomycin, FCCP, rotenone, and AA, and load each compound into corresponding injection port of all 24 ports of the sensor cartridge (Sect. 2.3, item 4, Table 2, **Note 9**).
11. Start mitostress assay. The first step prior to the actual measurements of OCR and ECAR involves cartridge calibration.
12. When calibration is complete, replace the utility plate with the microplate that contains neurons and start the assay. It takes about 2.5 h to complete one assay.
13. When the assay is complete, cells are washed and lysed with the ice-cold RIPA buffer (Sect. 2.3, item 5). The total protein content in each well is estimated using Bio-Rad Dc protein assay and a plate reader.

Data analysis

14. Export OCR and ECAR data into an Excel spreadsheet. Normalize OCR and ECAR values from each well to the respective total protein concentration. Plot the normalized OCR and/or ECAR values against time to observe changes in each well.

4 Data Interpretation

The significant advantage of the mitostress assay is that all major parameters related to mitochondrial function and bioenergetics can be obtained in intact neurons in a single experiment. An example of the experiment where primary neurons were treated with metformin, a mild inhibitor of the mitochondrial complex I [34] (Fig. 9a), and a description of all major bioenergetic parameters that could be gathered (Fig. 9b) are provided below.

1. *Basal respiration rate* represents level of neuronal respiration prior to the beginning of the mitostress assay and the addition of any of the inhibitors/uncouplers (Fig. 9b, Basal). Fluctuations

Table 2
The dilution scheme and stock concentrations of the reagents injected during the mitostress assay

Port	Compound	Volume/well, μl	Injection vol., μl	Stock conc.	Final conc.	Dilution factor
Starting volume/well		675	–	–	–	
A	Oligomycin	750	75	10 μg/ml	1 μg/ml	10
B	FCCP	825	75	11 μM	1 μM	11
C	Rotenone + AA	900	75	12 μM	1 μM	12

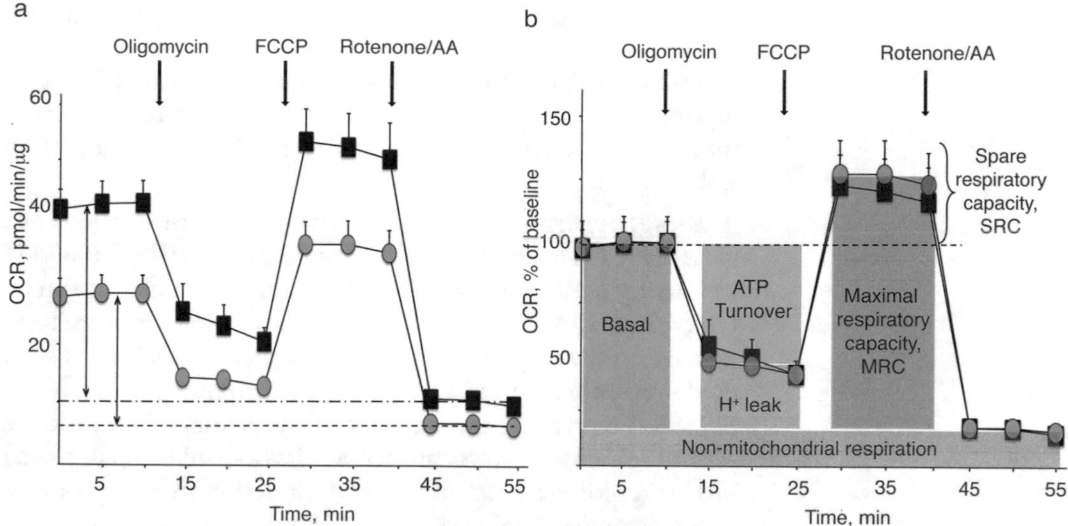

Fig. 9 Mitostress test in primary cortical neurons treated with metformin. (**a**) Mitostress test and OCR measured in wild type cortical neurons without treatment (*squares*) and after incubation with 1 μM metformin for 24 h (*circles*). Data represent actual values. Note a reduction in basal OCR indicated by arrows. (**b**) Parameters of cellular energetics obtained using mitostress assay in experiments conducted in (**a**). Data are presented as a percent of a baseline

in basal respiration rate reflect a response to changes in the environment or treatment with pharmacological compounds or toxicants. For example, treatment with 1 μM of metformin 24 h prior to measurements resulted in a substantial reduction in basal OCR in primary cortical neurons (Fig. 9a). This is in a good agreement with the known property of metformin to transiently inhibit mitochondrial complex I activity.

2. The mitochondrial *ATP turnover rate* is based on the amount of oxygen consumption linked to ATP production and represents the ability of neurons to synthesize ATP. It is determined based on a decrease in OCR in response to oligomycin compared to basal respiration rate (Fig. 9b, ATP turnover). For example, an increase in basal respiration and a decrease in ATP turnover rate in response to particular treatment indicate a failure of mitochondria to meet energy demands. It should be noted that this parameter does not represent the total amount of ATP produced in the cell [2].
3. *Proton leak*. The remaining level of respiration after the addition of oligomycin is not coupled to ATP production and indicates the rate of proton leak across the mitochondrial membrane in situ (Fig. 9, H^+ leak). This respiration is controlled mainly by uncoupling and is insensitive to the effect of oligomycin. A large increase in proton leak and basal respiration rate in response to treatment may indicate decreased

mitochondrial efficiency in meeting increased energy demands. On contrary, decrease in proton leak is a measure of better coupling efficiency that could be used to monitor efficacy of experimental therapeutics. In case of metformin, at concentration tested, there were no changes in ATP turnover or proton leak (Fig. 9b).

4. *Maximal respiratory capacity (MRC) and spare respiratory capacity (SRC)*. Addition of uncouplers such as FCCP allows estimating MRC and SRC, the indicators of cellular energetic limits (Fig. 9, MRC, SRC). Reduction in MRC may indicate damage to the respiratory chain. However, other factors such as altered substrate oxidation may contribute to lower MRC [2]. SRC represents the capacity of mitochondria to produce ATP in response to stress or increased workload [35]. Reduced SRC can indicate mitochondrial dysfunction that may not be apparent under basal conditions. Alternatively, cells with high SRC have a greater ability to sustain stress, which could be used as a measure of improved mitochondrial function and a marker of therapeutic efficacy. As could be seen from Fig. 9b, metformin does not alter MRC but does not significantly increase SRC either. Thus, in this experiment, mild reduction of complex I activity does not affect other parameters of brain energetics.
5. *Non-mitochondrial respiration* represents aerobic respiration of the cell independent from mitochondria. It is measured after an addition of rotenone/AA (Fig. 9b). The non-mitochondrial OCR most often remains unchanged [2] and needs to be subtracted from all other rates for accurate estimation of mitochondrial-related respiration.
6. *Mitochondrial coupling efficiency* is the ratio between ATP turnover and basal respiration. This parameter normally changes with energy demand and experimental conditions but is usually between 70% and 90% in intact cells [36–39]. A drop in coupling efficiency indicates increased proton leak through uncoupling where energy is dissipated as heat.
7. The *respiratory control ratio (RCR)* is the ratio between MRC and proton leak, which indicates the tightness of coupling between respiration and ATP synthesis. A high RCR implies that mitochondria have greater ability to respond quickly to high-energy demands despite idling at low rate of respiration. However, similar to coupling efficiency, the RCR value constantly changes according to cellular environment and does not have a specific range of reference values.
8. *In conclusion*, application of a Seahorse XF24 Extracellular Flux Analyzer allows estimation of energetics in primary intact neurons from wild-type and genetically modified mice

aiding deeper understanding of the disease mechanisms. Data obtained in the mitostress assay described in this chapter could be used to monitor efficacy of therapeutic approaches in respect to mitochondrial function and cellular energetics.

5 Notes

1. Neurons will not adhere to the plastic tissue culture dish without supporting substrate such as PO. This is an essential step to ensure the formation of the monolayer and good spreading of neurons in a well. This procedure needs to be conducted in a sterile tissue culture hood. We routinely use PO to obtain robust and well-dispersed cultures of cortical and striatal neurons. However, other substrates including poly-L-lysine or polyethyleneimine could be used.
2. We plate neurons and keep them for 3 days in serum(+) media to establish healthy cultures. However, serum(+) media also support survival and proliferation of glial cells. Since respiration in neuronal and glial cells differ significantly, it is essential to ensure glia-free cultures. Substitution of serum(+) plating media with Neurobasal serum(−) media promotes neuronal survival but inhibits proliferation of dividing supporting cells (also *see* **Note 3**).
3. If cells are extracted from the embryos at gestational day E17, neuronal cultures usually do not have many glial cells and do not require AraC treatment. However, if neurons were plated from the older embryos (E18–19), the amount of glia on the plate may be substantial requiring AraC treatment.
4. We found that plating in the low volume for an hour ensures that neurons spread evenly and adhere well across the bottom of the plate (Fig. 3a).
5. During mixing, media equilibrate with atmosphere oxygen, which allows neurons to recover from previous measurements. If cells exhibit high aerobic respiration and need longer time for recovery, the length of time for mixing may be increased.
6. If experiment cannot be carried out on the next day after the cartridge was equilibrated, it could be wrapped in Parafilm to prevent drying out and stored at 4 °C for up to 72 h.
7. When adding or removing media to/from the wells with cultured neurons, we recommend tilting the dish and adding media along the wall of the well to avoid any potential lifting of the cells. For the same reason, we strongly recommend using a transferring pipette to collect medium from the side of the well instead of a vacuum device.

8. In cases where treatment or genetic manipulation leads to high respiration and fast acidification of XF assay media affecting ECAR measurements, it is advisable to reduce incubation time to ~30 min.
9. Note that stock concentration of each compound is different to account for the fact that each injection increases the volume in the well. Use XF assay media to dilute compound stocks.

Acknowledgment

We thank Dr. L. Storjohann for providing data for Fig. 6. This research was supported by grant from NIEHS R01ES020715 (to E.T.). Its contents are solely the responsibility of the authors and do not necessarily represent the official views of the NIH.

References

1. Vacanti NM, Divakaruni AS, Green CR, Parker SJ, Henry RR, Ciaraldi TP, Murphy AN, Metallo CM (2014) Regulation of substrate utilization by the mitochondrial pyruvate carrier. Mol Cell 56:425–435
2. Brand MD, Nicholls DG (2011) Assessing mitochondrial dysfunction in cells. Biochem J 435:297–312
3. Huttemann M, Lee I, Pecinova A, Pecina P, Przyklenk K, Doan JW (2008) Regulation of oxidative phosphorylation, the mitochondrial membrane potential, and their role in human disease. J Bioenerg Biomembr 40:445–456
4. Mitchell P, Moyle J (1969) Estimation of membrane potential and pH difference across the cristae membrane of rat liver mitochondria. Eur J Biochem 7:471–484
5. Nicholls DG (1974) The influence of respiration and ATP hydrolysis on the proton-electrochemical gradient across the inner membrane of rat-liver mitochondria as determined by ion distribution. Eur J Biochem 50:305–315
6. Brand MD, Chien LF, Ainscow EK, Rolfe DFS, Porter RK (1994) The causes and functions of mitochondrial proton leak. BBA-Bioenergetics 1187:132–139
7. Rolfe DFS, Brand MD (1997) The physiological significance of mitochondrial proton leak in animal cells and tissues. Biosci Rep 17:9–16
8. Nelson DL, Cox MM (2013) Lehninger Principles of Biochemistry. W.H. Freeman and Company, New York, NY
9. Jastroch M, Divakaruni AS, Mookerjee S, Treberg JR, Brand MD (2010) Mitochondrial proton and electron leaks. Essays Biochem 47:53–67
10. Okun JG, Lummen P, Brandt U (1999) Three classes of inhibitors share a common binding domain in mitochondrial complex I (NADH: ubiquinone oxidoreductase). J Biol Chem 274:2625–2630
11. Sun F, Huo X, Zhai YJ, Wang AJ, Xu JX, Su D, Bartlam M, Rao ZH (2005) Crystal structure of mitochondrial respiratory membrane protein complex II. Cell 121:1043–1057
12. Lai B, Zhang L, Dong L-Y, Zhu Y-H, Sun F-Y, Zheng P (2005) Inhibition of Qi site of mitochondrial complex III with antimycin A decreases persistent and transient sodium currents via reactive oxygen species and protein kinase C in rat hippocampal CA1 cells. Exp Neurol 194:484–494
13. Thierbach G, Reichenbach H (1981) Myxothiazol, a new antibiotic interfering with respiration. Antimicrob Agents Chemother 19:504–507
14. Bergmann F, Keller BU (2004) Impact of mitochondrial inhibition on excitability and cytosolic Ca(2+) levels in brainstem motoneurones from mouse. J Physiol 555:45–59
15. Penefsky HS (1985) Mechanism of inhibition of mitochondrial adenosine triphosphatase by dicyclohexylcarbodiimide and oligomycin: relationship to ATP synthesis. Proc Natl Acad Sci U S A 82:1589–1593
16. Juthberg SKA, Brismar T (1997) Effect of metabolic inhibitors on membrane potential and ion conductance of rat astrocytes. Cell Mol Neurobiol 17:367–377

17. Tretter L, Chinopoulos C, AdamVizi V (1997) Enhanced depolarization-evoked calcium signal and reduced ATP/ADP ratio are unrelated events induced by oxidative stress in synaptosomes. J Neurochem 69:2529–2537
18. Loomis WF, Lipmann F (1948) Reversible inhibition of the coupling between phosphorylation and oxidation. J Biol Chem 173: 807–808
19. Quaranta M, Borisov S, Klimant I (2012) Indicators for optical oxygen sensors. Bioanal Rev 4:115–157
20. Clark LC, Wolf R, Granger D, Taylor Z (1953) Continuous recording of blood oxygen tensions by polarography. J Appl Physiol 6:189–193
21. Zhang J, Nuebel E, Wisidagama DRR, Setoguchi K, Hong JS, Van Horn CM, Imam SS, Vergnes L, Malone CS, Koehler CM, Teitell MA (2012) Measuring energy metabolism in cultured cells, including human pluripotent stem cells and differentiated cells. Nat Protoc 7(6):1068–1085. doi:10.1038/nprot.2012.1048
22. Ceroni P, Lebedev AY, Marchi E, Yuan M, Esipova TV, Bergamini G, Wilson DF, Busch TM, Vinogradov SA (2011) Evaluation of phototoxicity of dendritic porphyrin-based phosphorescent oxygen probes: an in vitro study. Photochem Photobiol Sci 10:1056–1065
23. Owicki JC, Wallace Parce J (1992) Biosensors based on the energy metabolism of living cells: the physical chemistry and cell biology of extracellular acidification. Biosens Bioelectron 7:255–272
24. http://www.seahorsebio.com/.
25. Xun Z, Rivera-Sanchez S, Ayala-Penã S, Lim J, Budworth H, Skoda EM, Robbins PD, Niedernhofer LJ, Wipf P, McMurray CT (2012) Targeting of XJB-5-131 to mitochondria suppresses oxidative DNA damage and motor decline in a mouse model of Huntington's disease. Cell Rep 2:1137–1142
26. Walls KC, Coskun P, Gallegos-Perez JL, Zadourian N, Freude K, Rasool S, Blurton-Jones M, Green KN, LaFerla FM (2012) Swedish Alzheimer mutation induces mitochondrial dysfunction mediated by HSP60 mislocalization of amyloid precursor protein (APP) and beta-amyloid. J Biol Chem 287:30317–30327
27. Siuda J, Jasinska-Myga B, Boczarska-Jedynak M, Opala G, Fiesel FC, Moussaud-Lamodière EL, Scarffe LA, Dawson VL, Ross OA, Springer W, Dawson TM, Wszolek ZK (2014) Early-onset Parkinson's disease due to PINK1 p.Q456X mutation – clinical and functional study. Parkinsonism Relat Disord 20:1274–1278
28. Lange ML, Zeng Y, Knight A, Windebank A, Trushina E (2012) Comprehensive method for culturing embryonic dorsal root ganglion neurons for Seahorse Extracellular Flux XF24 Analysis. Front Neurol 3:175–182
29. Horak Z, Kolarik J, Sipek M, Hynek V, Vecerka F (1998) Gas permeability and mechanical properties of polystyrene-polypropylene blends. J Appl Polym Sci 69:2615–2623
30. Gerencser AA, Neilson A, Choi SW, Edman U, Yadava N, Oh RJ, Ferrick DA, Nicholls DG, Brand MD (2009) Quantitative microplate-based respirometry with correction for oxygen diffusion. Anal Chem 81:6868–6878
31. Brewer GJ, Torricelli JR (2007) Isolation and culture of adult neurons and neurospheres. Nat Protoc 2:1490–1498
32. Banker G, Goslin K (1998) Culturing nerve cells. London, Cambridge
33. Trushina E, Nemutlu E, Zhang S, Christensen T, Camp J, Mesa J, Siddiqui A, Tamura Y, Sesaki H, Wengenack TM, Dzeja PP, Poduslo JF (2012) Defects in mitochondrial dynamics and metabolomic signatures of evolving energetic stress in mouse models of familial Alzheimer's disease. PLoS One 7:e32737
34. Viollet B, Guigas B, Sanz Garcia N, Leclerc J, Foretz M, Andreelli F (2012) Cellular and molecular mechanisms of metformin: an overview. Clin Sci 122:253–270
35. Desler C, Hansen TL, Frederiksen JB, Marcker ML, Singh KK, Juel Rasmussen L (2012) Is there a link between mitochondrial reserve respiratory capacity and aging? J Aging Res 2012:192503–192503
36. Nobes CD, Brown GC, Olive PN, Brand MD (1990) Non-ohmic proton conductance of the mitochondrial inner membrane in hepatocytes. J Biol Chem 265:12903–12909
37. Brand MD, Harper ME, Taylor HC (1993) Control of the effective P/O ratio of oxidative phosphorylation in liver mitochondria and hepatocytes. Biochem J 291:739–748
38. Jekabsons MB, Nicholls DG (2004) In situ respiration and bioenergetic status of mitochondria in primary cerebellar granule neuronal cultures exposed continuously to glutamate. J Biol Chem 279:32989–33000
39. Amo T, Yadava N, Oh R, Nicholls DG, Brand MD (2008) Experimental assessment of bioenergetic differences caused by the common European mitochondrial DNA haplogroups H and T. Gene 411:69–76

Chapter 6

Measuring ATP in Axons with FRET

Lauren Y. Shields, Bryce A. Mendelsohn, and Ken Nakamura

Abstract

Synaptic transmission is an energetically demanding process that consumes much of the brain's energy. In many neurodegenerative diseases, such as Parkinson's disease (PD), Alzheimer's disease (AD), and Huntington's disease (HD), synapses are lost at early stages, which may be caused by an inability to meet this high energy demand. Until recently, however, tools were unable to measure energy levels in synapses, which limited the understanding of the physiological and pathological requirements for ATP at the nerve terminal. Here described is an approach to measure ATP specifically in axons and at the synapse with a FRET-based ATP sensor. This approach involves imaging ATP levels in live cultures of primary neurons, specifically examining ATP production by respiration and glycolysis, based on the level of neural activity. This approach also allows assessment of different cell types for their relative and required levels of synaptic energy.

Key words ATP, FRET, Synapse, Axon, Energy, Mitochondria, Respiration, Glycolysis

1 Introduction

1.1 Background

The brain only accounts for 2.5% of the body's mass, but it uses 20% of the body's resting energy [1]. Much of this energy is believed to be used by axons and their synapses to support and maintain the ionic gradients needed to initiate and propagate action potentials [2]. It also supports synaptic vesicle cycling, which has particularly high energy requirements [3, 4]. Most of the energy used by neurons is produced by aerobic respiration [2]; however, some energy also derives from glycolysis, especially in cell culture models, where glucose levels (typically ~25–30 mM) are far higher than in the brain (~1–1.5 mM [5, 6]). Indeed, the availability of substrates largely determines the extent of reliance on respiration versus glycolysis in cell lines (Fig. 1) and in neurons [4].

In many neurodegenerative diseases, synapses are highly vulnerable, which may be caused by their high energy requirements and concurrent energy failure [7]. Indeed, axons are lost before the cell body in a range of neurodegenerative diseases where

Stefan Strack and Yuriy M. Usachev (eds.), *Techniques to Investigate Mitochondrial Function in Neurons*, Neuromethods, vol. 123, DOI 10.1007/978-1-4939-6890-9_6, © Springer Science+Business Media LLC 2017

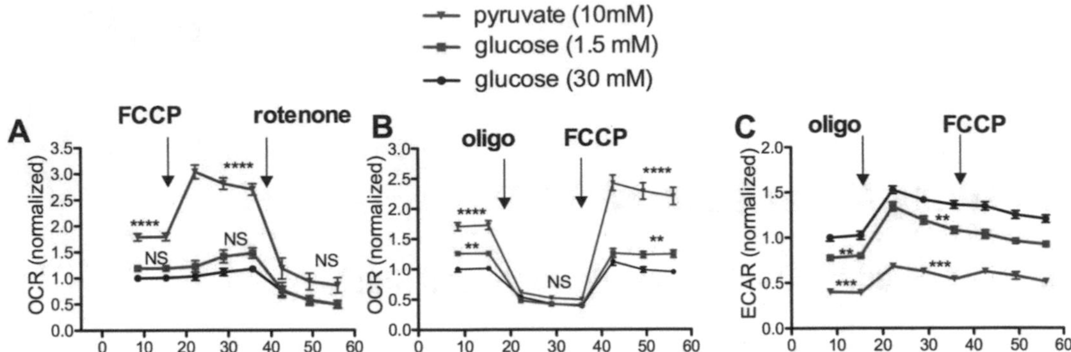

Fig. 1 Substrate availability regulates the proportion of ATP derived from aerobic and anaerobic respiration. The relative rates of aerobic respiration (oxygen consumption rate, OCR) and anaerobic respiration (extracellular acidification rate, ECAR) were estimated in HeLa cells with a 96-well Seahorse Extracellular Flux Analyzer. *Arrows* indicate addition of oligomycin (1 μM) to block ATP synthase, FCCP (1 μM) to uncouple mitochondria and assess maximal respiration, or rotenone (1 μM) to block all electron flow through complex I. (**a** , **b**) OCR was greatest in medium containing pyruvate (10 mM) only, and lowest in standard high-glucose medium (30 mM). (**c**) In contrast, ECAR was greatest in high-glucose. ** $p < 0.01$, *** $p < 0.001$, **** $p < 0.0001$, NS is not significant versus high glucose (30 mM) for both time points before the next drug addition by two way Anova with repeated measures and Bonferroni multiple comparisons post hoc test. Data shows mean ± SEM, $n = 3$ separate experiments with 6–8 wells/group for each experiment

mitochondria are disrupted, including PD, AD, and HD, as well as in animal models of mitochondrial dysfunction [7, 8].

In some cases, mitochondrial dysfunction appears to cause neurodegeneration. For example, mutations in the mitochondrial protein PTEN-induced putative kinase 1 (PINK1) produce neurodegeneration in familial PD. In most cases, however, it is unknown if mitochondrial dysfunction causes or contributes to neurodegeneration, or whether it is simply a consequence of the degenerative process. It is also unclear what aspects of mitochondrial dysfunction drive degeneration. A central function of mitochondria is to produce ATP. However, although mitochondria are required for aerobic respiration, the energetic requirements of neurons can also be supported (at least in part) by glycolysis. Mitochondria also have other functions, including calcium buffering, reactive oxygen species (ROS) production, apoptosis, and neurotransmitter and lipid metabolism [9–11].

The ability of mitochondria to support the energetic needs of neurons could be compromised in several ways. First, energy production might be compromised by an intrinsic deficit in mitochondrial function, as may occur with mitochondrial disorders involving defects in the respiratory chain (e.g., Leigh's disease). It could also be affected by a decrease in the mass and/or a disruption of the normal distribution of mitochondria within neuronal processes, especially axons. Indeed, mitochondrial movement and turnover in the axon are affected by proteins involved in neurodegeneration, such as PINK1 and Parkin in PD and tau and amyloid-beta in AD [8, 12, 13]. However, these effects have not yet been directly connected to a loss of energy, and it is unknown whether this loss occurs at the synapse or cell body, or both. With previous approaches, researchers have been unable to determine if energy decreases specifically at synapses and, if so, whether it causes neuronal dysfunction or degeneration. To a large extent, this limitation has been because the primary approaches to study energy failure have lacked specificity for cell type and subcellular compartment. Additionally, the approach of isolating synaptic compartments by synaptosome preparations requires neurons to be broken apart, and thus it cannot fully recapitulate the biology of firing intact cells. Furthermore, although mitochondrial membrane potential provides important information about mitochondrial function, it does not directly measure (or always change in parallel with) energy output or ATP levels [14].

Multiple substrates can support neuronal ATP levels [3, 5, 15]. By directly measuring ATP, researchers might better understand the neuron's preferred and physiologic energetic substrates. Contrasting theories posit that neurons directly metabolize glucose [16, 17] or preferentially metabolize lactate provided via the astrocyte neuron–lactate shuttle [18]. By examining ATP levels at the synapse in the presence of different substrates and/or while blocking specific metabolic processes, researchers can begin to clarify which mechanisms support synaptic transmission.

1.2 Introduction to FRET

Fluorescence resonance energy transfer (FRET), also known as Förster resonance energy transfer, capitalizes on the transfer of energy from one fluorescent protein (the donor) to another (the acceptor) to probe the spatial and temporal characteristics of biological processes. This energy transfer occurs when the donor is excited and emits within the acceptor's excitation spectrum, which, in turn, excites the acceptor, causing it to emit fluorescence [19].

Imamura et al. developed a FRET-based ATP sensor by linking cyan fluorescent protein (CFP) and Venus (a variant of yellow fluorescent protein (YFP)) with an ATP-binding domain cloned from a bacterial F_0F_1-ATP synthase (Fig. 2) [20]. In the unbound conformation, CFP and Venus are not in close contact, and the FRET levels remain low; however, when the sensor binds ATP, CFP and Venus are brought closer together, creating more efficient energy transfer and higher FRET levels. This sensor has several advantages: it selects for ATP versus ADP and AMP; it resists changes in pH over physiological ranges; and, importantly, it has a dead mutant control that can reveal unanticipated artifacts that may affect the fluorescent signal [20]. There are multiple versions of this FRET sensor, distinguished by their varying affinities for ATP and subcellular localizations. The cytosolic sensor

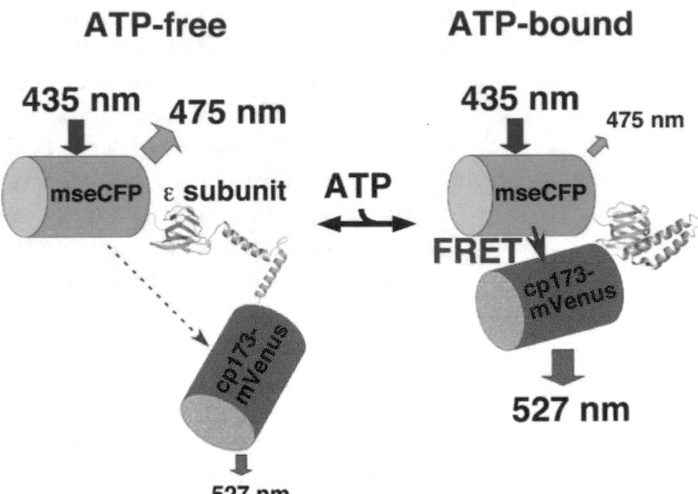

Fig. 2 FRET-based ATP probe. Variants of cyan fluorescent protein (CFP; mseCFP) and yellow fluorescent protein (YFP; cp173-mVenus) were connected by the ε subunit of *Bacillus subtilis* F_0F_1-ATP synthase. In the ATP-free form, the extended conformation leads to low FRET efficiency. In the ATP-bound form, the two fluorescent proteins are drawn closer together, increasing FRET efficiency. Adapted from Imamura et al., 2009 [20] with permission from *Proceedings of the National Academy of Sciences*

discussed here—AT1.03^{YEMK}—has a dissociation constant (K_d) of 1.2 mM at 37 °C [20], which was calculated using recombinant FRET sensor protein.

2 Materials

2.1 Neuronal Cultures

1. *Neuronal cultures.* Cultures can be prepared from various brain regions as previously described [21]. For hippocampal cultures, hippocampi are dissected in 37 °C Hank's BSS supplemented with glucose (20 mM) and HEPES (10 mM) (HBSS++). Neurons are grown in neuronal media consisting of Earle's Minimum Essential Medium supplemented with 5% fetal bovine serum (FBS), 21 mM glucose, 1% GlutaMAX, 2% B27 supplement (Gibco 17504-044), and 0.1% serum extender (Fisher 355006) [22].
2. *Coverslips and cloning rings.* Neurons are cultured on coverslips (round, 25 mm diameter, #1.5 thickness; Warner Instruments 64–0715) inside of cloning rings (8 × 8 mm; Fisher 955221).
3. *Vacuum Grease.* Rings are secured and sealed to coverslips with grease (Fisher 541718).

4. *Poly-L-lysine hydrobromide.* Coverslips are coated with 1 mg/mL Poly-L-lysine hydrobromide (PLL; Sigma P2636) prepared in borax solution (2.48% boric acid, 3.8% borax in dH_2O; pH 8.5) to facilitate neuronal attachment.
5. *Imaging chamber.* For live imaging, coverslips are mounted inside an imaging chamber (RC-21 BRFS chamber; Warner 640226).

2.2 Expression of Sensor

1. *Plasmids.* The ATP-FRET sensor ($AT1.03^{YEMK}$) is subcloned into the pCAGGS vector downstream of the chicken actin promoter [23]. For controls, single Venus and CFP sequences are also separately subcloned into two pCAGGS vectors downstream of the chicken actin promoter, as discussed below. DNA is prepared by endotoxin-free maxiprep purification.
2. *Transfection reagents.* Neuronal cultures are transfected with plasmids using the Amaxa Basic Neuron SCN Nucleofector Kit (Lonza VSPI-1003).

2.3 Imaging of ATP at the Synapse

1. *Microscope.* Neurons are imaged with a Nikon Ti-E inverted microscope with a 40× or 60× objective, or similarly equipped epifluorescent microscope.
2. *Imaging solutions.* Cells are imaged in Tyrode's imaging buffer (Table 1) with glucose (30 mM) and/or pyruvate (10 mM) at pH 7.4.
3. *Oligomycin A.* Oligomycin inhibits ATP synthase and is used to inhibit mitochondrial respiration. Oligomycin (Sigma-Aldrich 75351) is prepared as a 10 mM stock in DMSO and typically used at 3 μM.
4. *Rotenone.* Rotenone inhibits complex I and is used to inhibit mitochondrial respiration. Rotenone (Sigma-Aldrich r8875) is

Table 1
Normal Tyrode's with glucose and/or pyruvate

Stock	For 1 L	Final concentration, mM
NaCl	7.4 g	126.5
KCl	0.18 g	2.5
$CaCl_2$ (1 M stock)	2 mL	2
$MgCl_2$ (1 M stock)	2 mL	2
HEPES	2.38 g	10
Glucose	5.4 g	30
Pyruvate	1.1 g	10

prepared as a 500 μM stock in DMSO and typically used at 500 nM.

5. *2-Deoxyglucose*. 2-deoxyglucose (2DG) competitively inhibits hexokinase, the first step of glycolysis, and is used to inhibit glycolysis. 2DG (Sigma-Aldrich D8375) is prepared as a 500 mM stock in milliQ water and typically used at 5–10 mM, though the degree of inhibition of glycolysis will also depend on the glucose concentration.
6. *Iodoacetate*. Iodoacetate (IAA) irreversibly inhibits glyceraldehyde-3-phosphate dehydrogenase (GAPDH) and is used to inhibit glycolysis. IAA (Sigma-Aldrich I2512) is prepared as a 1 M stock in DMSO and typically used at 1 mM.

2.4 Analysis

MetaMorph software. MetaMorph software (version 7.7.7.0) is used to identify regions of interest and analyze fluorescence intensity. After processing, mathematical calculations are performed as described below.

3 Methods

3.1 Neuronal Cultures

Neuronal culture has already been discussed thoroughly in previous protocols [21, 23]. Briefly, for live imaging, rat or mouse neurons are cultured inside cloning rings attached to prepared coverslips. Before culturing, coverslips and cloning rings are cleaned by momentary immersion in 100% ethanol and then immediately in 100% acetone, flame-dried, and autoclaved. Cloning rings are mounted in the center of coverslips with a sparse, but sufficient amount of vacuum grease to prevent any leakage of liquid from inside the coverslip. Coverslips and cloning rings are sterilized under ultraviolet light for a minimum of 15 min.

Coverslips are then coated with PLL to facilitate neuronal attachment (Sect. 2.1, item 4). To do this, 70 μL of PLL solution is added per cloning ring, and coverslips/cloning rings are incubated at 37 °C for a minimum of 45 min. Cloning rings are then rinsed with water 4×.

For hippocampal cultures, p0 rat or mouse pups are sacrificed, their brains isolated, and their hippocampi quickly separated from the meninges and dissected in HBSS++ pre-warmed to 37 °C. The tissue is then treated with 0.25 mM trypsin, washed 4× with HBSS++ (Sect. 2.1, item 1), and triturated with p1000 in 750 μL neuronal media. After adding an additional 3.75 mL neuronal media, cells are further dissociated by filtering through a 40 μm sterile filter before counting.

3.2 Expression of Sensor

Per transfection, 0.6 μg of the ATP sensor is used to transfect 600,000 cells. Cells are spun down at 1050 rpm for 6 min, and neuronal media is removed. Cells are then resuspended in 20 μL AMAXA solution, transferred to the transfection cuvette per AMAXA protocol, and electroporated in an AMAXA Nucleofector 2 (Lonza) using the electroporation protocol described by SCN Basic Neuro Program 1 (http://bio.lonza.com/fileadmin/groups/marketing/Downloads/Protocols/Generated/Optimized_Protocol_261.pdf). Cells then recover in 480 μL neuronal media at 37 °C for 10 min and are plated inside cloning rings at an appropriate density (48,000–60,000 cells per cloning ring). Media is completely replaced with 200 μL fresh neuronal media, warmed to 37 °C, 1.5 h after plating. Three to four days post-transfection, 50 μL media is replaced with 50 μL fresh, neuronal glial-inhibitor warmed to 37 °C at a concentration of 50 μM, giving a final concentration of 12.5 μM (FUDR, Sigma F0503; uridine, Sigma U6381) to minimize glial overgrowth.

3.3 Imaging ATP at the Synapse

Typically, cultures are allowed to develop mature hippocampal synapses for 10–14 days in vitro [24, 25] before live imaging. An example imaging protocol is provided in Fig. 3 and involves a few specific steps described below.

1. *Imaging*. Sequential images are obtained at each imaging point in the FRET (λ_{ex} = 430/24 nm, λ_{em} = 535/30 nm), CFP (λ_{ex} = 430/24 nm, λ_{em} = 470/24 nm), and YFP channels (λ_{ex} = 500/20 nm, λ_{em} = 535/30 nm) (stars, Fig. 3) with an ET ECFP/EYFP filter set (Chroma). Exposure length is 150 ms per channel, the minimum needed to consistently obtain a signal-to-noise ratio that is clearly discernible from background. Images are captured every 4 min to limit exposure (see bleaching discussed in troubleshooting), except immediately before and after stimulation, to capture any acute changes in ATP levels caused by neuronal transmission.
2. *Perfusion*. To switch buffers during imaging, a perfusion valve–control system (VC-8, Warner Instruments) can be used and controlled by MetaMorph software. This helps to precisely change substrates or add drugs during live imaging. In Fig. 3, neurons are imaged in pyruvate until minute 9 and then pyruvate + 2DG + IAA (arrow) for the remaining time.
3. *Stimulation*. To simulate neuronal firing and augment ATP requirements, cultures can be stimulated with an A385 current isolator and a SYS-A310 accupulser signal generator (World Precision Instruments). Stimulating cultures at 10 Hz for 60 s promotes the preferential release of vesicles in the recycling pool [22]. Stimulation at 30 Hz for 3 s causes preferential release of vesicles in the readily releasable pool [26]. Cultures are stimulated once when only pyruvate is available, and again when glycolysis is fully blocked with 2DG and IAA (arrowheads, Fig. 3).

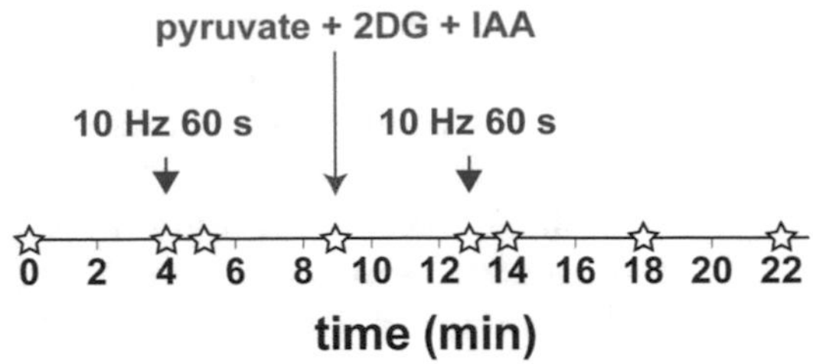

Fig. 3 Example of imaging sequence for FRET protocol. Neurons are imaged live in Tyrode's (Table 1) for 22 min, with images taken every 4 min and before and after stimulation (*stars*). Two stimulations are performed at 10 Hz for 60 s (*arrowheads*). Neurons are first perfused in pyruvate for 9 min, and then perfused in pyruvate + 2-deoxyglucose (2DG) and iodoacetate (IAA). Imaging sequences should be tailored to each experiment. This sequence is used to capture the neuronal reliance on mitochondrial-ATP during stimulation. Interventions other than stimulation are also possible

4. *Substrates.* Neurons are imaged in pyruvate to assess mitochondrial-derived ATP (Fig. 3). Neurons can also be imaged in glucose with or without pyruvate to promote both glycolysis and respiration.
5. *Inhibition of metabolic pathways.* To distinguish whether glycolysis or mitochondria contribute ATP, in addition to limiting which substrates are available (i.e., glucose for glycolysis and respiration, pyruvate for respiration alone), each of these metabolic pathways can be inhibited. In Fig. 3, glycolysis is inhibited by adding 2DG and IAA and mitochondrial-derived ATP is measured. Alternatively, neurons could be given glucose plus oligomycin or rotenone to inhibit respiration and allow glycolytic-derived ATP to be measured. Drugs are added acutely to minimize compensatory changes or nonspecific consequences of cellular toxicity.
6. *Dead or fusion mutant (control).* To control for artifacts that may non-specifically impact the FRET signal, use a dead or fusion mutant of the FRET sensor. The dead mutant ($AT1.03^{R122K/R126K}$) has mutations that eliminate its affinity for ATP [20], meaning that any changes in the FRET signal are independent of ATP. Similarly, a fused FRET sensor in which the donor and acceptor fluorophores are directly linked can also be used as a control for cellular or other experimental factors that may influence the FRET independent of ATP [4].
7. *Bleed-through corrections (recommended control).* Additional transfections with the donor and acceptor alone are required to control for bleed-through between FRET channels. Specifically, bleed-through occurs when the emission of the donor fluorophore is detected in the FRET image, and/or if the excitation wavelengths used to excite the donor also excite

the acceptor fluorophore [27]. Although these corrections are frequently omitted in the literature, they are recommended. The magnitude of this correction depends on the specific FRET pair and experimental paradigm. Using the CFP-Venus ATP sensor, corrections generally do not change the direction of trends seen in the final results, but do augment trends. To perform these corrections, neurons are transfected with either CFP (as discussed in plasmids) or Venus alone. Imaging is then performed in the same channels as discussed above (i.e., FRET, CFP, and Venus). BT_{CFP}(donor bleed-through) and BT_{Venus}(direct excitation of the acceptor) are calculated by determining the ratios of $I_{FRET}/\ I_{CFP}$ and $I_{FRET}/\ I_{Venus}$, respectively, where I_X is the background-corrected fluorescence intensity (Avg $\text{Intensity}_{\text{channel}}$ – Avg $\text{Intensity}_{\text{background in same channel}}$). ImageJ has created a plug-in to assist in bleed-through corrections called PixFRET, which is thoroughly explained in its User Guide [28]. MetaMorph also contains instructions for calculating bleed-through. Importantly, if background fluorescence is subtracted from the FRET images, it must also be subtracted from single-fluorophore images prior to calculating bleed-through coefficients. Also, bleed-through coefficients are dependent not only on the specific fluorophores, but also on the excitation and emission filters, and other aspects of the imaging setup, and thus must be recalculated if these parameters are changed.

8. *Expression* versus *FRET (control)*. Ideally, FRET measurements are independent of fluorophore concentration, but it is important to establish this for each experimental system. To validate the data, create a scatter plot of FRET levels versus a simultaneously expressed fluorescent marker, such as the intensity of mCherry-synaptophysin or the acceptor fluorescence itself [29]. After performing linear regression, there should be no correlation between the two within the range of values studied, indicating that differing FRET values are not the result of different expression levels. Additionally, tracking relative changes in FRET in a single cell throughout a time course mitigates the impact that the initial FRET value has on the results.

3.4 Sensor Calibration

Calibration of the ATP-FRET sensor permits FRET measurements to be converted into absolute concentrations of ATP (low millimolar range for mammalian cells). This is accomplished by adding exogenous ATP in defined concentrations to permeabilized sensor-expressing cells and measuring the FRET signal at each ATP concentration. Our lab has achieved reproducible results using 1 nM XF-PMP reagent (Seahorse Bioscience, #102504-100) and the buffer described by Tanaka et al. [30] (140 mM KCl, 6 mM NaCl, 1 mM $MgCl_2$, 0.465 mM $CaCl_2$, 2 mM EGTA, and 12.5 mM HEPES, pH 7.0), using HeLa cells, which remain intact following permeabilization, and 0–5 mM exogenous ATP. However, it's important to recognize that permeabilization buffer can never precisely mimic the cytosolic milieu, and thus the resulting calibration provides only an estimate of the true ATP level in an intact cell. To maximize the accuracy of such calibrations, experimental and calibration experiments should be performed in the same temperature and with the same microscopy settings. Ideally the same cell type would also be used, although we have found that the permeabilization process severely disrupts the morphology of synaptic boutons [4].

Permeabilization. Incubate cells in XF-PMP in buffer for 4 min, then gently wash and incubate in buffer without ATP for an additional 30 min. Replace buffer with MgATP (Sigma A9187, dissolved in buffer and pH-adjusted) in defined concentrations and measure FRET 5–10 min after the addition of ATP (Fig. 4) [4].

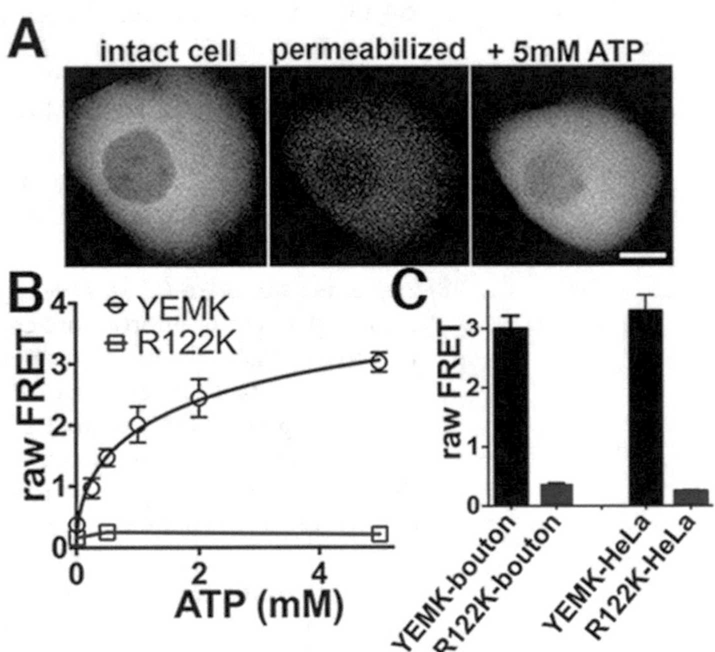

Fig. 4 Calibration of the ATP-FRET sensor. (**a**) Upon permeabilization, nearly all FRET signal is lost from the HeLa cell, indicating that ATP has diffused out. Addition of exogenous ATP then increases the FRET signal. Scale bar = 5 μm (**b**) Calibration curve in HeLa cells using 0, 0.25, 0.5, 1, 2, and 5 mM MgATP with the ATP sensor (YEMK, *black circles*) and the mutant sensor that does not bind ATP (R122K, *red squares*). Error bars are ±SD. (**c**) comparison of raw FRET of ATP FRET sensor and dead mutant in intact HeLa cells and hippocampal boutons. This research was originally published in The Journal of Biological Chemistry. Divya Pathak, Lauren Shields, Bryce A. Mendelsohn, Dominik Haddad, Wei Lin, Akos A. Gerencser, Hwajin Kim, Martin D. Brand, Robert H. Edwards and Ken Nakamura (2015) The role of mitochondrially derived ATP in synaptic vesicle recycling. © the American Society for Biochemistry and Molecular Biology

Other labs have reported the use of streptolysin-O or alpha-hemolysin as permeabilization reagents with different buffers in analogous experiments [3, 30].

3.5 Analysis of FRET Images

1. *Measurement of fluorescence intensity.* To calculate FRET values, the average intensity of fluorescence in each image must be measured at the synapse, in each channel, at each time point. As depicted in Fig. 5, boutons should be selected based on Venus and morphology (located in a string of boutons) at the first imaged time point. Synaptic identity can be confirmed by co-transfection with a fluorescent synaptic marker, such as mCherry-synaptophysin. For analysis, a circular region of interest (ROI) is used that is ≈4–5 μm in diameter, allowing for a small amount of bouton movement. The bouton should remain inside the ROI throughout the imaging protocol. Exclude boutons from analysis for any of the following prospectively defined reasons:
   (a) Low expression of the ATP sensor usually most obvious in the CFP channel. Expression should be sufficiently above background to reliably and sensitively observe CFP fluorescence.

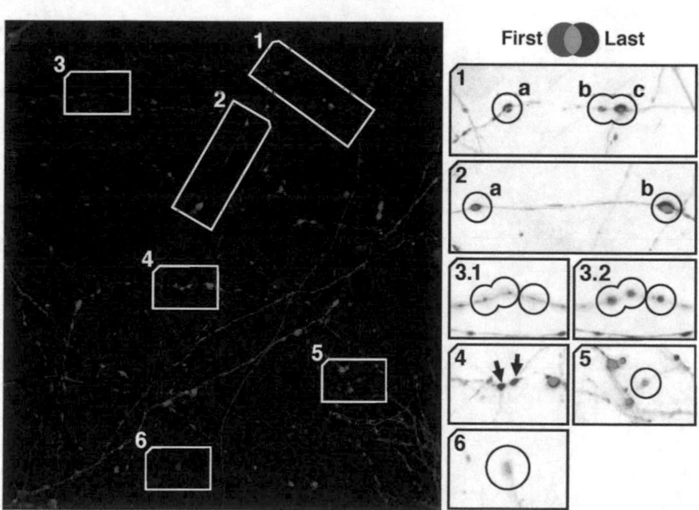

Fig. 5 Analysis of FRET Images. To analyze FRET values, boutons should be selected using the Venus channel based on expression and location in strings (*1a–c* , *2a, b*). Axonal identity can be further confirmed based on mCherry-synaptophysin (not shown). The position of boutons during the first (*magenta*) and last (*blue*) frames should be compared. Areas of the bouton that overlap in both the first and last frames are depicted in *cyan* . Boutons should be excluded if they undergo swelling during imaging (*3.1* and *3.2*), move (*4*), have low expression (*5*), or are out of focus (*6*)

(b) Movement throughout the time course such that the bouton no longer remains entirely in the ROI.

(c) Swelling of the bouton (which typically precedes synaptic degeneration or disappearance of the axon or cell attached to the bouton).

(d) A shift in focus such that the bouton is no longer in focus at the end of the time course.

A minimum of ten boutons should remain selected per field. An additional ROI is selected for the background. Regions are transferred from the Venus stack to the CFP and FRET image stacks and saved. Using *Region Measurements* under the *Measure* heading, MetaMorph can analyze the average intensity of each ROI across all planes in the stack (i.e., all time points). Data can be exported to an Excel document and saved for further calculations.

Analysis can also be performed on cell bodies as described above, using an ROI the same size as for boutons and selecting a background ROI. If there are varying levels of focus, image the cell body and boutons in two separate fields, if needed (though they can be done simultaneously in the same run using the Stage Position option in MetaMorph).

2. *Calculation of FRET.* The FRET/donor ratio is calculated for each bouton as previously described [31], where FRET = (I_{FRET} − I_{CFP}*BT_{CFP} − I_{Venus}*BT_{Venus}) / I_{CFP}, such that I_X is the background-corrected fluorescence intensity (Avg $Intensity_{FRET}$ – Avg $Intensity_{background}$) measured in a given channel at each time point. As noted above, bleed-through corrections are recommended (Sect. 3.3, step 7). These calculations can be performed with Microsoft Excel or an automated script. For relative changes in FRET, values are normalized to the first time point.

4 Expected Results and Interpretation

1. *ATP levels.* Results from the protocol in Fig. 3 are depicted in Fig. 6 for hippocampal cultures from floxed Drp1 (dynamin-related protein 1, the central mitochondrial fission protein) mice transfected with a control vector (pCAGGS alone) or cre recombinase (Drp1KO) [15]. Notably, the initial stimulation produced only a transient drop in ATP levels. However, when neurons were forced to rely entirely on their mitochondria for ATP (by blocking all glycolysis with 2DG and IAA), ATP levels dropped to a far greater extent in the Drp1KO boutons than in controls, indicating that the Drp1KO mitochondria were unable to maintain basal ATP levels in the bouton.

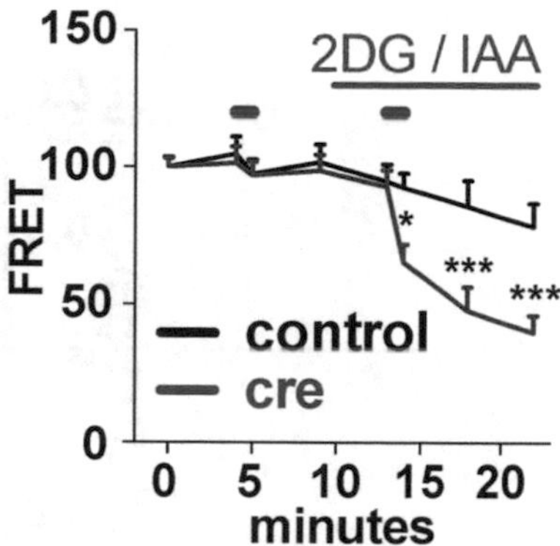

Fig. 6 Dysfunctional mitochondria cannot maintain ATP when glycolysis is blocked. Primary hippocampal cultures with floxed Drp1 (dynamin-related protein 1, the central mitochondrial fission protein) were co-transfected with an ATP-based FRET sensor (AT1.03^{YEMK}) [20] and either a control vector or cre (Drp1KO). Without glycolysis (blocked with 2-deoxyglucose (2DG) and iodoacetate (IAA) (*orange horizontal bar*)), ATP levels significantly decreased at the synapse of Drp1KO neurons after stimulation at 10 Hz for 60 s (*blue horizontal bars*). Adapted from Shields et al. 2015 [15] with permission from *Cell Death and Disease*

2. *Functional correlate.* VGLUT1-pHluorin provides a correlate for the functional consequences of a given decrease in ATP. This method allows visualization of synaptic-vesicle cycling in individual synaptic vesicles based on the pH sensitivity of pHluorin [4, 23]. When vesicles are exocytosed, fluorescence increases, which is quenched upon endocytosis and reacidification. As depicted in Fig. 7, when glycolysis was blocked and ATP levels dropped, Drp1KO (cre) neurons could no longer undergo endocytosis and re-acidify vesicles.

5 Notes/Troubleshooting

5.1 Poor Neuronal Health

Poor neuronal health is indicated by low numbers of surviving neurons with poor development or loss of neurites. More subtle deficits in neuronal health are suggested by an impaired capacity for synaptic vesicle cycling. Beaudoin et al. thoroughly describes how to obtain healthy mouse neuronal cultures [21]. In particular, the speed of dissection and cell preparation for plating (particularly the dissection step) strongly impacts neuronal health. Trituration can also impact neuronal health if done too vigorously or too many times. Waiting too long to change the neuronal media after the initial plating also leads to poor health.

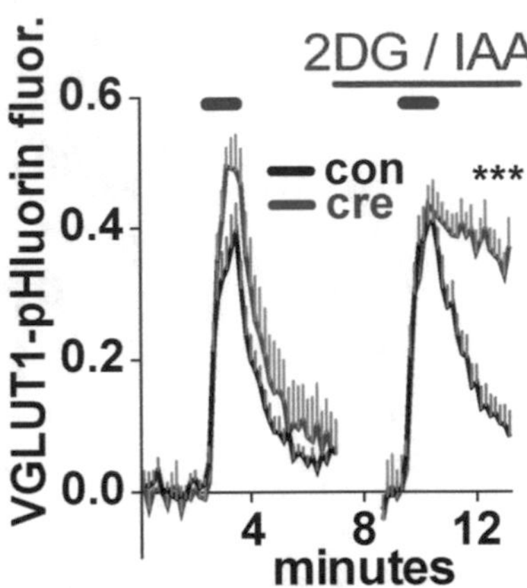

Fig. 7 Synaptic-vesicle cycling is blocked at low ATP levels. A VGLUT1-pHluorin reporter with a pH-sensitive GFP was targeted to the lumen of synaptic vesicles to measure synaptic-vesicle cycling in individual synapses [23]. Compared to control (con), Drp1KO (cre) markedly impaired endocytosis when glycolysis was blocked with 2-deoxyglucose (2DG) and iodoacetate (IAA), and cultures were stimulated at 10 Hz for 60 s (*blue bar*). Adapted from Shields et al. 2015 [15] with permission from *Cell Death and Disease*

As transfection can often cause some toxicity, neuronal health can also be supported by co-plating transfected neurons with additional untransfected neurons (5000–8000 untransfected cells per cloning ring).

5.2 Poor Expression

Poor expression of the sensor, particularly in the CFP channel, can produce unreliable FRET values. To improve expression, confirm the concentration and purity of plasmids, and consider increasing the amount of DNA transfected. DNA amounts need to be titrated, because too much DNA will eventually increase neuronal death. In addition, ensure that DNA is purified using low or endotoxin-free protocols.

5.3 Movement of Boutons During Imaging

If a significant proportion of boutons move beyond the range of the ROI and, therefore, must be excluded, there are a few options for correction. This movement may be caused by temperature changes during imaging. To avoid this effect, ensure that the solutions, coverslip, and imaging chamber are maintained at the same temperature throughout imaging. It is also important to ensure that the imaging chamber is fully stabilized on the scope and that there are no issues with erroneous stage movements. Shortening runs can also help decrease movement that occurs incrementally over time. If available, engaging perfect focus can minimize changes in the focal plane.

If movement cannot be entirely prevented, images can be corrected post hoc with ImageJ plugins such as stackReg and TurboReg, or the Auto Align function in MetaMorph, under the *Apps* heading.

5.4 Unstable Baseline FRET

An unstable baseline may occur at the beginning of imaging, often seen as a jump in FRET values. To avoid this increase and stabilize the baseline, capture three images in each channel (nine images total) before beginning the time course.

5.5 Photobleaching and Phototoxicity

Photobleaching can produce erroneous FRET values [32]. To minimize these problems, decrease the number and/or frequency of images, decrease the exposure time in each channel to the minimum required for a clear signal, and/or use a sensitive camera with optimal excitation and emission filters. Parallel runs on the dead mutant or direct fusion control can be used to directly evaluate whether photobleaching occurs.

5.6 Other Causes of Artifactual FRET Values

A common cause of unexpectedly low FRET levels is the presence of residual mitochondrial toxins in the perfusion tubing, which can be resolved by more thoroughly rinsing the tubing with buffer between runs.

5.7 Alternative Sensors

Alternative versions of the ATP sensor can be used for their increased brightness (e.g., to combat low expression), different affinities, or unique spectral properties to allow simultaneous imaging with other fluorophores [20, 33–36]. Additionally, both an ATP/ADP sensor and luciferase-based sensor are alternate approaches [3, 37].

6 Summary

This approach with FRET allows for measurements of ATP at the synapse. In particular, it distinguishes between ATP levels at the cell body versus synapse to provide valuable information about pathology when synapses are particularly vulnerable. Additionally, it differentiates between mitochondrial- and glycolytic-derived ATP to increase understanding of how neurons rely on ATP from different substrates. Finally, it can discriminate between glial and neuronal metabolism in mixed cultures, while maintaining intact neurons and synapses that can recapitulate energetic conditions that exist while firing. By measuring ATP levels specifically at the synapse, researchers in the neuronal field can better understand the energetic requirements at the synapse and, perhaps, why they are uniquely vulnerable in neurodegenerative diseases.

Acknowledgments

This work was supported by a Burroughs-Wellcome Fund Award (K.N.), NIH (1RO1NS091902, K.N.), Pediatric Scientist Development Program Award (K12-HD000850, B.M.), a Hillblom fellowship (L.S.), and a NSF Fellowship (L.S.). We thank Crystal Herron for editorial assistance and Giovanni Maki for graphical design.

References

1. Solokoff L (1960) The metabolism of the central nervous system in vivo. In: Field J, Magoun H, Hall V (eds) Handbook of physiology. American Physiological Society, Washington, DC, pp 1843–1864
2. Harris JJ, Jolivet R, Attwell D (2012) Synaptic energy use and supply. Neuron 75(5): 762–777
3. Rangaraju V, Calloway N, Ryan TA (2014) Activity-driven local ATP synthesis is required for synaptic function. Cell 156(4):825–835
4. Pathak D, Shields LY, Mendelsohn BA, Haddad D, Lin W, Gerencser AA, Kim H, Brand MD, Edwards RH, Nakamura K (2015) The role of mitochondrially derived ATP in synaptic vesicle recycling. J Biol Chem 290(37):22325–22336
5. McNay EC, Fries TM, Gold PE (2000) Decreases in rat extracellular hippocampal glucose concentration associated with cognitive demand during a spatial task. Proc Natl Acad Sci U S A 97(6):2881–2885
6. Rex A, Bert B, Fink H, Voigt JP (2009) Stimulus-dependent changes of extracellular glucose in the rat hippocampus determined by in vivo microdialysis. Physiol Behav 98(4): 467–473
7. Pathak D, Berthet A, Nakamura K (2013) Energy failure-does it contribute to neurodegeneration? Ann Neurol 74(4):506–516
8. Berthet A, Margolis EB, Zhang J, Hsieh I, Zhang J, Hnasko TS, Ahmad J, Edwards RH, Sesaki H, Huang EJ, Nakamura K (2014) Loss of mitochondrial fission depletes axonal mitochondria in midbrain dopamine neurons. J Neurosci 34(43):14304–14317
9. Rizzuto R, De Stefani D, Raffaello A, Mammucari C (2012) Mitochondria as sensors and regulators of calcium signalling. Nat Rev Mol Cell Biol 13(9):566–578
10. Nunnari J, Suomalainen A (2012) Mitochondria: in sickness and in health. Cell 148(6):1145–1159
11. Waagepetersen HS, Sonnewald U, Schousboe A (2003) Compartmentation of glutamine, glutamate, and GABA metabolism in neurons and astrocytes: functional implications. Neuroscientist 9(5):398–403
12. Vossel KA, Zhang K, Brodbeck J, Daub AC, Sharma P, Finkbeiner S, Cui B, Mucke L (2010) Tau reduction prevents Aβ-induced defects in axonal transport. Science 330(6001):198–198
13. Sheng Z-H, Cai Q (2012) Mitochondrial transport in neurons: impact on synaptic homeostasis and neurodegeneration. Nat Rev Neurosci 13:77–93
14. Perry SW, Norman JP, Barbieri J, Brown EB, Gelbard HA (2011) Mitochondrial membrane potential probes and the proton gradient: a practical usage guide. Biotechniques 50(2): 98–115
15. Shields LY, Kim H, Zhu L, Haddad D, Berthet A, Pathak D, Lam M, Ponnusamy R, Diaz-Ramirez LG, Gill TM, Sesaki H, Mucke L, Nakamura K (2015) Dynamin-related protein 1 is required for normal mitochondrial bioenergetic and synaptic function in CA1 hippocampal neurons. Cell Death Dis 6:e1725
16. Patel AB, Lai JC, Chowdhury GM, Hyder F, Rothman DL, Shulman RG, Behar KL (2014) Direct evidence for activity-dependent glucose phosphorylation in neurons with implications for the astrocyte-to-neuron lactate shuttle. Proc Natl Acad Sci U S A 111(14):5385–5390
17. Lundgaard I, Li B, Xie L, Kang H, Sanggaard S, Haswell JD, Sun W, Goldman S, Blekot S, Nielsen M, Takano T, Deane R, Nedergaard M (2015) Direct neuronal glucose uptake heralds activity-dependent increases in cerebral metabolism. Nat Commun 6:6807
18. Pellerin L, Magistretti PJ (2012) Sweet sixteen for ANLS. J Cereb Blood Flow Metab 32(7):1152–1166
19. Snapp EL, Hegde RS (2006) Rational design and evaluation of FRET experiments to mea-

sure protein proximities in cells. Curr Protoc Cell Biol Chapter 17:Unit 17.9

20. Imamura H, Huynh Nhat KP, Togawa H, Saito K, Iino R, Kato-Yamada Y, Nagai T, Noji H (2009) Visualization of ATP levels inside single living cells with fluorescence resonance energy transfer-based genetically encoded indicators. Proc Natl Acad Sci U S A 106(37): 15651–15656
21. Beaudoin GM 3rd, Lee SH, Singh D, Yuan Y, Ng YG, Reichardt LF, Arikkath J (2012) Culturing pyramidal neurons from the early postnatal mouse hippocampus and cortex. Nat Protoc 7(9):1741–1754
22. Nemani VM, Lu W, Berge V, Nakamura K, Onoa B, Lee MK, Chaudhry FA, Nicoll RA, Edwards RH (2010) Increased expression of alpha-synuclein reduces neurotransmitter release by inhibiting synaptic vesicle reclustering after endocytosis. Neuron 65(1):66–79
23. Voglmaier SM, Kam K, Yang H, Fortin DL, Hua Z, Nicoll RA, Edwards RH (2006) Distinct endocytic pathways control the rate and extent of synaptic vesicle protein recycling. Neuron 51(1):71–84
24. Banker GA, Cowan WM (1979) Further observations on hippocampal neurons in dispersed cell culture. J Comp Neurol 187(3):469–493
25. Basarsky TA, Parpura V, Haydon PG (1994) Hippocampal synaptogenesis in cell culture: developmental time course of synapse formation, calcium influx, and synaptic protein distribution. J Neurosci 14(11 Pt 1):6402–6411
26. Pyle JL, Kavalali ET, Piedras-Renteria ES, Tsien RW (2000) Rapid reuse of readily releasable pool vesicles at hippocampal synapses. Neuron 28(1):221–231
27. Vogel SS, Thaler C, Koushik SV (2006) Fanciful FRET. Sci STKE 2006(331):re2
28. Feige JN, Sage D, Wahli W, Desvergne B, Gelman L (2005) PixFRET, an ImageJ plug-in for FRET calculation that can accommodate variations in spectral bleed-throughs. Microsc Res Tech 68(1):51–58
29. Nakamura K, Nemani VM, Wallender EK, Kaehlcke K, Ott M, Edwards RH (2008) Optical reporters for the conformation of alpha-synuclein reveal a specific interaction with mitochondria. J Neurosci 28(47): 12305–12317
30. Tanaka T, Nagashima K, Inagaki N, Kioka H, Takashima S, Fukuoka H, Noji H, Kakizuka A, Imamura H (2014) Glucose-stimulated single pancreatic islets sustain increased cytosolic ATP levels during initial Ca2+ influx and subsequent Ca2+ oscillations. J Biol Chem 289(4):2205–2216
31. Xia Z, Liu Y (2001) Reliable and global measurement of fluorescence resonance energy transfer using fluorescence microscopes. Biophys J 81(4):2395–2402
32. Zal T, Gascoigne NR (2004) Photobleaching-corrected FRET efficiency imaging of live cells. Biophys J 86(6):3923–3939
33. Yaginuma H, Kawai S, Tabata KV, Tomiyama K, Kakizuka A, Komatsuzaki T, Noji H, Imamura H (2014) Diversity in ATP concentrations in a single bacterial cell population revealed by quantitative single-cell imaging. Sci Rep 4:6522
34. Tsuyama T, Kishikawa J, Han YW, Harada Y, Tsubouchi A, Noji H, Kakizuka A, Yokoyama K, Uemura T, Imamura H (2013) In vivo fluorescent adenosine 5′-triphosphate (ATP) imaging of *Drosophila melanogaster* and *Caenorhabditis elegans* by using a genetically encoded fluorescent ATP biosensor optimized for low temperatures. Anal Chem 85(16): 7889–7896
35. Kotera I, Iwasaki T, Imamura H, Noji H, Nagai T (2010) Reversible dimerization of *Aequorea victoria* fluorescent proteins increases the dynamic range of FRET-based indicators. ACS Chem Biol 5(2):215–222
36. Nakano M, Imamura H, Nagai T, Noji H (2011) Ca(2)(+) regulation of mitochondrial ATP synthesis visualized at the single cell level. ACS Chem Biol 6(7):709–715
37. Sun T, Qiao H, Pan PY, Chen Y, Sheng ZH (2013) Motile axonal mitochondria contribute to the variability of presynaptic strength. Cell Rep 4(3):413–419

Chapter 7

Characterizing Metabolic States Using Fluorescence Lifetime Imaging Microscopy (FLIM) of NAD(P)H

Thomas S. Blacker and Michael R. Duchen

Abstract

The intrinsic fluorescence of the metabolic cofactors NADH and NADPH has been used as an indicator of the metabolic state of living tissues for more than 50 years. Combining this approach with modern developments in laser-scanning microscopy, pulsed laser sources, and time-resolved fluorescence detection allows measurement of the fluorescence lifetime of NAD(P)H within cells and tissues using fluorescence lifetime imaging microscopy (FLIM). As the fluorescence lifetime of a molecule is extremely sensitive to its local environment at the nanoscale, NAD(P)H FLIM can provide information on the metabolic state of complex biological preparations at a molecular level. Here, we describe experimental and theoretical considerations to the application of the technique in the study of neuroscience.

Key words NAD, NADP, Redox signaling, Metabolism, Microscopy, Fluorescence lifetime, Autofluorescence, Live tissue, In vivo

1 Background and Method Overview

Cellular autofluorescence has been used as an in situ reporter of redox state since the 1960s, when Britton Chance and his colleagues showed that light of wavelength 366 nm incident upon living tissue caused emission with a spectral maximum at 460 nm [1]. Removing the oxygen supply to the tissue caused the intensity of this emission to increase, returning back to its initial value when normoxia was restored. The molecule present in the tissue responsible for this effect was shown to be nicotinamide adenine dinucleotide (NAD), absorbing in the near-ultraviolet portion of the electromagnetic spectrum and emitting blue light as fluorescence.

NAD is a key electron carrier, responsible for ferrying reducing equivalents between the catabolic redox reactions that convert the chemical potential energy in the bonds of substrate molecules into the usable "currency" of adenosine triphosphate (ATP) [2]. For example, in glycolysis, the anaerobic respiration of glucose taking place in the cytosol, the transfer of a hydride ion from the oxidation

Stefan Strack and Yuriy M. Usachev (eds.), *Techniques to Investigate Mitochondrial Function in Neurons*, Neuromethods, vol. 123, DOI 10.1007/978-1-4939-6890-9_7, © Springer Science+Business Media LLC 2017

of glyceraldehyde 3-phosphate converts the oxidized form, NAD^+, into the reduced NADH. Conversely, lactate dehydrogenase reduces pyruvate, the end product of glycolysis, into lactate, with associated oxidation of NADH to NAD^+, allowing the supply of ATP by glycolysis to proceed.

NAD is also central to mitochondrial aerobic respiration, where oxidation of pyruvate produced by glycolysis yields approximately 28 further ATP molecules to the two produced from a single molecule of glucose anaerobically [3]. Here, NADH carries reducing equivalents from the sequential substrate oxidation taking place in the citric acid cycle to complex I of the inner membrane electron transport chain where it is oxidized to NAD^+. Passage of the donated electrons along the respiratory chain to complex IV is coupled to the pumping of protons from the mitochondrial matrix into the intermembrane space, establishing an electrochemical gradient across the membrane to provide the energy source for ATP synthesis [4]. The NAD^+/NADH ratio therefore plays a key role in both anaerobic and aerobic ATP production. Thus, the redox states of the cytosolic and mitochondrial NAD pools are key regulators of cell metabolism.

The redox ratio of the mitochondrial NAD pool is determined by the balance between oxidation by complex I and reduction by the citric acid cycle [5]. The intrinsic photophysics of NAD can be exploited in order to quantify this redox balance in living cells and tissues. While the nicotinamide moiety of NAD^+ absorbs light of wavelength 220 ± 10 nm, reduction of this group to form NADH shifts the absorption peak to 340 ± 30 nm [6]. Following excitation, NADH emits fluorescence at 460 ± 50 nm [7], which can be measured in living samples by fluorometry or fluorescence microscopy. Comparison of the NADH fluorescence intensity under resting conditions to that measured following maximal oxidation in response to the uncoupler carbonyl cyanide 4-(trifluoromethoxy) phenylhydrazone (FCCP) and maximal reduction following inhibition of complex I by rotenone, allows the redox ratio of the mitochondrial NAD pool to be estimated [5], shown schematically in Fig. 1.

While this assay of NADH fluorescence intensity provides a tentative assessment of the mitochondrial redox ratio, it discards any cytosolic contributions to the change in signal, and its overall accuracy is limited by two experimental assumptions. Firstly, the method ignores any changes in the fluorescence quantum yield of mitochondrial NADH caused by alterations in the biochemical environment of the metabolic cofactors under the applied treatments. The fluorescence quantum yield of a molecule quantifies the ratio of photons emitted as fluorescence to photons absorbed. Isolated in solution, NADH possesses a quantum yield of only 2 % [8], around 50 times lower than most synthetic fluorescent dyes, where quantum yields of 50–100 % are typical [9]. Inside the cell,

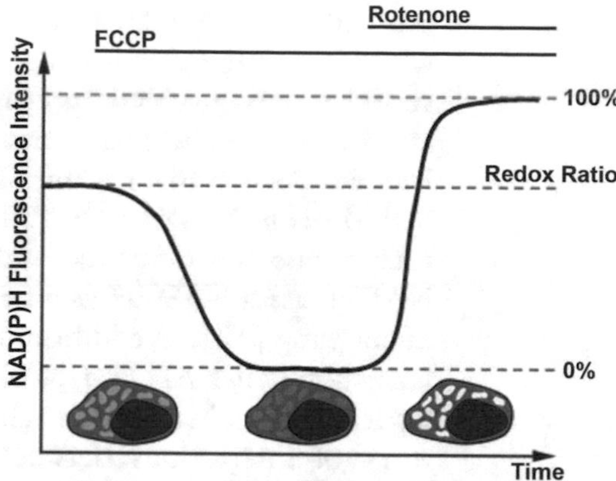

Fig. 1 NAD(P)H fluorescence intensity assay. The intensity in the presence of FCCP and rotenone defines the dynamic range upon which to compare the resting fluorescence level, allowing the resting redox ratio of the cells to be inferred

the binding of NADH to an enzyme increases its fluorescence quantum yield by five to ten times by restricting the conformational freedom of the nicotinamide ring from which excitation is localized [10]. As electron transport chain inhibition and uncoupling can alter the proportion of bound NADH inside the cell [11, 12], accounting for the associated changes in quantum yield is necessary in order to increase the accuracy of metabolic characterization using NADH fluorescence.

Secondly, the described assay ignores any redox changes in the NADP pool, in addition to the NAD pool, in response to the applied treatments. NADP is spectrally identical to NAD but functionally distinct, binding to different sets of enzymes to act as an electron carrier for anabolic reactions such as lipid and nucleotide synthesis and maintenance of the glutathione defence against reactive oxygen species (ROS) [13]. In the brain, the total NAD pool exceeds that of NADP by around 11 to one [14]. However, its role in anabolism requires the NADP pool to be maintained in a largely reduced state, in contrast to the largely oxidized state of the catabolic cofactor NAD. Thus, the reduced populations of the two pools are of similar magnitude [15]. Any NADH imaging experiment will therefore detect a mixture of fluorescence arising from both NADH and NADPH. The combined signal is therefore designated NAD(P)H [5].

If the NADPH contribution to the total NAD(P)H signal was constant, its presence would have no bearing on the results obtained from the dynamic FCCP/rotenone assay. However, although NAD and NADP regulate separate subsets of the reactions of metabolism, their primary reduction pathways are highly interconnected. For example, in the cytosol, rather than directly entering the NADH-producing glycolysis, glucose 6-phosphate can enter the NADPH-producing pentose phosphate pathway where it is oxidized to ribulose 5-phosphate and subsequently converted into the glycolytic intermediates glyceraldehyde 3-phosphate and fructose 6-phosphate, leading to further NADH production. In the mitochondria, the mitochondrial transhydrogenase transfers hydride from NADH to $NADP^+$, powered by translocation of protons across the inner membrane [16]. Reduction of NAD^+ to NADH using NADPH as an electron donor is also possible upon uncoupling [17]. In addition, the redox states of the cytosolic and mitochondrial NAD and NADP pools are linked by the malate–aspartate and citrate–α-ketoglutarate shuttles [18, 19]. This network of interactions between the distinct pools confirm that NADPH cannot be regarded as a static background against which dynamic redox changes in the NAD pool take place.

The development of fluorescence lifetime imaging microscopy (FLIM) provides increased molecular-level information from NAD(P)H fluorescence imaging experiments. FLIM measures the fluorescence lifetime of a molecule at each pixel of an image. The fluorescence lifetime reflects the average amount of time a molecule spends in the excited state before fluorescence emission and is proportional to the fluorescence quantum yield. As such, the average fluorescence lifetime of free NAD(P)H increases from 0.4 ns to values between 0.6 and 7 ns upon binding to different enzymes [20, 21]. Indeed, we have recently observed greatly increased fluorescence lifetimes associated with high levels of bound NADPH [12]. This enzyme-specificity and detection of quantum yield variations demonstrates that NAD(P)H FLIM has the potential to circumvent the limitations of intensity-based NAD(P)H fluorescence assays discussed previously. The technique has therefore attracted interest as a method to quantitatively characterize differences in metabolic state in a variety of complex tissue types [12, 22–25].

In this chapter, we detail the experimental approaches required to obtain images of the lifetime distribution of NAD(P)H fluorescence in living cells and tissues. We will also discuss the theoretical considerations necessary to perform the data analysis process and provide an exemplar final dataset. We emphasize that, although the technique may sound complex, it requires only a simple and relatively inexpensive modification to any existing two photon confocal imaging system. This chapter is intended to be platform-neutral, acting as a guide to the subtle technicalities of NAD(P)H FLIM, to be consulted alongside the literature provided with the specific hardware available to the reader.

2 Equipment and Materials

While so-called "frequency domain" methods for measurement of fluorescence lifetimes exist, based on observing phase shifts between illumination and emission [26], NAD(P)H FLIM is more usually performed using the "time domain" time-correlated single photon counting (TCSPC) technique, due to the weak fluorescence signals involved [27]. In contrast to standard laser-scanning microscopy, in TCSPC FLIM, a pulsed laser source is used to repeatedly excite the sample as the beam is raster scanned. The laser intensity is kept sufficiently low to ensure that no more than one fluorescence photon is emitted per pulse. A detector sensitive to single photons registers the emission of fluorescence and timing electronics calculate the time delay between the arrival of the fluorescence photon and the incidence of the pulse. This time delay is recorded in a histogram associated with the pixel location from where the fluorescence was emitted. After many excitation–emission events, the histogram at each pixel will resemble the time-dependent exponential decay of the excited-state population of target molecules at that location in the cell. Subsequent fitting to an appropriate model using data analysis software reveals the fluorescence lifetimes present at that position. The measured lifetimes are then either displayed as color-coded images or average values are presented after being calculated across regions-of-interest (ROIs) in the samples. A schematic overview of this experimental procedure is given in Fig. 2.

Most existing confocal microscopy systems can be modified to permit FLIM. Commercial suppliers of the sensitive PMT detectors and associated counting electronics required include Becker & Hickl and PicoQuant. The counting electronics can now be contained on a PCI card to be housed in a standard desktop PC, such as that used to control the microscope itself. Both companies also provide the necessary data acquisition and analysis software alongside the hardware. Additionally, third-party FLIM analysis packages are now available, such as TRI2 [28], Globals [29] and OMERO FLIMFit [30]. The requirement for a pulsed laser can often involve significant expense, so FLIM is frequently supplied as an add-on to a multiphoton microscope, where the necessary modelocked Ti:Sapphire laser serves as the excitation source for both techniques and the sensitive external FLIM detector is swapped in for the detector used for standard two photon imaging when required. The typical ~12.5 ns dwell period between the ~80 MHz pulses of a Ti:Sapphire oscillator is also highly suitable for capturing the NAD(P)H fluorescence decay in its entirety.

As with all living-sample microscopy techniques, provision for maintaining the physiology of the cells or tissues at the microscope stage is paramount. For example, temperature-controlled stages, gas supply and perfusion systems may all be required to maintain a complex in vivo model. Additionally, as FLIM images typically require acquisition times of the order of minutes, it is crucial that the sample remains still in order to maintain the spatial integrity of the resultant image. Care must also be taken that no measures acting to preserve the sample at the microscope introduce external

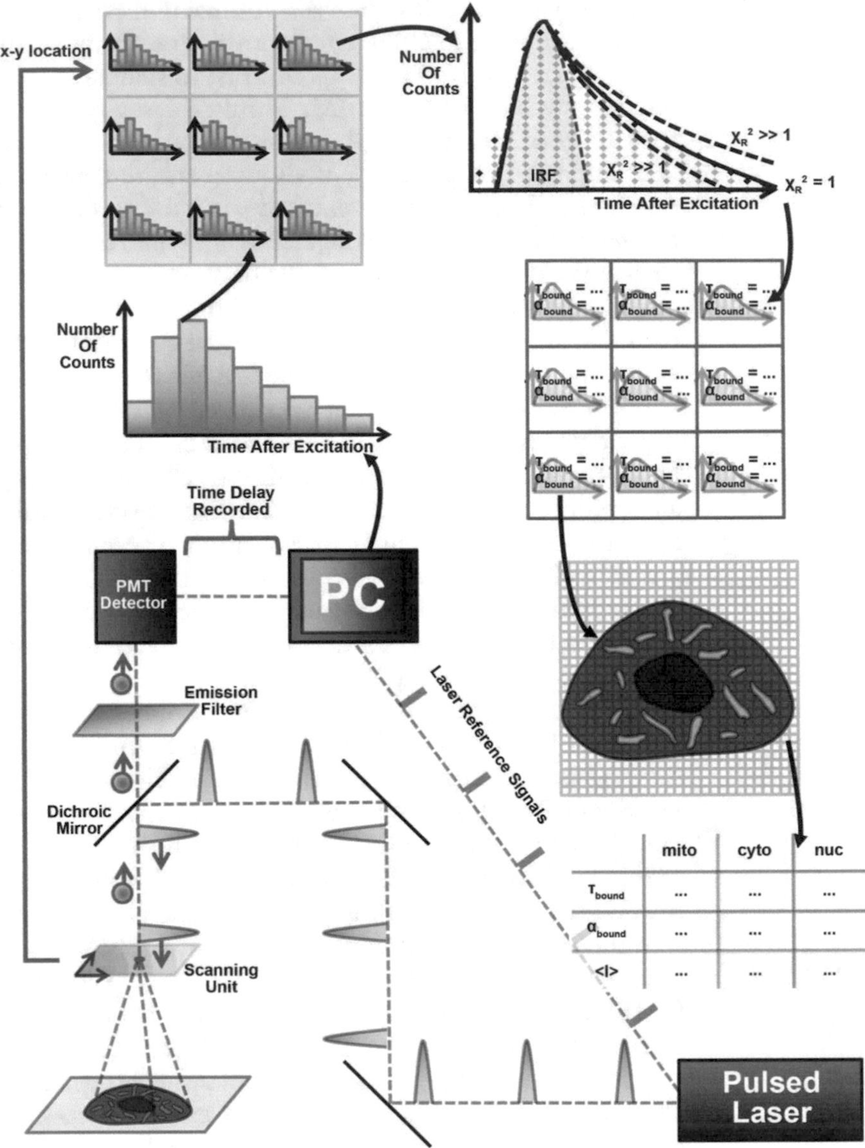

Fig. 2 The TCSPC FLIM method. A pulsed laser allows the rapidly repeated excitation of the sample as the laser is scanned across it. This builds fluorescence decay recordings at each pixel of the image on a photon-by-photon basis. Fitting a biexponential decay model at each pixel allows the average lifetime parameters in different ROIs to be calculated

fluorophores that will interfere with the NAD(P)H signal, such as adhesives and media pH indicators.

As the advantage of in vivo techniques such as NAD(P)H FLIM lies in their non-destructive application to complex multicellular preparations, an upright microscope with water-dipping objectives is preferred. This allows a variety of sample types to be imaged, ranging from basic neuron–astrocyte cocultures on glass coverslips to hippocampal brain slices and cranial-window studies.

Regardless of the complexity of the apparatus required to maintain the biological model at the microscope stage, if the setup permits live imaging using standard fluorescence microscopy, no further sample-preparation measures are required in order to attempt NAD(P)H FLIM.

3 Methods

3.1 Experimental Procedures

1. Following mounting of the sample on the microscope stage, tune the excitation laser to the required wavelength. The ubiquity of Ti:Sapphire lasers for multiphoton microscopy means this is typically performed using two-photon excitation at 720 nm. Justifications and alternatives are discussed in Note 1.
2. Select an appropriate major dichroic to direct the incident illumination towards the sample and the emitted fluorescence (410–510 nm) towards the FLIM detector. In the light path between dichroic and detector, place filters to isolate NAD(P)H fluorescence. In the Zeiss LSM510, we make use of the 460 ± 25 nm bandpass filter to select the region surrounding the 460 nm emission peak. As the signal is weak, a fairly broad bandwidth is preferable; more liberal filtering will permit enhanced signal levels but increase the risk of contamination by other sources of autofluorescence, discussed further in Note 2.
3. Given the extended collection times required for high quality fluorescence decay data, photoinduced damage to the living tissue is a major concern in NAD(P)H FLIM. These phenomena are discussed in more detail in Note 3. Experimental variables to be considered to minimize photodamage include incident laser power, repetition rate, excitation wavelength, scanning speed, image magnification and acquisition time. While specific settings will depend on the type of preparation being imaged, as a rule of thumb we choose the fastest scanning speed using the lowest laser power that permits sufficient photons to be counted (100–500 in the dimmest pixel of interest, see Note 4) in an acquisition period the length of which does not induce damage. For standard cell culture studies, this amounts to pixel dwell times of 1.6 μs, laser powers at the back of the objective of ~10 mW and 2 min acquisition.
4. Once a routine imaging protocol has been established based on the considerations above, repeat images should be taken for each sample being analyzed in order to account for biological variation. For example, under each experimental condition, multiple tissue preparations or cell culture dishes should be imaged, and multiple regions of each sample if possible. This permits a rigorous comparison of metabolic states to be made in the data analysis stage.

5. Upon completion of the imaging session, fluorescence lifetime parameters are extracted from the data using least-squares fitting in dedicated FLIM analysis software. This can be a lengthy process, given the computational complexity of carrying out this procedure at every pixel. For NAD(P)H FLIM of the type described here, the model fit to the data takes the form of a biexponential decay,

$$I(t) = I(0)[(1-\alpha_{\text{bound}})\exp(-\frac{t}{\tau_{\text{free}}}) + \alpha_{\text{bound}}\exp(-\frac{t}{\tau_{\text{bound}}})] \quad (7.1)$$

where τ_{free} and τ_{bound} are the average lifetimes of the free and bound NAD(P)H populations respectively and α_{bound} is the fraction of bound NAD(P)H. While the population of molecules at each location in the cell will undoubtedly possess more than two distinct fluorescence lifetimes, given the large variety of NAD(P)H-binding enzymes available, this model typically minimizes the χ^2 fitting statistic for signal levels obtainable from living tissues. Further discussion of these considerations is included in Note 5.

6. Following binning to increase the number of counts in the darkest pixel of interest to ~4000 (see Note 4), the pixel-by-pixel biexponential fitting algorithm can be initiated. For enhanced quality of fit and increased accuracy of lifetime determination, a custom instrument response function (IRF) should ideally be defined, the measurement of which is described in Note 6. However, fitting using a computer-generated IRF based on the rise time of the fluorescence decay or discounting the rise from the fit altogether (tail-fitting) is considered sufficient for internal comparison between the datasets generated in an individual experiment. It is important that fit settings are kept constant across the images of an individual experiment to ensure that any differences in fluorescence parameters between samples can be confidently assigned as real and not an artefact of the fitting process.

7. After fitting has been performed in each dataset, ROIs should be defined in each image from which mean values of τ_{bound} and α_{bound} are then extracted. For example, separate ROIs may be drawn for neurons and astrocytes in a brain slice sample to compare the metabolic states of these two distinct cell types, or cytosolic and nuclear ROIs could be used to compare NAD(P) compartmentalization. The importance of taking means across ROIs, as opposed recording parameter values from individual pixels, is discussed in Note 7.

8. The NAD(P)H fluorescence intensity information provided by standard imaging approaches is also contained within the FLIM datasets by summation of the photon counts at each pixel. If a particular data analysis software does not provide the

capability to measure the average intensity across a ROI, raw matrices of photon counts can be exported. Custom algorithms can then be written to generate intensity images from this data. Advanced analysis approaches such as this are discussed in Note 8.

9. After extracting mean parameter values for each ROI in each image, the results should be collated in order to calculate average values of τ_{bound}, α_{bound} and fluorescence intensity $<I_{total}>$ for each ROI type under each experimental condition. The final dataset takes the form of these mean values with uncertainties given by the standard error of the results obtained for each ROI type. Statistically significant differences in each parameter between experimental conditions are then ascertained with appropriate statistical tests.

3.2 Example Study

To illustrate a typical NAD(P)H FLIM experiment, we present a simple study to demonstrate the impact of perturbations to the NADP pool on the NAD(P)H fluorescence decay parameters. As a key function of the NADP pool is in maintaining the reduction of the cell's glutathione ROS defence, we performed NAD(P)H FLIM on cells cultured in the presence and absence of buthionine sulfoximine (BSO). This compound inhibits glutamate–cysteine ligase, and as such disrupts the synthesis of glutathione [31]. Thus an upstream effect on the NADP pool may be expected.

HeLa cells were grown in Advanced Dulbecco's Modified Eagle's Medium (DMEM) supplemented with 10 % fetal bovine serum, 2 mM GlutaMAX, 100 U ml^{-1} penicillin and 100 μg ml^{-1} streptomycin (all from Life Technologies). Cells were cultured as monolayers in sterile 75 cm^2 tissue culture flasks (all plasticware from Thermo Scientific Nunc) in a 37 °C, 5 % CO_2 incubator. 72 h prior to imaging, cells were seeded in sterile 6-well plates (20,000 cells per well) containing 22 mm glass coverslips. After allowing the cells to settle overnight, 100 μM BSO was added to half the wells and the cells were imaged 48 h later.

NAD(P)H FLIM was performed on an upright LSM 510 microscope (Carl Zeiss) with a 1.0 N.A. 40× water-dipping objective using a 650-nm short-pass dichroic and 460 ± 25 nm emission filter. At the microscope stage, coverslips were maintained at 37 °C mounted in a metal ring constructed in our workshops, for which the cover slip forms the base, and bathed in DMEM solution buffered by 10 mM HEPES. Two-photon excitation was provided by a Chameleon (Coherent) Ultra Ti:Sapphire laser tuned to 720 nm, with on-sample powers kept constant below 10 mW. Emission events were registered by an external detector (HPM-100, Becker & Hickl) attached to a commercial time-correlated single photon counting electronics module (SPC-830, Becker & Hickl) contained on a PCI board in a desktop computer. Scanning was performed continuously for 2 min with a pixel dwell time of 1.6 μs.

Three coverslips grown in the presence of BSO were imaged along with three untreated control coverslips. Images from three separate regions of each coverslip were taken, giving a total of nine images for each condition. Photon counts at each pixel were 5 × 5 binned and biexponential fluorescence decays were fit using SPCImage v3.08 by the iterative reconvolution method, with an IRF measured using a potassium dihydrogen phosphate (KDP) crystal (see Note 6). Matrices of the fit parameters α_{bound}, τ_{bound} and the total photons recorded at each pixel were exported from SPCImage and converted into grayscale images using MATLAB (The Mathworks) for further analysis using ImageJ (NIH, see Note 8). Extranuclear regions of interest were drawn for each image and the mean parameter values recorded. These were then averaged over all images for each treatment, giving the final lifetime parameter measurements and standard error uncertainties. Statistically significant differences in α_{bound} and τ_{bound} ($P < 0.05$) were judged using a Student's *t*-test and in normalized fluorescence intensity using a Wilcoxon signed-rank test.

The results of this standard NAD(P)H FLIM experiment are summarized in Fig. 3. Here it can be seen that the metabolic state induced by BSO treatment can be characterized by a decreased enzyme-bound fluorescence lifetime (τ_{bound}=2.85 ± 0.05 ns vs. 2.61 ± 0.02 ns, P = 0.0004), decreased NAD(P)H fluorescence intensity (I = 1.0 ± 0.09 au vs. 0.78 ± 0.03, P = 0.03) and unchanged bound NAD(P)H fraction (α_{bound} = 0.221 ± 0.006), relative to untreated conditions. Previous work has correlated changes in α_{bound} with changes in the balance of ATP production by oxidative and glycolytic means [22]. Combining this interpretation with the results obtained here therefore suggests that BSO treatment causes no change in catabolic flux. However, in our recent work [12] we observed that interpretation of this parameter may not be so straightforward. Changes in α_{bound} seemed to be dependent on the

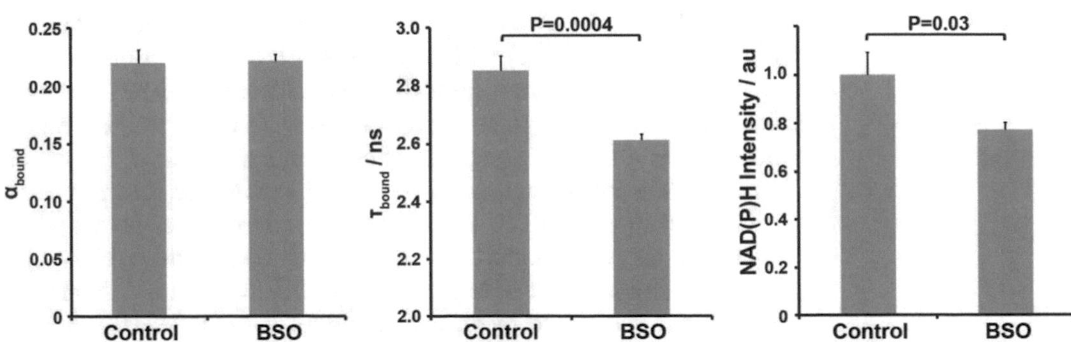

Fig. 3 Results of NAD(P)H FLIM investigation into BSO treatment in HeLa cells. Disrupting glutathione synthesis using BSO left the fraction of bound NAD(P)H unchanged, but the enzyme-bound fluorescence lifetime and the fluorescence intensity both decreased

mechanism of perturbation to the aerobic–anaerobic metabolic balance and its associated effect on the total redox state of the NAD(P) population.

Our recent work has also linked changes in τ_{bound} to changes in the relative concentrations of bound NADPH and bound NADH [12]. Making the assumption that these two species possessed finite and distinct fluorescence lifetimes of 4.4 and 1.5 ns respectively led to expressions for NADH and NADPH concentrations of,

$$[\mathrm{NADH}] = k\left(\frac{4.4 - \tau_{\mathrm{bound}}(ns)}{2.9}\right)\frac{I_{\mathrm{total}}}{(1-\alpha_{\mathrm{bound}})\tau_{\mathrm{free}} + \alpha_{\mathrm{bound}}\tau_{\mathrm{bound}}} \quad (7.2)$$

$$[\mathrm{NADPH}] = k\left(\frac{\tau_{\mathrm{bound}}(ns) - 1.5}{2.9}\right)\frac{I_{\mathrm{total}}}{(1-\alpha_{\mathrm{bound}})\tau_{\mathrm{free}} + \alpha_{\mathrm{bound}}\tau_{\mathrm{bound}}} \quad (7.3)$$

Further work is undoubtedly required to test the validity of this simplified model in different cell types. However, applying Eqs. (7.2) and (7.3) to each image acquired in this experiment, setting k such that the untreated concentration of NADH is 1 au, implied that BSO treatment left the NADH concentration unchanged but the NADPH population decreased from 0.89 ± 0.07 au to 0.58 ± 0.03 au ($P = 0.01$, Wilcoxon signed-rank test). This is consistent with increased oxidation of the NADP pool in order to maintain reduced glutathione levels in the face of a depleted total glutathione population.

4 Notes

1. Choice of excitation wavelength.

   Two-photon excitation is typically employed in NAD(P)H FLIM studies as mode-locked Ti:Sapphire lasers are frequently available in biological imaging facilities to cater for advanced multiphoton microscopy techniques. The two-photon absorption spectrum of NADH peaks at 700 nm with an absolute cross section of approximately 4 GM [32], yet a variety of wavelengths have previously been used for two-photon excitation of NAD(P)H in live cells and tissues, ranging from 700 nm [12] to 800 nm [33]. While shorter wavelengths will provide more efficient excitation and subsequently decreased acquisition times, longer wavelengths have previously been associated with decreased probability of photodamage [34]. We find 720 nm to be a suitable compromise between these factors.

   The availability of computer-controlled, second-harmonic generating optical parametric oscillators and pulsed diode lasers operating in the near-ultraviolet is making single-photon excitation a more viable prospect for NAD(P)H FLIM. This would have the advantage of increased signal levels due to the inclusion of out of focus fluorescence at the expense of axial resolution. However, quartz optics may be necessary to minimize losses in the excitation beam path. Regardless of the excitation source, it is crucial to consider emission filtering in parallel with the incident wavelength, as contaminating sources of autofluorescence will be more prominent in different spectral regions.

2. Autofluorescence and emission filtering.
   The potential for other autofluorophores to contaminate the weak NAD(P)H signal make emission filtering crucial to the success of an NAD(P)H FLIM experiment. Zipfel et al. (2003) previously investigated the two-photon induced fluorescence properties of a number of endogenous fluorophores, including NADH, folic acid, vitamins A and D, pyridoxine and flavin [32]. Of these, the spectral characteristics of folic acid and vitamins A and D were most similar to those of NAD(P)H, with two-photon absorption maxima at ~700 nm and emission maxima around 460 nm. These metabolites are present in much lower concentrations than NAD(P)H [35], and weak absorption and low quantum yields ensure their contribution in the NAD(P)H spectral window is negligible. However, the two-photon action cross section of flavin was shown to be at least 10-times greater than any other autofluorescent compound across a wide spectral range. As this emits at redder wavelengths to that of NAD(P)H, the most crucial spectral region to efficiently block during NAD(P)H FLIM measurements is 530 ± 30 nm.

3. Causes and identification of photodamage.
   The low fluorescence quantum yields of fluorophores intrinsic to biological tissue indicate that the bulk of excitation events in the absence of external dyes do not result in the emission of a fluorescence photon. The fate of the majority of energy absorbed by the tissue therefore lies in excited-state photochemical reactions and localized heating due to vibrational relaxation. As such, photodamage is of significant concern in autofluorescence imaging. Viability of cultured cells following imaging can be assessed by a trypan blue exclusion test [36]. However, such a method is insensitive to subtle changes in cell physiology induced by the incident light as plasma membrane rupture is indicative of complete cell death [37]. Photodamage in NAD(P)H imaging has previously been linked with the onset of photobleaching [34]. As such, the principal means by which damage can be avoided is by ensuring that the imaging settings result in constant fluorescence count rates over the course of acquisition. Count rate monitoring is a standard feature of most TCSPC electronics.

4. Signal-to-noise considerations.
   Data acquired using the TCSPC technique follows Poisson statistics, leading to the signal-to-noise ratio increasing as the square root of the signal. As such, the more signal is acquired, the more accurately the lifetimes present can be extracted. Using computational modeling, we have previously found that ~4000 fluorescence counts in a measured decay is a benchmark for a precise reflection of the true mean NAD(P)H lifetimes present with biexponential fitting, assuming the final result is taken as an average of the lifetime parameter across a region of interest of the image [12]. For TCSPC data digitized in 256 time bins, this corresponds to ~200 counts in the peak bin for a typical intracellular NAD(P)H decay. In theory, a FLIM image should therefore be acquired until the dimmest region of the sample reaches 4000 counts. However, in practice, this limit will not be reached in a reasonable acquisition time. Thus, this threshold is often reached by "binning" the counts from neighboring pixels in the data analysis process, sacrificing spatial resolution to gain temporal resolution. Pixels with 100 to 500 photon counts will therefore approximately reach the threshold of 4000 counts using binning levels of between 3 × 3 and 7 × 7.

5. Biexponential fitting.
   Least-squares fitting of TCSPC data, such as described here, aims to vary the parameters of a defined decay model in order to minimize the χ_R^2 (reduced chi-squared) value of the fit. The χ_R^2 statistic describes the extent to which deviations of the measured data to the model can be ascribed to the expected level of noise. As such, if the difference between the fluorescence predicted by the model at a given time delay after excitation and the measured data is greater than that expected by Poisson statistics ($\sqrt{N}$, where N is the number of photon counts) a high χ_R^2 value results. A χ_R^2 of 1 is considered a perfect agreement between model and data. As a rule of thumb, decay components should be added until addition of a further component causes no subsequent decrease in χ_R^2 . Increasing the complexity of the model beyond this point typically results in significant fitting errors [38].

   A minimized χ_R^2 does not rule out more complex decay models describing the true underlying fluorescence decay processes occurring. However, such higher-order processes are obscured by the noise inherent in photon-by-photon counting. This is the case in NAD(P)H FLIM, where a distribution of fluorescence lifetimes corresponding to NAD(P)H bound to a large variety of different enzymes would be expected. However, maintaining the integrity of the sample results in relatively low signal levels, masking these additional components. As such, canonical NAD(P)H FLIM studies typically extract and analyze the behavior of only the two lifetimes τ_{free} and τ_{bound} [33, 39, 40].

6. Custom IRF measurement.
The largely random amplification processes taking place in the sensitive detectors used for TCSPC result in a distribution of possible delay times between fluorescence emission and the photon being registered by the counting electronics. The shape of this distribution is known as the instrument response function (IRF). This finite time response causes the measured TCSPC histogram at each pixel to resemble not the pure multiexponential decay of the NAD(P)H species present at that location but a convolution of this decay with the IRF. FLIM analysis software will therefore provide the facilities to deconvolute these two functions, typically by the iterative reconvolution method. This requires a measurement of the IRF of the FLIM system being used, obtained by performing a fluorescence decay measurement on a sample of infinitely short lifetime. The measured histogram will then reflect the IRF of the system.
To perform this measurement, we use the second harmonic generation (SHG) signal of potassium dihydrogen phosphate (KDP) crystals with the Ti:Sapphire laser set to 920 nm; twice the central wavelength of the emission filters. Such crystals can be produced by leaving a molar solution of KDP to evaporate overnight on a glass coverslip. For the most accurate representation of the IRF, all experimental configurations should be maintained from the NAD(P)H FLIM experiments. As such, the SHG approach is not ideal as it requires the incident illumination to be at a longer wavelength than used experimentally. Additionally, it is not possible to perform IRF measurements on FLIM apparatus with single-photon excitation in this way. Gold nanorods, possessing wide absorption and emission spectra and an ultrashort lifetime, have therefore been suggested as a more suitable target [41].

7. Necessity of ROI parameter averages.
The stochastic nature of the fluorescence process ensures that repeated lifetime measurements on the same fluorescent sample will report a range of different decay parameters. While the spread of this distribution of measurements will be almost negligible for high-quality TCSPC recordings taken from fluorophores in solution, this effect will be significant in recordings made from live cells where Poisson noise is more prominent. For example, we previously simulated NAD(P)H FLIM data and observed standard deviations of ±0.5 ns for τ_{bound} with ~5000 counts in the decay, despite a constant underlying lifetime of 3.5 ns in each of the 100 datasets [12]. This uncertainty decreased with increasing photon counts. However, the

mean of the reported values remained at 3.5 ns across all signal levels. As such, accurate measurements of the parameters α_{bound} and τ_{bound} can be extracted from NAD(P)H FLIM images by taking means of each parameter across a ROI in which the metabolic state can be assumed to be constant, such as the cytosolic, mitochondrial and nuclear compartments in a cell culture study. This process therefore represents a pseudo-bootstrap approach to assessing the precision of the NAD(P)H FLIM measurement, by taking the standard deviation of the fluorescence decay parameters across the ROI.

8. Optimizing analysis procedures.
   The process of data analysis in a FLIM experiment is frequently more time-consuming than the acquisition of the data itself. Batch processing of the pixel-by-pixel least-squares fitting procedure is a desirable feature in any FLIM analysis software. Additionally, at the experimental planning stage, all possible steps to reduce human input to ROI definition are strongly encouraged. For example, we frequently use tetramethylrhodamine methyl ester (TMRM) or MitoTracker dyes to aid subcellular segmentation into cytosolic and mitochondrial ROIs. In tissue studies, cell-specific dyes can be applied in order to segment individual cell types, such as sulforhodamine B for the specific staining of astrocytes in whole brains [42]. Ensuring that the emission from these secondary fluorophores is strongly separated from the NAD(P)H signal is crucial, particularly as they are likely to possess significantly higher absorption cross-sections and quantum yields than the autofluorescence.

   While least-squares fitting of large fluorescence decay datasets is typically performed most efficiently in dedicated FLIM analysis software, we frequently find that subsequent processing steps are more straightforward in established software for standard confocal imaging such as ImageJ, particularly if large numbers of images have been acquired. As such software is generally designed purely for the analysis of fluorescence intensity images, tables of the fluorescence decay fit parameters must be exported and converted into grayscale images in which the intensity value at each pixel encodes the value of α_{bound} or τ_{bound} there. Due to its design for handling both matrices and image processing, we find MATLAB the most suitable tool for performing this conversion. With the aid of intensity images produced in a similar manner using the total photon counts at each pixel, the image analysis software of choice can then be used to define ROIs and extract mean lifetime parameters from them, with intensity values converted back to lifetime parameters using the inverse of the scaling performed to produce the image.

5 Conclusion

The high sensitivity of the fluorescence lifetime of a molecule to its local environment ensures that NAD(P)H FLIM can add substantial information to the already-established techniques of autofluorescence intensity measurement. Owing to its lack of reliance on external dyes and pharmacological calibrations, alongside its compatibility with multiphoton excitation, NAD(P)H permits quantitative comparison of the metabolic states of living tissue samples, including complex biological preparations such as cranial windows and hippocampal slices.

To extend the technique away from comparative measurements of variations in lifetimes and decay component weightings, efforts are now under way to link the values of the fluorescence decay parameters to underlying biochemical quantities, such as the total NAD(P) redox state [11] and the balance between NADH an NADPH [12]. As the physiological significance of changes in the fluorescence decay parameters are further understood, along with improvements in analysis procedures and FLIM technology, NAD(P)H FLIM appears to be a promising addition to the arsenal of experimental techniques to study metabolic function in both neuroscience and beyond.

References

1. Chance B, Cohen P, Jobsis F, Schoener B (1962) Intracellular oxidation-reduction states in vivo. Science 137:499–508
2. Ying W (2006) NAD+ and NADH in cellular functions and cell death. Front Biosci 11: 3129–3148
3. Osellame LD, Blacker TS, Duchen MR (2012) Cellular and molecular mechanisms of mitochondrial function. Best Pract Res Clin Endocrinol Metab 26:711–723
4. Mitchell P (1961) Coupling of phosphorylation to electron and hydrogen transfer by a chemi-osmotic type of mechanism. Nature 191:144–148
5. Duchen MR, Surin A, Jacobson J (2003) Imaging mitochondrial function in intact cells. Methods Enzymol 361:353–389
6. De Ruyck JJ et al (2007) Towards the understanding of the absorption spectra of NAD(P)H/NAD(P)+ as a common indicator of dehydrogenase enzymatic activity. Chem Phys Lett 450:119–122
7. Patterson GH, Knobel SM, Arkhammar P, Thastrup O, Piston DW (2000) Separation of the glucose-stimulated cytoplasmic and mitochondrial NAD(P)H responses in pancreatic islet beta cells. Proc Natl Acad Sci U S A 97: 5203–5207
8. Scott TG, Spencer RD, Leonard NJ, Weber G (1970) Synthetic spectroscopic models related to coenzymes and base pairs. V. Emission properties of NADH. Studies of fluorescence lifetimes and quantum efficiencies of NADH, AcPyADH, [reduced acetylpyridineadenine dinucleotide] and simplified synthetic models. J Am Chem Soc 92:687–695
9. Resch-Genger U, Grabolle M, Cavaliere-Jaricot S, Nitschke R, Nann T (2008) Quantum dots versus organic dyes as fluorescent labels. Nat Methods 5:763–775
10. Blacker TS, Marsh RJ, Duchen MR, Bain AJ (2013) Activated barrier crossing dynamics in the non-radiative decay of NADH and NADPH. Chem Phys 422:184–194

11. Bird DK et al (2005) Metabolic mapping of MCF10A human breast cells via multiphoton fluorescence lifetime imaging of the coenzyme NADH. Cancer Res 65:8766–8773
12. Blacker TS et al (2014) Separating NADH and NADPH fluorescence in live cells and tissues using FLIM. Nat Commun 5:3936
13. Ying W (2008) NAD+/NADH and NADP+/NADPH in cellular functions and cell death: regulation and biological consequences. Antioxid Redox Signal 10:179–206
14. Sies H (1982) Metabolic compartmentation. Academic Press, New York
15. Pollak N, Niere M, Ziegler M (2007) NAD kinase levels control the NADPH concentration in human cells. J Biol Chem 282:33562–33571
16. Huxley L, Quirk PG, Cotton NPJ, White SA, Jackson JB (2011) The specificity of proton-translocating transhydrogenase for nicotinamide nucleotides. Biochim Biophys Acta Bioenerg 1807:85–94
17. Enander K, Rydstrom J, Rydstrom J (1982) Energy-linked nicotinamide nucleotide transhydrogenase. Kinetics and regulation of purified and reconstituted transhydrogenase from beef heart mitochondria. J Biol Chem 257: 14760–14766
18. LaNoue KF, Williamson JR (1971) Interrelationships between malate-aspartate shuttle and citric acid cycle in rat heart mitochondria. Metabolism 20:119–140
19. Wallace DC (2012) Mitochondria and cancer. Nat Rev Cancer 12:685–698
20. Gafni A, Brand L (1976) Fluorescence decay studies of reduced nicotinamide adenine dinucleotide in solution and bound to liver alcohol dehydrogenase. Biochemistry 15:3165–3171
21. Vishwasrao HD, Heikal AA, Kasischke KA, Webb WW (2005) Conformational dependence of intracellular NADH on metabolic state revealed by associated fluorescence anisotropy. J Biol Chem 280:25119–25126
22. Stringari C et al (2012) Metabolic trajectory of cellular differentiation in small intestine by phasor fluorescence lifetime microscopy of NADH. Sci Rep 2
23. Yaseen MA et al (2013) In vivo imaging of cerebral energy metabolism with two-photon fluorescence lifetime microscopy of NADH. Biomed Opt Express 4:307–321
24. Shah AT et al (2014) Optical metabolic imaging of treatment response in human head and neck squamous cell carcinoma. PLoS One 9:e90746
25. Zholudeva LV, Ward KG, Nichols MG, Smith HJ (2015) Gentamicin differentially alters cellular metabolism of cochlear hair cells as revealed by NAD(P)H fluorescence lifetime imaging. J Biomed Opt 20:051032
26. Zhang Q, Piston DW, Goodman RH (2002) Regulation of corepressor function by nuclear NADH. Science 295:1895–1897
27. Becker W et al (2004) Fluorescence lifetime imaging by time-correlated single-photon counting. Microsc Res Tech 63:58–66
28. Barber PR et al (2009) Multiphoton time-domain fluorescence lifetime imaging microscopy: practical application to protein-protein interactions using global analysis. J R Soc Interface 6:S93–S105
29. Digman MA, Caiolfa VR, Zamai M, Gratton E (2008) The phasor approach to fluorescence lifetime imaging analysis. Biophys J 94:L14–L16
30. Allan C et al (2012) OMERO: flexible, model-driven data management for experimental biology. Nat Methods 9:245–253
31. Keelan J, Allen NJ, Antcliffe D, Pal S, Duchen MR (2001) Quantitative imaging of glutathione in hippocampal neurons and glia in culture using monochlorobimane. J Neurosci Res 66:873–884
32. Zipfel WR et al (2003) Live tissue intrinsic emission microscopy using multiphoton-excited native fluorescence and second harmonic generation. Proc Natl Acad Sci U S A 100:7075–7080
33. Skala MC et al (2007) In vivo multiphoton microscopy of NADH and FAD redox states, fluorescence lifetimes, and cellular morphology in precancerous epithelia. Proc Natl Acad Sci 104:19494–19499
34. Tiede LM, Nichols MG (2006) Photobleaching of reduced nicotinamide adenine dinucleotide and the development of highly fluorescent lesions in rat basophilic leukemia cells during multiphoton microscopy. Photochem Photobiol 82:656–664
35. Wishart DS et al (2013) HMDB 3.0—the human metabolome database in 2013. Nucleic Acids Res. 41:D801–D807
36. Strober W (2001) Trypan blue exclusion test of cell viability. Curr Protoc Immunol. Appendix 3, Appendix 3B
37. Brauchle E, Thude S, Brucker SY, Schenke-Layland K (2014) Cell death stages in single apoptotic and necrotic cells monitored by Raman microspectroscopy. Sci Rep 4:4698
38. James DR, Ware WR (1985) A fallacy in the interpretation of fluorescence decay parameters. Chem Phys Lett 120:455–459
39. Wang HW et al (2008) Differentiation of apoptosis from necrosis by dynamic changes of

reduced nicotinamide adenine dinucleotide fluorescence lifetime in live cells. J Biomed Opt 13:9

40. Yu Q, Heikal AA (2009) Two-photon autofluorescence dynamics imaging reveals sensitivity of intracellular NADH concentration and conformation to cell physiology at the single-cell level. J Photochem Photobiol B Biol 95:46–57

41. Talbot CB et al (2011) Application of ultrafast gold luminescence to measuring the instrument response function for multispectral multiphoton fluorescence lifetime imaging. Opt Express 19:13848–13861

42. Appaix F et al (2012) Specific in vivo staining of astrocytes in the whole brain after intravenous injection of sulforhodamine dyes. PLoS One 7:e35169

Chapter 8

Techniques for Simultaneous Mitochondrial and Cytosolic Ca^{2+} Imaging in Neurons

Jacob E. Rysted, Zhihong Lin, and Yuriy M. Usachev

Abstract

Mitochondria efficiently buffer Ca^{2+} influx during excitation, which limits the amplitude of cytosolic Ca^{2+} rise, and then slowly release Ca^{2+} back into the cytosol thereby extending the duration of cytosolic Ca^{2+} response. This mitochondrial Ca^{2+} cycling helps shape Ca^{2+} transients and regulates Ca^{2+}-dependent functions in neurons such as excitability, synaptic plasticity, bioenergetics, and survival. Therefore identifying the molecular components of mitochondrial Ca^{2+} transport in neurons and defining their pharmacological and functional properties presents an important and challenging task for neuroscientists. Fulfilling this task requires a set of tools for simple and reliable measurement of Ca^{2+} concentration inside and outside mitochondria, i.e., within the mitochondrial matrix and cytosol, respectively. In this chapter we describe instrumentation and techniques for simultaneous measurements of mitochondrial and cytosolic Ca^{2+} concentration in central and peripheral neurons by using synthetic (e.g., Fura-2) and genetic (e.g., mito-R-GECO1) fluorescent Ca^{2+} indicators. We include detailed protocols for preparing primary cultures of hippocampal and dorsal root ganglion (DRG) sensory neurons, gene transfer into these cells and the use of fluorescent microscopy and patch clamp for comprehensive characterization of Ca^{2+} fluxes across the mitochondrial and plasma membranes.

Key words Ca^{2+}, Fura-2, Fluorescent microscopy, Genetic transfection, Genetically encoded Ca^{2+} indicator (GECI), Mito-R-GECO1, Mito-LAR-GECO1.2, Patch clamp

1 Introduction

1.1 Mitochondrial Ca^{2+} Transport in Neurons

Mitochondria play a prominent role in Ca^{2+} transport and signaling in neurons. The organelle efficiently buffers Ca^{2+} influx during neuronal excitation via the Ca^{2+} uniporter, and then slowly releases Ca^{2+} back into the cytosol via Na^+/Ca^{2+} and H^+/Ca^{2+} exchangers, which limits the amplitude and prolongs duration of $[Ca^{2+}]_i$ responses [1–6]. Consequently, mitochondria have a tremendous ability to regulate many Ca^{2+}-dependent functions in neurons including excitability, synaptic transmission, gene expression, and survival [2, 7–13]. Ca^{2+} transported into the mitochondrial matrix stimulates Ca^{2+}-dependent dehydrogenases and boosts ATP production to meet increased energy demand during excitation [14–16].

Stefan Strack and Yuriy M. Usachev (eds.), *Techniques to Investigate Mitochondrial Function in Neurons*, Neuromethods, vol. 123, DOI 10.1007/978-1-4939-6890-9_8, © Springer Science+Business Media LLC 2017

However, an excessive Ca^{2+} load within mitochondria can trigger neurotoxic processes that contribute to neuronal death following stroke and in several neurodegenerative disorders such as Alzheimer's, Parkinson's, and Huntington's diseases [17–22].

Ca^{2+} shuttling between the cytosol and the mitochondrial matrix involves crossing of the outer and inner mitochondrial membranes, OMM and IMM, respectively. OMM is considered to be Ca^{2+} permeable, whereas IMM contains a collection of specialized Ca^{2+} transporter molecules that enable Ca^{2+} flow across the IMM in a regulated fashion [23, 24]. Although the physiological characteristics of mitochondrial Ca^{2+} transport have been known for several decades, the molecules that mediate the transport across the IMM remained enigmatic until recently. In 2004, Kirichok and colleagues demonstrated that mitochondrial Ca^{2+} uniporter functions as a Ca^{2+}-activated and Ca^{2+}-selective ion channel [25]. In 2011, two groups identified MCU (mitochondrial calcium uniporter, also known as CCDC109A) as a pore-forming subunit of the complex [26, 27]. Recent nuclear magnetic resonance and electron microscopy data suggest that MCU subunits form a pentamer structure [28]. Opening of the channel enables robust Ca^{2+} flow from the intermembrane space into the matrix that is driven by a large membrane potential (~150–180 mV) across the IMM. MCU has a paralog, MCUb (also known as CCDC109B) whose expression in most tissues is much lower than that of MCU [29]. MCU and MCUb physically interact, likely forming heteromultimers in the IMM [29, 30]. It has been proposed that MCUb negatively regulates mitochondrial Ca^{2+} uptake [29], although the exact role of MCUb remains a subject of debate [28].

In addition, several auxiliary subunits of the MCU complex have been recently discovered. These include mitochondria Ca^{2+} uptake isoforms MICU1, MICU2, and MICU3 [31, 32], essential MCU regulator (EMRE) [30], and MCU regulator 1 (MCUR1) [33, 34]. MICU1 and MICU2 are responsible for Ca^{2+}-dependent gating of the MCU channel [35, 36]. EMRE is a transmembrane IMM protein that is required for MCU Ca^{2+} permeation and its MICU1-dependent regulation [37] (but see Vais et al. [38]). MCUR1 has been proposed to function as a scaffold factor for the MCU complex assembly [34]. However, two groups have disputed the MCU-MCUR1 connection and attributed other MCU-independent functions to MCUR1 [39, 40].

Under physiological conditions, Ca^{2+} extrusion from mitochondria in neurons is mediated primarily by the mitochondrial Na^+/Ca^{2+} exchanger [6, 23, 41]. It has been proposed that the molecular identity of this transporter is NCLX ($Na^+/Li^+/Ca^{2+}$ exchanger), also known as SLC8b1 [42]. Another molecule, Letm1, has been shown to function as a Ca^{2+}/H^+ exchanger in the IMM [43–45] (but see [46, 47]). Letm1 can move Ca^{2+} in and out of mitochondria, depending on the H^+ and Ca^{2+} gradients across the IMM, as well as on the mitochondrial membrane potential.

Despite significant progress in recent years, the molecular era of mitochondrial Ca^{2+} transport research is in its early stage. We still know little about the roles of the listed molecules in specific organ systems. The nervous system is not an exception. Recent advancements in the development of fluorescent genetically encoded Ca^{2+} indicators (GECI) that can be specifically targeted to mitochondria, have greatly improved our ability to examine molecular and functional properties of mitochondrial Ca^{2+} transport in neurons and in other cell types.

1.2 Mitochondrial Ca^{2+} Indicators

Before the era of genetic Ca^{2+} indicators, a synthetic fluorescent Ca^{2+} probe Rhod-2 and its analogs with lower Ca^{2+} affinities, e.g., Rhod-FF and Rhod-5N, were used by a number of investigators to monitor Ca^{2+} concentration within the mitochondrial matrix ($[Ca^{2+}]_{mt}$) [48–50]. Although originally developed for measuring cytosolic Ca^{2+} concentration ($[Ca^{2+}]_{cyt}$), Rhod-2 was found to strongly accumulate in mitochondria due to its positive charge. The latter property made Rhod-2 a useful tool for studying mitochondrial Ca^{2+} signaling in a variety of cell types, which led to significant progress in characterizing mitochondrial Ca^{2+} transport in live cells. An important limitation of this approach is that Rhod-2 (and other Rhod indicators) is also present in the cytosol, and therefore the cytosolic Rhod-2 fluorescence almost inevitably contaminates the Rhod-2 signal. An impact of this cytosolic contamination is determined by the dye partitioning between the mitochondria and cytosolic compartments and also depends on a specific dye loading protocol as well as cell type used. It also requires additional nontrivial steps to unload the dye from the cytosol by either permeabilizing the cells or by using whole-cell patch clamp dialysis of Rhod-2 dyes from the cytosol [50]. In addition, Rhod-2 was reported to affect the mitochondrial morphology and membrane potential [49], as well as to compartmentalize in other organelles such as lysosomes [51].

The first genetic mitochondrial Ca^{2+} indicator was developed in 1992 by Rizzuto and colleagues [52]. It was based on a Ca^{2+}-sensitive bioluminescent protein aequorin and was targeted to mitochondria by fusing it with the N-terminal targeting sequence of cytochrome c oxidase (COX) subunit VIII. Although the mitochondria-targeted aequorin (mtAEQ) enabled unequivocal monitoring of mitochondrial Ca^{2+} signaling in live cells, its low quantum yield and weak signal made it nearly impossible to use mtAEQ for monitoring $[Ca^{2+}]_{mt}$ at a single-cell level. A requirement for a chromophore, coelenterazine, to generate a functional photoprotein and irreversible Ca^{2+}-dependent consumption of mtAEQ further complicate its usage [53, 54].

A major progress in mitochondrial Ca^{2+} imaging was made with the invention of fluorescent genetic Ca^{2+} indicators based on the green fluorescent protein (GFP) and calmodulin in late 1990th–early 2000th by Roger Tsien and colleagues [55, 56] as well as by several other groups [57–59]. These new indicators could be specifically targeted to mitochondria (and other organelles) and have a much greater quantum efficiency, which makes them suitable for monitoring mitochondrial Ca^{2+} signals in individual cells with high spatial and temporal resolution. These indicators can be broadly divided into two types. Indicators of the first type, called Cameleons, are based on Förster resonance energy transfer (FRET) and consist of two fluorescent proteins, a shorter wavelength donor (e.g., ECFP = enhanced cyan fluorescent protein) and a longer wavelength acceptor (e.g., EYFP = enhanced yellow fluorescent protein or cp-Venus) [55, 60, 61]. The donor and acceptor proteins are linked via calmodulin (CaM) that serves as a Ca^{2+} sensor, and the CaM-binding peptide M13 that is derived from the myosin light chain kinase. Ca^{2+} binding to CaM results in the conformational change that enhances FRET between the donor and acceptor proteins, which can be detected by fluorescent microscopy. The second type of indicators is represented by GECIs such as Pericam [58] and GCaMP [59]. They are based on a single permutated GFP-derived protein that is flanked by CaM and M13 peptide. Ca^{2+} binding to CaM leads to an M13-CaM interaction and conformation change, which ultimately causes an increase or decrease in emitted fluorescence [60, 61].

Over the last 10–15 years, these originally developed GECI have been substantially improved through structure-guided remodeling as well as through a directed protein evolution in an *E. coli*-based system combined with large-scale screening and optimization. This led to generation of a broad mutlicolor palette of GECIs consisted of dozens of fluorescent indicators that differ in their Ca^{2+} affinity, quantum yield, dynamic range, and spectral and kinetic characteristics [62–69]. Many of these GECIs were targeted to mitochondria by attaching one or more copies of cytochrome c oxygenase subunit VIII targeting sequence. The most commonly used mitochondrial GECIs (mito-GECIs) are listed in Table 1. The choice of an appropriate mitochondrial Ca^{2+} indicator is dictated by its spectral properties (absorption/excitation and emission wavelengths), Ca^{2+} affinity (K_d, determines the range of $[Ca^{2+}]_{mt}$ changes that will be reliably reported by a mito-GECI), dynamic range (fold-change in fluorescence intensity when indicator transitions from a Ca^{2+}-free to a Ca^{2+}-bound state), Hill coefficient/constant (determines the steepness of fluorescence intensity change with $[Ca^{2+}]_{mt}$ increase; indicators with Hill constant ~1 have a near linear response curve; indicators with larger Hill constant, e.g., ~2–3, have a less gradual, more abrupt change in fluorescence once Ca^{2+} concentration approaches the K_d of the

Table 1
Properties of genetically encoded mitochondrial Ca^{2+} indicators (mito-GECIs)

Indicator	Ex (nm)	Em (nm)	Dynamic range	K_d (μM)	Hill constant	References
mtPericam (ratiometric)	418/494	511	10	1.7	1.1	[2, 43, 58, 80, 95]
mtInvPericam (inverse)	490	513	6.7	0.2	1	[33, 58, 96]
4mtD3cpv	436[a]	475/535	6.1	0.6		[62, 97–99]
mito-GCaMP2	487	511	5	0.15	3.8	[99–101]
mito-GCaMP5G	500/400	515	32.7	0.46	2.5	[61, 102, 103]
mito-R-GECO1	561	589	16	0.48	2.06	[63]
mito-LAR-GECO1.2	557	584	8.7	12.0	1.4	[65]
mito-GEM-GECO1	390	455/511	110	0.34	2.9	[40, 63]
CEPIA2mt	498	512	5.1	0.16	2.5	[67]
CEPIA3mt	498	512	5.0	11.0	1.5	[67]
CEPIA4mt	498	512	4.9	56.0	1.7	[67]

Ex excitation wavelength, *Em* emission wavelength, K_d Ca^{2+} dissociation constant. Ex and Em peaks are given for Ca^{2+}-bound forms of the indicators

[a]FRET optimized excitation wavelength is shown for 4mtD3cpv

indicator). An additional important factor to consider is whether a mito-GECI can be used in a dual-wavelength ratiometric mode, which is the case with 4mtD3cpv and mito-GEM-GECO1. A significant advantage of the ratiometric mode used for $[Ca^{2+}]_{mt}$ measurements is that it is less sensitive to the concentration (expression) of an indicator, its photobleaching and preparation movement artifacts. In addition, ratiometric indicators are easier to calibrate and use for determining absolute $[Ca^{2+}]_{mt}$ values. On a flip side, ratiometric indicators typically require a more sophisticated instrumentation. Several recent excellent reviews provide comprehensive discussion of theoretical and practical aspects of GECI use in various biological applications [54, 60, 61, 70, 71].

1.3 Ca^{2+} Indicators for Simultaneous Measurements of $[Ca^{2+}]_{mt}$ and $[Ca^{2+}]_{cyt}$ in Neurons

Mitochondrial Ca^{2+} transport shuttles Ca^{2+} between the two cellular compartments, the cytosol and the mitochondrial matrix. Therefore, comprehensive examination of the transport requires simultaneous monitoring of Ca^{2+} concentration in both compartments. In addition, $[Ca^{2+}]_{cyt}$ monitoring provides an important control for potential off-target effects of molecular and pharmacological manipulations directed at mitochondrial Ca^{2+} transporters. For example, in neurons if a putative mitochondria targeting drug X inadvertently inhibits voltage-gated Ca^{2+} influx (but not mitochondrial Ca^{2+} uptake), this could result in a decreased stimulus-evoked $[Ca^{2+}]_{mt}$ elevation. In the absence of concurrent $[Ca^{2+}]_{cyt}$ monitoring one can wrongly conclude that drug X inhibits its mitochondrial Ca^{2+} uptake. Such data misinterpretation can be avoided if Ca^{2+} concentration is simultaneously measured both in the cytosol and mitochondria.

For cytosolic Ca^{2+} measurements, a synthetic fluorescent Ca^{2+} indicator, Fura-2, and its analogs remain some of the best (if not the best) in the field for a number of reasons [54, 72, 73]. First, Fura-2 is a dual-wavelength indicator, i.e., Ca^{2+} binding to Fura-2 causes its absorption/excitation spectrum shift to the left, producing the opposite changes in fluorescence intensity, i.e., increase for λ_{Ex} = 340 nm and decrease for λ_{Ex} = 380 nm. This property allows employment of the indicator in the ratiometric mode, which minimizes the sensitivity of Fura-2-based measurements to the dye concentration, photo-bleaching, and movement artifacts. Second, Fura-2 has a high quantum yield and is resistant to photo-bleaching. Third, the indicator is highly sensitive to Ca^{2+} (K_d~275 nM at room temperature; [74]) and is highly selective for Ca^{2+} relative to other metal ions. Fourth, the available acetoxymethyl (AM) form (esterified, membrane-permeable) of the indicator simplifies its loading into intact cells. Fifth, lower affinity Fura-2 analogs are available, such as Fura-4F (K_d = 770 nM), Fura-FF (K_d = 5500 nM), and Mag-Fura-2 (K_d = 25,000 nM). These additional Fura indicators have spectral properties that are very similar to that of Fura-2, which enables one to expand the dynamic range of $[Ca^{2+}]_{cyt}$ measurements using essentially the same instrumentation, and like Fura-2 are available in AM form for simple loading into intact cells (The Molecular Probes Handbook; 4th edition). Therefore, we aimed at identifying a genetic mitochondrial Ca^{2+} indicator that will be compatible with Fura-2 and other Fura indicators, thus enabling simultaneous $[Ca^{2+}]_{mt}$ and $[Ca^{2+}]_{cyt}$ measurements.

After extensive experimentation with various mito-GECIs, we identified red-shifted mitochondrial Ca^{2+} indicators mito-R-GECO1 and mito-LAR-GECO1.2 ([63, 65] and Table 1) as excellent indicators for combined imaging with Fura-2 or Fura-FF in neurons (Figs. 1 and 2). Mito-R-GECO1 has a high-to-intermediate Ca^{2+} affinity (K_d = 0.48 μM) [63], whereas mito-LAR-GECO1 is a low-affinity Ca^{2+} indicator (K_d= 12 μM) [65]. These indicators have several advantages over other mito-GECIs used for the same purpose (Table 1). First, there is no significant spectral overlap between Fura-2 (Fura-FF) and mito-R-GECO1 (mito-LAR-GECO1.2) indicators. In contrast, the excitation spectra of several other mitochondrial indicators, such as 4mtD3cpv, mito-GEM-GECO1, and mtPericam, substantially overlap with the excitation spectrum of Fura-2 (Table 1). Second, the red-shifted excitation of mito-R-GECO1 (mito-LAR-GECO1.2) reduces cell autofluorescence and phototoxicity as compared to indicators with blue-shifted excitation light. Third, both indicators have high quantum yield [63, 65],

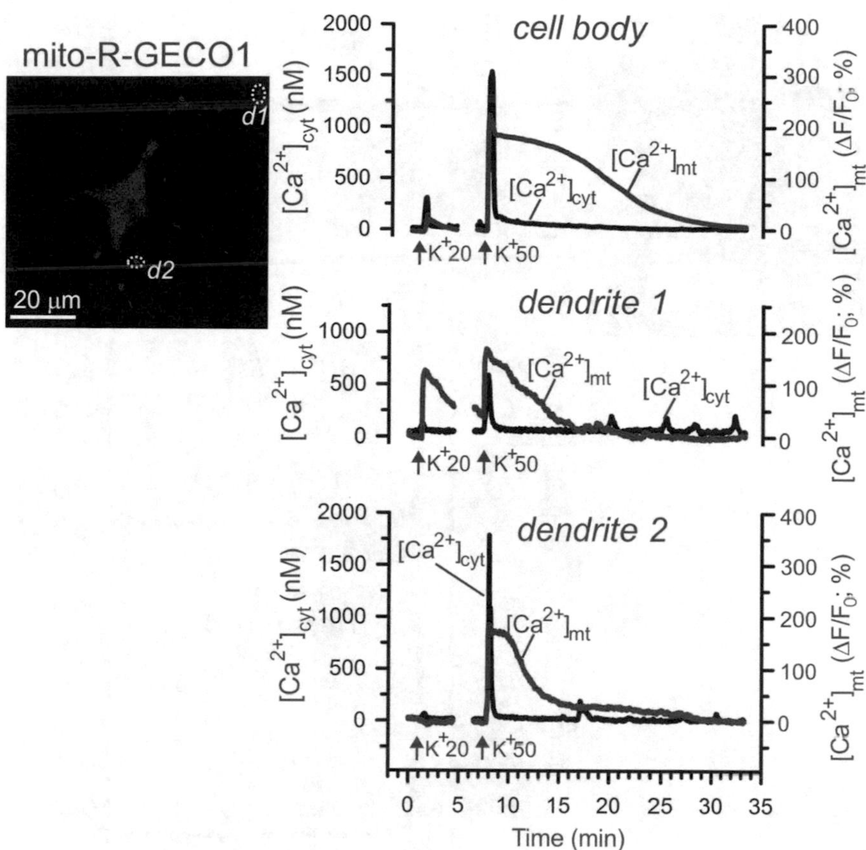

Fig. 1 Simultaneous monitoring of mitochondrial ($[Ca^{2+}]_{mt}$) and cytosolic ($[Ca^{2+}]_{cyt}$) Ca^{2+} concentrations in hippocampal neurons using mito-R-GECO1 and Fura-2, respectively. Image shows cellular distribution of mito-R-GECO1 (λ_{Ex} = 550 nm) in a cultured hippocampal neuron. The graphs show depolarization (20 or 50 mM KCl, 30 s)-induced changes in $[Ca^{2+}]_{mt}$ (*red*) and $[Ca^{2+}]_{cyt}$ (*black*) in the cell body (*top plot*) and individual mitochondria within small dendritic segments d1 = dendrite1 (*middle plot*) and d2 = dendrite 2 (*bottom plot*). All three plots represent the same experiment, and demonstrate heterogeneity among $[Ca^{2+}]_{mt}$ and $[Ca^{2+}]_{cyt}$ responses recorded from various regions. *White dashed lines* on the image denote regions of interest within the dendrites. A selective inhibitor of voltage-gated Na^+ channels, tetrodotoxin (200 nM), was added to the extracellular solutions to prevent action potential firing

which enables $[Ca^{2+}]_{mt}$ measurement at a single-cell and even single-organelle level (Figs. 1 and 2). Fourth, these indicators are characterized by high dynamic ranges, e.g., 16 for mito-R-GECO1 and 8.7 for mito-LAR-GECO1.2 (Table 1). Fifth, both indicators have Hill coefficient ~1 (2.06 for mito-R-GECO1 and 1.4 for mito-LAR-GECO1.2), which make them relatively linear reporters of $[Ca^{2+}]_{mt}$ within broad ranges below and near of their respective K_d. Much larger Hill coefficients (2.5–3.8) such as those reported for mito-GCaMP2, mito-GCaMP5G, mito-GEM-GECO1, and CEPIA2mt (Table 1) are associated with highly nonlinear response curves, which substantially complicates $[Ca^{2+}]_{mt}$ quantification based on fluorescence intensity change of these indicators [70]. A very high

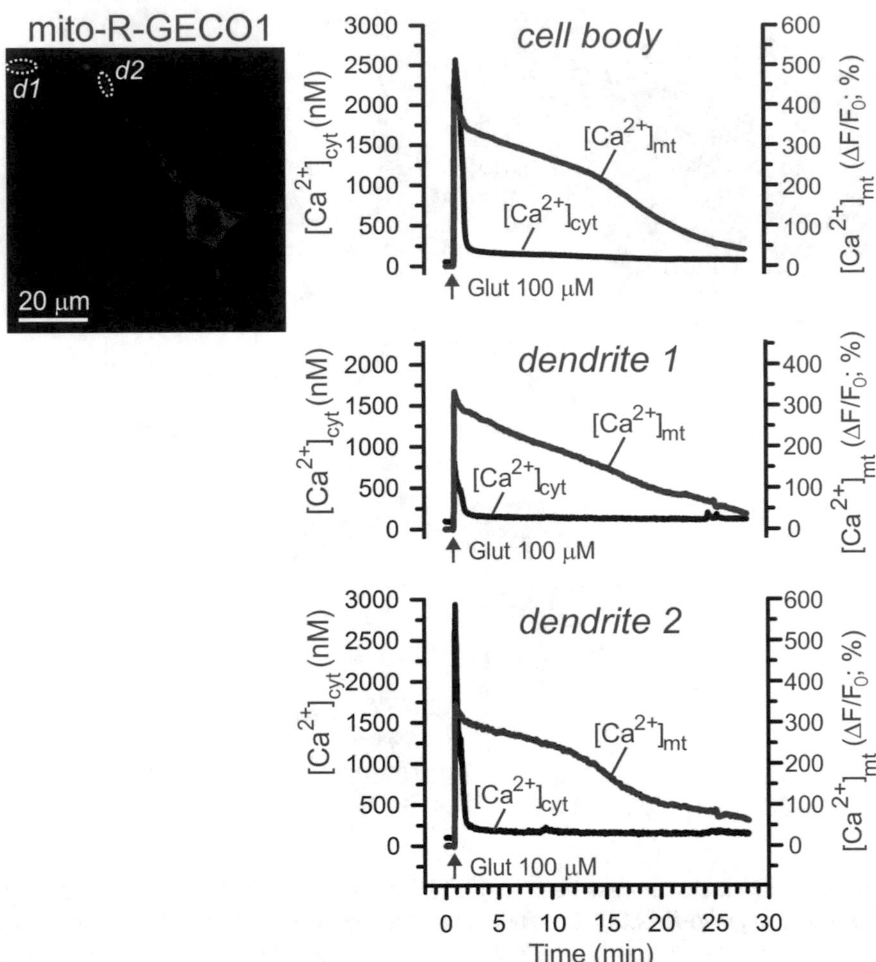

Fig. 2 Glutamate-induced $[Ca^{2+}]_{mt}$ and $[Ca^{2+}]_{cyt}$ changes were simultaneously recorded in a hippocampal neuron using mito-R-GECO1 (*red*) and Fura-2 (*black*), respectively. Image shows distribution of mito-R-GECO1 fluorescence ($\lambda_{Ex} = 550$ nm) in a cultured transfected hippocampal neuron. Graphs show simultaneous recordings of $[Ca^{2+}]_{mt}$ and $[Ca^{2+}]_{cyt}$ from three different regions, cell body (*top plot*), dendritic segment (d1 = dendrite 1; *middle plot*) and dendritic segment 2 (d2 = dendrite 2; *bottom plot*). The time of glutamate 100 μM application (30 s) is shown by *arrow*. The perfusion solutions also contained 200 nM of tetrodotoxin (to block voltage-gated Na^+ channels and action potential firing) and 10 μM of the glutamate NMDA receptor co-agonist glycine. Dendritic regions of interest are shown by *white dashed lines* on the image

Hill coefficient of 3.8 for mito-GCaMP2 (Table 1) makes it a nearly all-or-none sensor of $[Ca^{2+}]_{mt}$ responses.

A choice of a specific pair of indicators, such as Fura-2 and mito-R-GECO1; Fura-2 and mito-LAR-GECO1.2; or Fura-FF and mito-LAR-GECO1.2, depends on the expected range of $[Ca^{2+}]_{mt}$ and $[Ca^{2+}]_{cyt}$ changes in a given cell type under specific experimental conditions. We have previously published implementation of the Fura-2 and mito-LAR-GECO1.2 pair for combined $[Ca^{2+}]_{mt}$ and $[Ca^{2+}]_{cyt}$ monitoring in neurons [65]. In this chapter, we focus on a combination of the high-affinity types of cytosolic and mitochondrial indicators, Fura-2 and mito-R-GECO1. However, given the spectral similarities with their lower affinity analogs, Fura-FF and mito-LAR-GECO1, the overall methods and instrumentation will be very similar for the other possible combinations of Fura and mito-(LA)R-GECO probes.

2 Materials and Equipment

In this section we describe materials that are required for preparing and transfecting mouse hippocampal and DRG neurons, as well as imaging equipment that we use for simultaneous monitoring $[Ca^{2+}]_{cyt}$ and $[Ca^{2+}]_{mt}$ in neurons.

2.1 General Equipment and Materials

1. Sterile laminar flow dissection hood.
2. Sterile laminar flow culture hood.
3. Dissection microscope
4. Benchtop centrifuge.
5. 5 and 10 % CO_2 cell culture incubators.
6. Light microscope and hemacytometer for counting cells.
7. 6-well cell culture plates.
8. Sterile 25 mm glass coverslips (Fisher Scientific, Cat. #12-545-86).
9. Sterile 8 mm cloning cylinders (Sigma, Cat. #CLS31668).
10. Sterile scissors and forceps.
11. Sterile surgical blades (#10 and #15T; Cincinnati Surgical Co.).
12. 0.2 μm syringe filters (Acrodisc Cat.# 4427T, 13 mm diameter).

2.2 Hippocampal Culture Reagents

1. Culture Medium **H1**.
   - 500 ml Neurobasal-A (Gibco/Thermofisher, Cat.# 10888-022).
   - 1.25 ml of 200 mM L-glutamine (Gibco Cat.# 25030-149).
   - 5 ml of 1 M HEPES (pH 7.3, adjusted with NaOH).
   - 2.5 ml Pen–Strep (penicillin–streptomycin mixture; Gibco Cat.# 15140-122) (exclude for Lipofectamine transfection).
2. Culture Medium **H2**.
   - 90 ml of Minimal Essential Medium (MEM; Gibco Cat.#11095-080).
   - 10 ml of heat inactivated horse serum (HIHS; Gibco Cat.#26050-070).
   - 48 μl of 1.25 % insulin (in 0.01 M HCl; Sigma Cat.# I-5500).

- 0.5 ml Pen–Strep (penicillin–streptomycin mixture; Gibco Cat.# 15140-122).
- pH = 7.35.

3. Culture supplement B27 (Gibco Cat.# 17504-044).
4. Trypsin Solution. Trypsin (Sigma Cat.# T-7409) solution (3 ml; 1 mg/ml in culture medium H1); must be sterilized by filtration via a 0.2 μm syringe filter tip (Acrodisc, 13 mm diameter).
5. Sterile DPBS (Dulbecco's phosphate buffered solution; Gibco, Cat. #14190-144).
6. Poly-L-ornithine solution. Dilute 20 mg poly-L-ornithine (Sigma P-4538) in 100 ml of 150 mM boric buffer (pH 9.0).
7. Laminin (Gibco; Cat.#23017-015; 0.05 mg/ml solution in sterile DPBS).

2.3 Dorsal Root Ganglia (DRG) Culture Reagents

1. Culture Medium **D1** (500 ml).
   - Dulbecco's modified Eagle medium (DMEM; Sigma, Cat.#D-5523) 5 g dissolved in 487.5 ml of distilled water.
   - 10 ml of 1 M HEPES (pH 7.4, adjusted with NaOH).
   - 2.5 ml Pen–Strep (penicillin–streptomycin mixture; Gibco, Cat.# 15140-122).
2. Culture Medium **D2** (500 ml).
   - Dulbecco's modified Eagle medium (DMEM; Sigma, Cat.#D-5523) 5 g dissolved in 450 ml of culture grade purified distilled water.
   - Sodium bicarbonate ($NaHCO_3$) 1.85 g.
   - Adjust pH to 7.35 using 1 M HCl (or if necessary with NaOH).
   - 2.5 ml Pen–Strep (penicillin–streptomycin mixture; Gibco, Cat.# 15140-122).
3. DRG Complete Culture Medium (100 ml).
   - Culture Medium D2 90 ml.
   - Heat-inactivated horse serum (HIHS; Gibco, Cat.#26050-070) 5 ml.
   - Fetal bovine serum (FBS; Gibco, Cat.# 26140-079) 5 ml.
   - Nerve growth factor (NGF; AdD Serotec, Cat.# PMP04Z) 50 ng/ml.
   - Insulin (Sigma, Cat.# I-5500) 6 μg/ml.
   - The solution must be prepared fresh and should not be stored at 4 °C longer than 2–3 days due to rapid deterioration of the sera, NGF, and insulin (see Note 1).
4. Collagenase A (Roche, Cat.# 10103578001).
5. Pronase E (Amresco, Cat.# E629-1G).

2.4 Transfection Reagents and Equipment

1. Lipofectamine 2000 (hippocampal neurons; Invitrogen, Cat.# 11668019).
2. Opti-MEM I transfection medium for hippocampal neurons (Gibco, Cat.# 31985-070).
3. Transfection Medium **H3**.
   - 500 ml Neurobasal-A (Gibco/Thermofisher, Cat.# 10888-022).
   - 1.25 ml of 200 mM L-glutamine (Gibco Cat.# 25030-149).
   - 5 ml of 1 M HEPES (pH 7.3, adjusted with NaOH).
4. Amaxa Nucleofector II (Lonza; Illinois) is an electroporator that we use for transfecting DRG neurons.
5. Amaxa Mouse Neuron Nucleofector Kit (Lonza, Cat.# VPG-1001) contains transfection cuvettes and required solutions optimized for transfecting mouse cultured neurons.
6. cDNA plasmids such as mito-R-GECO1 (Addgene Cat.# 46021), mito-LAR-GECO1.2; (Addgene Cat.# 61245) or others, amplified and purified using endotoxin-free Qiagen EndoFree Plasmid Maxi kit (Qiagen Cat.#12362).

2.5 Buffers for Ca^{2+} Imaging Experiments

1. *HEPES buffered Hanks Salt Solution (HH Buffer).*
   140 mM NaCl, 5 mM KCl, 1.3 mM $CaCl_2$, 0.4 mM $MgSO_4$, 0.5 mM $MgCl_2$, 0.4 mM KH_2PO_4, 0.6 mM Na_2HPO_4, 3 mM $NaHCO_3$, 10 mM glucose, 10 mM HEPES, 1 ml Phenol Red, pH 7.4, 310 mOsm/kg with sucrose.
2. *50 mM KCl HEPES buffered Hanks Salt Solution (K+50).*
   95 mM NaCl, 50 mM KCl, 1.3 mM $CaCl_2$, 0.4 mM $MgSO_4$, 0.5 mM $MgCl_2$, 0.4 mM KH_2PO_4, 0.6 mM Na_2HPO_4, 3 mM $NaHCO_3$, 10 mM glucose, 10 mM HEPES, 1 ml phenol red, pH 7.4, 310 mOsm/kg with sucrose.
3. *Ca2+ Free HEPES buffered Hanks Salt Solution (Ca2+ Free HH Buffer).*
   140 mM NaCl, 5 mM KCl, 0.5 mM $MgCl_2$, 0.4 mM $MgSO_4$, 0.4 mM KH_2PO_4, 0.6 mM Na_2HPO_4, 3 mM $NaHCO_3$, 10 mM glucose, 10 mM HEPES, 1 ml phenol red, pH 7.4, 310 mOsm/kg with sucrose, and 1 mM EGTA.
4. Fura-2/AM (Invitrogen Cat.#F-1201). Prepare a 2 mM stock solution of Fura-2/AM dissolved in DMSO supplemented with 20 % Pluronic F-127 (Invitrogen Cat.# P-6866), aliquot and store at −20 °C until use.
5. Fura-FF/AM (Setareh-Biotech Cat.# 6546). Prepare a 2 mM stock solution of Fura-FF/AM dissolved in DMSO supplemented with 20 % Pluronic F-127 (Invitrogen Cat.# P-6866), aliquot and store at −20 °C until use.

2.6 Intracellular and Extracellular Buffers for Patch Clamp Recordings

1. *Ca2+ currents intracellular buffer.*
   125 mM Cs gluconate, 10 mM NaCl, 3 mM Mg-ATP, 1 mM $MgCl_2$, 0.5 mM GTP, 100 μM Fura-2 (Invitrogen Cat.# F-1200), 10 mM HEPES, pH 7.25 (adjusted with CsOH), 290 mOsm/kg (adjusted with sucrose).
2. *Ca2+ currents extracellular buffer.*
   115 mM CholineCl, 30 mM TEACl, 1.3 mM $CaCl_2$, 1 mM $MgCl_2$, 10 mM glucose, 10 mM HEPES and 1 μM tetrodotoxin, pH 7.4 (adjusted with TEAOH), 310 mOsm/kg (adjusted with sucrose).
3. *Current-clamp intracellular buffer (for monitoring electrical activity).*
   125 mM K gluconate, 10 mM NaCl, 3 mM Mg-ATP, 1 mM $MgCl_2$, 0.5 mM GTP, 100 μM Fura-2 (Invitrogen Cat.# F-1200), 10 mM HEPES, pH 7.25 (adjusted with KOH), 290 mOsm/kg (adjusted with sucrose).
4. *Current-clamp extracellular buffer.*
   HH buffer (Sect. 2.5, item 1).

2.7 Fluorescent Imaging System

Our imaging system is built by using an inverted Olympus IX-71 microscope (Olympus, Japan) and a TILL Photonics CCD-camera based imaging system (TILL Photonics/FEI, Germany/USA) (Fig. 3). It enables high temporal (up to 4 ms) and spatial (pixel = 166 nm) resolution imaging with continuous excitation spectrum capability in the 320–680 nm range. Although we focus on a specific configuration that we have successfully used for a variety of research applications, systems/components with similar characteristics and capabilities are also available from a number of other manufacturers. Several excellent reviews and a book provide comprehensive discussion of instrumentation for Ca^{2+} imaging in vivo and in vitro [54, 75–77].

1. *Microscope.*
   Inverted IX-71 microscope (Olympus, Japan) equipped with high numerical apertures oil immersion Olympus objectives 40× (NA = 1.35) and 60× (NA = 1.40), and dry Olympus objectives 10× (NA = 0.40) and 20× (NA = 0.75).
2. *TILL Photonics imaging system* (TILL Photonics/FEI, Germany/USA).
   - Polychrome V that houses a 150 W Xenon high stability lamp, power supply and a monochromator (scanner based diffraction grating) with the 320–680 nm wavelength range and 15 nm spectral bandwidth.
   - Interline transfer CCD camera TILL Imago VGA (640 × 480 pixels; TILL Photonics). Without binning, pixel size = 250 nm for 40× objective, and pixel size = 166 nm for 60× objective.

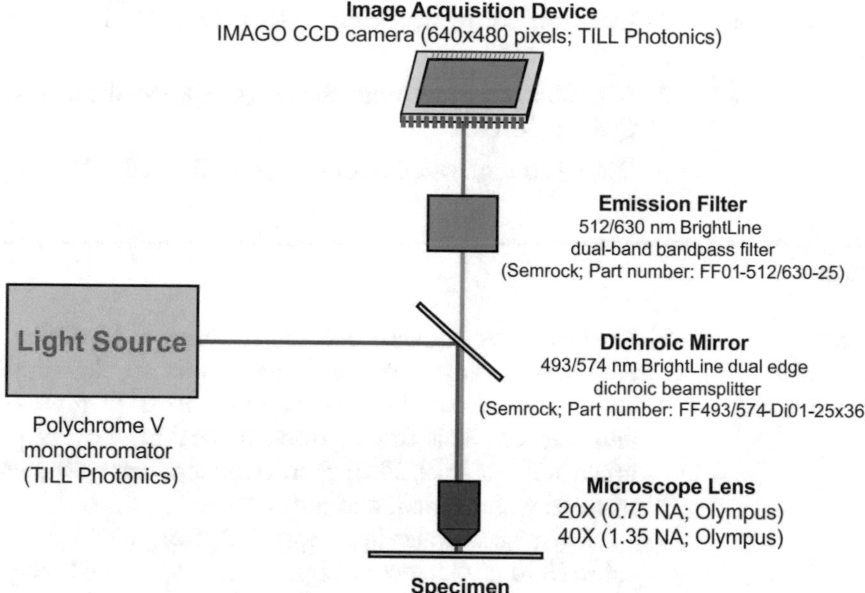

Fig. 3 Schematic of the imaging system for simultaneous measurement of mito-R-GECO1 and Fura-2 fluorescence

- ICU controller unit.
- TILLvisION 4.0.1.2 software (earlier version) or Live Acquisition (LA) software (later version) (TILL Photonics/ FEI).

3. *Filter set.*

   The filter set we use is optimized for combined Fura/mito-R-GECO1 and Fura/GCaMP/mito-R-GECO1 imaging (Fig. 3). Since selective excitation is achieved using a monochromator, no excitation filter is used.

   Dual band 493/574 nm BrightLine dichroic beamsplitter (Semrock, Cat.# FF493/574-Di01).

   Dual band 512/630 bandpath BrightLine emission filter (Semrock, Cat.# FF01-512/630).

2.8 Patch Clamp Equipment

1. Patch clamp amplifier Axopatch 200-B (Axon Instruments/ Molecular Devices).
2. Controller unit Digidata1440A (Axon Instruments/Molecular Devices).
3. Software pCLAMP 10 (Axon Instruments/Molecular Devices).
4. MP-225 micromanipulator (Sutter Instrument).
5. Micropipette puller P-87 (Sutter Instrument).
6. Glass capillary with filament for patch-pipette production (Narishige, Cat.# GD-1).

2.9 Field Stimulation Equipment

1. Electrical stimulator Grass S48 (Grass Medical Instruments; Quincy, MA).
2. Stimulus isolation unit SIU5 (Grass Medical Instruments; Quincy, MA).
3. Digital storage oscilloscope Siglent SDS 1052DL (Siglent).

3 Methods

3.1 Preparing Cultures of Mouse Hippocampal Neurons

1. Coverslip coating with poly-L-ornithine and laminin.
   All the procedures are performed under sterile conditions in a cell culture hood. Prior to coating, coverslips must be cleaned and washed. This can be done by soaking coverslips in 95 % ethanol for at least 24 h; then coverslips are rinsed with sterile distilled water, dried, and autoclaved.

   Coat glass coverslips (Sect. 2.1, item 8) with poly-L-ornithine (300 μl/each coverslip; Sect. 2.2, item 6), keep at room temperature for 1–2 h; wash twice with sterile distilled water (use vacuum suction for removing water). Add 50 μl of laminin to each coverslip (Sect. 2.2, item 7), incubate at 4 °C for at least 6 h (can be stored at 4 °C up to 3 months). Wash coverslips with sterile 1× DPBS (Sect. 2.2, item 5).
2. Place sterile cloning rings (Sect. 2.1, item 9) in the center of the coverslips in 6-well plates, and return the plates with coverslips to 5 % CO_2 incubator.
3. Dissect out the brain of a neonatal (P0-P1) mouse pup and place in chilled dissection media in a 35 mm petri dish. Dissect out the hippocampus then transfer it to a new petri dish with ice-chilled Culture Medium **H1** (Sect. 2.2, item 1).
4. Using a scalpel cut the hippocampus into small pieces and transfer to trypsin solution (Sect. 2.2, item 4) for 10 min.
5. After trypsin digestion, wash the cells two times with 4 ml of Culture Medium **H1** (Sect. 2.2, item 1) in 15 ml conical (Falcon) tubes. Use a new tube for each wash.
6. Collect and transfer cells to a new 15 ml tube with 1.8 ml of Culture Medium **H1** to mechanically dissociate the cells by triturating them with fire-polished Pasteur pipettes of increasingly smaller bore size (triturate 10–16 times per bore size pipette).
7. Culture the cells onto 25 mm glass coverslips coated with poly-L-ornithine/laminin in a 6-well plate by adding 150 μl of the cell suspension to cloning cylinders on the coverslips; then add 150 μl of Culture Medium **H2** (Sect. 2.2, item 2) to each cloning ring. Serum in Culture Medium **H2** will help to block the activity of trypsin if any residual amount remains after 2× wash (see above).

8. Incubate cells in 5 % CO_2 at 37 °C for 2 h, and then add 2 ml of Culture Medium **H1** supplemented with B27 (1 ml B27/50 ml Culture Medium **H1**) and remove the cloning cylinders carefully to avoid disturbing the cells.
9. Keep hippocampal cultures in 5 % CO_2 at 37 °C. Replace 50 % of culture medium with fresh Culture Medium **H1** supplemented with B27 every 3–4 days.

3.2 Transfection of Hippocampal Neurons

1. Hippocampal neurons are typically transfected after 6–7 days in culture/in vitro (6–7 DIV). The recipe is formulated for a single well of a 6-well plate.
2. *Tube A:* Add 4 μl of Lipofectamine 2000 (Sect. 2.4, item 1) to 0.5 ml of Opti-MEM I solution (Sect. 2.4, item 2) in a 5 ml plastic tube and gently mix, incubate for 5 min.
3. *Tube B:* Add 0.2–1.2 μg of plasmid DNA to 0.5 ml Opti-MEM I in another 5 ml tube and gently mix.
4. *Tube C:* Add content of tube B to tube A and mix gently; incubate for 20 min.
5. *Conditioned Medium:* For each well, transfer 1 ml of culture (conditioned) medium to a sterile 5 ml (or 15 ml tube if performed for multiple wells) and add 1 ml of fresh Culture Medium H2 supplemented with B27 (20 μl/ml), mix gently and store in 5 % CO_2 incubator.
6. For each well, gently aspirate the remaining culture medium (~1 ml) and replace it with the Tube C content (1 ml). Add 1 ml of Transfection Medium H3 (Sect. 2.4, item 3) supplemented with B27 (20 μl/ml). Incubate in 5 % CO_2 incubator at 37 °C for 2 h.
7. After 2 h, replace the transfection solution with Conditioned Medium (2 ml/well) described above.
8. Ca^{2+} imaging experiments are typically performed in 3–7 days after the transfection.

3.3 Culture and Transfection of DRG Neurons from Adult Mice

1. The protocol for coverslip cleaning and coating with poly-L-ornithine and laminin is the same protocol as for hippocampal cultures (Sect. 3.1, step 1). Wash coverslips with sterile 1× DPBS (Sect. 2.2, item 5) within ~1 h before plating DRG neurons.
2. Place sterile cloning rings (Sect. 2.1, item 9) in the center of the coverslips in 6-well plates, and return the plates with coverslips to 5 % CO_2 incubator.
3. For DRG neuronal culture we typically use 3 adult mice (6–10 week old) in order to collect a sufficient number of cells that is required for Amaxa-based transfection (nucleofection) protocol. Dissect out all of the DRG and place isolated ganglia into

an ice chilled Culture Medium **D1** (Sect. 2.3, item 1) in a 15 ml conical tube. Spin them down at 750 rpm for 5 min.

4. Use a 2 ml eppendorf tube to prepare Collagenase A (Sect. 2.3, item 4) digestion solution by dissolving 10 mg Collagenase A in 1 ml of Culture Medium **D1**. Sterilize the solution by filtering through a 0.2 μm Acrodisc filter into a new sterile 2 ml eppendorf tube.
5. Gently aspirate Culture Medium **D1** from DRG pellet (see step 3) in the 15 ml conical tube using 1 ml, and then 200 μl plastic pipette tip. Add the Collagenase A solution to the 15 ml tube, and then bring the volume to 5 ml with Culture Medium **D1** (2 mg/ml final concentration) and incubate at 37 °C for 20 min. In the middle of the digestion (in ~10 min), mix gently with a Pasteur pipette.
6. Aspirate gently the collagenase A solution using 1 ml, and then 200 μl plastic pipette tip. Be careful to avoid disturbance of DRG. Wash DRG by filling conical tube with 15 ml of Culture Medium **D1** and centrifuge at 750 rpm for 3 min.
7. Prepare Pronase E digestion solution by adding 5 mg of Pronase E (Sect. 2.3, item 5) to 1 ml of Culture Medium **D1** in a 2 ml eppendorf tube and sterilize as described above (Sect. 3.3, step 4).
8. After DRG are spun down, carefully aspirate media using pipettes with 1 ml, and then 200 μl pipette tip, and replace with pronase E solution, then fill volume to 5 ml with Culture Medium **D1** to a final concentration of 1 mg/ml of pronase E. Incubate in 5 % CO_2 at 37 °C for 10 min. In the middle of the digestion (in ~5 min), mix solution gently with a 1 ml plastic pipette (make sure that DRG are not caught into the pipette).
9. Carefully aspirate pronase E solution; then fill the tube to 15 ml with Culture Medium **D1** and centrifuge at 750 rpm for 3 min.
10. Carefully aspirate the media and then add 2 ml of Culture Medium **D1** and mechanically dissociate cells by triturating with fire-polished Pasteur pipettes of increasingly smaller size (triturate 15–20 times per pipette).
11. Centrifuge at 750 rpm for 5 min.
12. Prepare transfection solution by mixing 90 μl of mouse neuron nucleofector solution with 20 μl supplement (both are included in the Amaxa mouse nucleofector kit (Sect. 2.4, item 5).
13. Add plasmid DNA solution into a new eppendorf tube. We typically use 1–5 μg of DNA (or DNA mix), depending on a specific DNA. The solution volume should not exceed 5–7 μl. Mix 100 μl of the nucleofector solution with the plasmid DNA.

14. Transfer cell suspension to the tube with DNA–nucleofector solution mix, resuspend twice, and transfer to an Amaxa transfection cuvette included in the Amaxa mouse nucleofector kit (Sect. 2.4, item 5). Ensure that there are no bubbles inside the solution. Bubbles can be removed by tapping gently the cuvette.
15. Place cuvette into Amaxa Nucleofector II and transfect cells using G-013 protocol.
16. Transfer cells into a new eppendorf tube (2 ml) using a special plastic pipette included in the kit (Sect. 2.4, item 5). Fill up to 2 ml with DRG Complete Culture Medium (Sect. 2.3, item 3) to the tube and incubate in a 10 % CO_2 incubator for 5–10 min (required for cell recovery following the transfection).
17. Plate cell suspension into cloning cylinders on poly-L-ornithine/laminin coated glass coverslips in a 6-well plate (~300 μl/coverslip).
18. Incubate in 10 % CO_2 at 37 °C for ~1–2 h. Then, add 2 ml of DRG Complete Culture Medium (Sect. 2.3, item 3) to each well and carefully remove cloning cylinders to avoid disturbing cells. We typically use the cells after 2–3 days in culture. However they can be used as early as 24 h after plating. Culturing cells for more than 4–5 days requires treating them with AraC (5 μM, 24 h) for preventing proliferation of non-neuronal cells.

3.4 Simultaneous Imaging of $[Ca^{2+}]_{mt}$ and $[Ca^{2+}]_{cyt}$ in Neurons Using Mito-R-GECO1 and Fura-2, Respectively

1. Hippocampal or DRG cultures transfected with mito-R-GECO1 are loaded with Fura-2 prior to the experiment by incubating the cells in 2 ml of HH buffer (Sect. 2.5, item 1; in a 35 mm petri dish) containing 2–4 μM of Fura-2/AM for 30 min at room temperature in the dark. Sufficient loading of hippocampal neurons typically requires 4 μM of Fura-2/AM, whereas a lower concentration of Fura-2/AM, 2–3 μM, is typically sufficient for loading DRG neurons.
2. After 30 min, coverslips with Fura-2 loaded neurons are mounted in a flow-through chamber. The chamber was originally designed and described by Thayer and colleagues [78].
3. A flow-through chamber is installed on a stage of an inverted fluorescent microscope IX-71 (Sect. 2.7, item 1), and cells are perfused for at least 20 min to wash out Fura-2/AM. Cell perfusion is controlled by a custom made gravity-driven 7-channel perfusion system.
4. Transfected cells are identified by distinct mito-R-GECO1 fluorescence (λ_{Ex} = 550 nm; Figs. 1 and 2). After regions of interest (ROIs) are chosen, simultaneous monitoring of Fura-2 and mito-R-GECO fluorescence is performed by alternately exciting fluorescence at 340, 380 and 550 nm. We typically

use an oil-immersion 40× objective (NA=1.35; Olympus; Fig. 3) and 2 × 2 pixel binning. The latter increases signal/ noise ratio, but reduces the spatial resolution (pixel size = 0.5 μm). A typical exposure time used for Fura-2 is 5 ms, and for mito-R-GECO1 it ranges from 10 to 40 ms, depending on mito-R-GECO1 expression.

5. Cells are imaged at a sampling rate of 0.5–20 Hz (for triplicates of 340/380/550 images; Fig. 3). High-speed image acquisition experiments (sampling rate of 10–20 Hz) are typically limited to a 30–60 s duration. In turn, for experiments that last 30–60 min, we limit the sampling frequency to 0.5–2 Hz to prevent photobleaching of the dyes and minimize the risk of phototoxic damage to the cells.
6. $[Ca^{2+}]_{cyt}$ and $[Ca^{2+}]_{mt}$ elevations can be easily evoked by depolarization (Fig. 1) or by cell type specific agonists, such as glutamate (Fig. 2), NMDA, ATP, capsaicin, and others. Depolarization of specific magnitude can be easily achieved by applying extracellular solution with elevated K^+ concentration, e.g., 50 mM KCl (K^+50; Sect. 2.5, item 2) for 10–30 s (Fig. 1). A magnitude of depolarization in response to a specific high K^+ solution can be estimated using the Nernst eq. [79] or measured electrophysiologically. For example, by using patch clamp recording, Wheeler et al. showed that K^+20, K^+40, and K^+60 applications induced depolarization of sympathetic neurons from the resting potential of −60 mV to −37, −19, or −9 mV, respectively [13]. We found that high K^+ solutions of increasing KCl concentration provide a very convenient and reliable tool to elicit $[Ca^{2+}]_{mt}$ and $[Ca^{2+}]_{cyt}$ elevations of incremental amplitude in neurons (Fig. 1) [65]. We also commonly use extracellular field stimulation and patch clamp recording for studying $[Ca^{2+}]_{mt}$ and $[Ca^{2+}]_{cyt}$ in neurons, as described below.

3.5 Extracellular Field Stimulation for Studying $[Ca^{2+}]_{mt}$ and $[Ca^{2+}]_{cyt}$ Responses in Neurons

Extracellular field stimulation enables evoking either single action potentials or trains of action potentials of desired duration and frequency. The method is relatively simple, reproducible and is easy to quantify. It enables physiologically relevant stimulation of intact neurons. For this method, field potentials are generated by passing current between two platinum electrodes via a Grass S48 stimulator and a stimulus isolation unit SIU5 (Sect. 2.9) as previously described [80]. Trains of 1 ms pulses are delivered at 1–20 Hz, as required by experimental design, and monitored using a Siglent 1052DL digital storage oscilloscope. The stimulus voltage threshold sufficient to elicit a detectable increase in $[Ca^{2+}]_{cyt}$ from the cell body or neuronal processes is determined before beginning of an experiment, and the stimulus voltage for further experimentation is typically set at 20 V higher than the threshold voltage. An increase of the stimulus voltage above the threshold does not lead to a change in the amplitude of the $[Ca^{2+}]_{cyt}$ response. The amplitude of $[Ca^{2+}]_{cyt}$ and $[Ca^{2+}]_{mt}$ elevations can be easily controlled by altering the field stimulation duration or frequency. One common application of this method is determination of $[Ca^{2+}]_{cyt}$ threshold that is required to trigger mitochondrial Ca^{2+} uptake (or $[Ca^{2+}]_{mt}$ elevation) in peripheral and central neurons. For this method, the stimulation frequency is typically set to 10 Hz, and the duration of stimulation (i.e., action potential number) is gradually increased until $[Ca^{2+}]_{cyt}$ rises sufficiently to trigger an increase in $[Ca^{2+}]_{mt}$.

3.6 Simultaneous Imaging of $[Ca^{2+}]_{mt}$ and $[Ca^{2+}]_{cyt}$ and Patch Clamp Recording

As pointed in the introduction, simultaneous monitoring of $[Ca^{2+}]_{cyt}$ provides an important control for potential off-target effects of genetic and pharmacological manipulations that directed at mitochondrial Ca^{2+} transport and $[Ca^{2+}]_{mt}$. However, cytosolic Ca^{2+} signaling itself can be markedly affected by altered activity of mitochondrial Ca^{2+} transporters. For example, inhibition of mitochondrial Ca^{2+} uptake in DRG neurons results in a marked increase in the amplitude and decrease in duration of depolarization-induced $[Ca^{2+}]_{cyt}$ elevations [2, 81]. These considerations emphasize the need for an additional level of control that can be achieved by monitoring Ca^{2+} influx through plasma membrane during depolarization. In this regard, patch clamp recording of voltage-gated Ca^{2+} currents (I_{Ca}) can be used to quantify total Ca^{2+} entering the cell during a depolarization pulse as time integral of I_{Ca} ($\int I_{Ca}dt$) [81, 82]. There are two additional advantage of using whole-cell patch clamp recording for simultaneous $[Ca^{2+}]_{mt}$ and $[Ca^{2+}]_{cyt}$ measurements. First, the composition of both extracellular and intracellular solutions, including intracellular ATP concentration, can be easily controlled. Second, Fura-2 (or Fura-FF or other Ca^{2+} indicator; Sect. 2.6, items 1 and 3) is loaded via a patch pipette only into the cell under investigation, which significantly reduces the background and allows to study very fine structures such as axons and axonal boutons (presynaptic compartments), as well as dendrites and dendritic spines (postsynaptic compartments) [2, 83, 84].

Figure 4 shows an example of simultaneous measurements of $[Ca^{2+}]_{mt}$, $[Ca^{2+}]_{cyt}$ and I_{Ca} in DRG neurons in control and under the conditions when mitochondrial Ca^{2+} uniporter was blocked by including its selective inhibitor Ru360 (10 μM) in intracellular solution. In these experiments, $[Ca^{2+}]_{cyt}$ and $[Ca^{2+}]_{mt}$ of increasing amplitudes were induced by depolarization pulses (from −60 to 0 mV) of incremental duration. Examination of the amplitudes of $[Ca^{2+}]_{cyt}$ and $[Ca^{2+}]_{mt}$ elevations as function of net Ca^{2+} influx ($\int I_{Ca}dt$; Fig. 4) revealed that $[Ca^{2+}]_{cyt}$ increase is very sensitive to a short depolarization and tends to plateau at higher magnitudes of Ca^{2+} influx, while the amplitude of $[Ca^{2+}]_{mt}$ response continues to grow. A nearly complete inhibition of $[Ca^{2+}]_{mt}$ detected in the presence of Ru360 provides an additional control for mito-R-GECO1

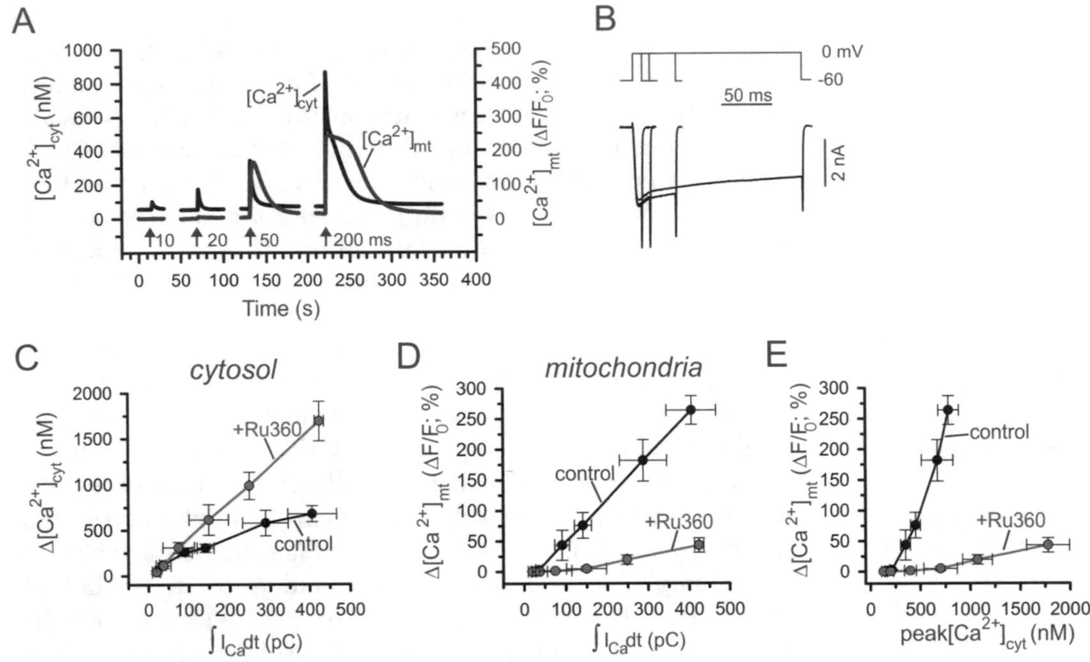

Fig. 4 Simultaneous measurements of $[Ca^{2+}]_{mt}$, $[Ca^{2+}]_{cyt}$, and Ca^{2+} currents (I_{Ca}) in DRG neurons. (**a**) $[Ca^{2+}]_{mt}$ (*red*) and $[Ca^{2+}]_{cyt}$ (*black*) were simultaneously monitored in DRG under whole-cell patch clamp conditions, voltage clamp. $[Ca^{2+}]_{mt}$ and $[Ca^{2+}]_{cyt}$ were evoked by depolarization pulses (−60 to 0 mV) of incremental duration: 10, 20, 50, and 200 ms (*arrows*). Ca^{2+} currents were isolated substituting intracellular K^+ with Cs^+ and Na^+, and extracellular Na^+ with choline and tetraethylammonium (TEA), supplemented with 1 μM tetrodotoxin. The full compositions of the intracellular and extracellular solutions are listed in Sect. 2.6, items 1 and 2. (**b**) The corresponding Ca^{2+} currents for the experiment in (**a**), are shown on an expanded time scale. (**c**, **d**) Analysis of $\Delta[Ca^{2+}]_{cyt}$ (**c**) and $\Delta[Ca^{2+}]_{mt}$ (**d**) as function of total Ca^{2+}/charge ($\int I_{Ca}dt$) transferred into DRG neurons during the voltage pulses as recorded in (**a**), under control conditions (*black*) and in the presence of the mitochondrial inhibitor Ru360 (10 μM) in patch pipette (*green*). The duration of depolarization pulses was 10, 20, 50, 100, 200, and 300 ms. Data for $\Delta[Ca^{2+}]_{cyt}$, $\Delta[Ca^{2+}]_{mt}$, and $\int I_{Ca}dt$ for a specific pulse duration were averaged among control cells ($n = 5$) and Ru360-treated cells ($n = 3$) and are presented as mean ± SEM. Note that inhibition of mitochondrial Ca^{2+} uptake led to a near twofold increase of the amplitude of $[Ca^{2+}]_{cyt}$ responses, suggesting a crucial role of mitochondria in buffering Ca^{2+} influx. (**e**) The same data as in (**c**) and (**d**) are used to plot $\Delta[Ca^{2+}]_{mt}$ as a function of peak $[Ca^{2+}]_{cyt}$ rise in the absence/control (*black*) and presence of 10 μM Ru360 in the intracellular solution (*green*), respectively

as a reliable reporter of mitochondrial Ca^{2+} dynamic (Fig. 4d, e). In turn, the observation that $[Ca^{2+}]_{cyt}$ responses are markedly increased in the presence of Ru360 (Fig. 4c) underscores a crucial role of mitochondrial Ca^{2+} uptake in buffering cytosolic Ca^{2+} in neurons.

A similar approach can be adapted for simultaneous monitoring of $[Ca^{2+}]_{mt}$, $[Ca^{2+}]_{cyt}$ and I_{Ca} for studying the role of mitochondria in buffering Ca^{2+} fluxes mediated by ligand-gated Ca^{2+} channels, such as TRP channels or glutamate NMDA receptors [85]. Finally, the current-clamp mode of whole-cell patch clamp recording can be used to study $[Ca^{2+}]_{mt}$ and $[Ca^{2+}]_{cyt}$ in response to action potentials that are simultaneously monitored.

3.7 Data Analysis and Calibration of Fluorescent Indicators

1. *[Ca2+]cyt analysis and Fura-2 calibration.*
Fura-2 is a dual excitation wavelength indicator, such that its excitation spectrum shifts upon Ca^{2+} binding [72]. As a result of the spectral shift, F_{340} nm fluorescence intensity increases whereas F_{380} nm intensity decreases when Ca^{2+} binds to Fura-2 (isobestic point is ~360 nm). Therefore, changes in Fura-2 fluorescence intensities can be easily converted to $[Ca^{2+}]_{cyt}$ by operating with a fluorescence ratio $R = F_{340}/F_{380}$ instead of using the intensity value F at a given wavelength. This is an important advantage because R does not depend on the dye concentration, illumination intensity or optical path length, whereas F depends on all of these parameters.

The following formula is used to calculate $[Ca^{2+}]_{cyt}$ based on $R = F_{340}/F_{380}$:

$$[Ca^{2+}]_{cyt} = K_d\beta(R - R_{\min})/(R_{\max} - R) \quad (1)$$

where R_{max} and R_{min} are F_{340}/F_{380} fluorescence ratios determined for Ca^{2+}-bound and Ca^{2+}-free forms of Fura-2, respectively; K_d is Ca^{2+} dissociation constant of Fura-2, and $\beta = F_{min,380}/F_{max,380}$, where $F_{max,380}$ and $F_{min,380}$ are fluorescence intensities for λ_{Ex}=380 nm determined for Ca^{2+}-bound and Ca^{2+}-free form of Fura-2, respectively [72]. Fluorescence intensity is corrected for background that is determined in an area that does not contain a cell.

R_{min}, R_{max} and β values are determined using ionomycin-based calibration procedure. Ionomycin is a Ca^{2+} ionophore that enables equilibrating Ca^{2+} concentration across the plasma membrane and intracellular membranes, such as those of mitochondria and endoplasmic reticulum [86]. Hence, for in situ calibration, transfected with mito-R-GECO1 and loaded with Fura-2 neurons are exposed to 10 μM ionomycin in a Ca^{2+}-free HH extracellular buffer (with 1 mM EGTA; Sect. 2.5, item 3) to determine R_{min} and $F_{min,380}$, which is followed by exposure to 10 μM ionomycin in an HH buffer containing 1.3 mM Ca^{2+} to determine R_{max} and $F_{max,380}$. Prolonged exposure to ionomycin can be toxic to neurons. Therefore, we use 2–3 brief (2 min) applications of 10 μM ionomycin to ensure its stepwise accumulation in the plasma and intracellular membranes until cytosolic and organellar Ca^{2+} concentrations equilibrate with extracellular Ca^{2+} concentration. For these brief applications, ionomycin can be applied directly into a perfusion chamber (while perfusion is stopped during the application), to prevent contamination of the perfusion system with ionomycin. Using this approach, the calibration values were determined as follows: R_{min} = 0.19, R_{max} = 2.98 and β = 6.77. The K_d value used for Fura-2 was 275 nM [74]. Image data are initially processed using TILLvisION 4.0.1.2 or Live Acquisition software (TILL

Photonics/FEI), followed by additional analysis using Sigma Plot 13 and GraphPad 6 software.

2. *[Ca2+]mt analysis and mito-R-GECO1 calibration.* Unlike Fura-2, mito-R-GECO1 is a single wavelength Ca^{2+} fluorescent indicator, i.e., it does not undergo a spectral shift upon Ca^{2+} binding. Therefore, the ratiometric formula (1) is no longer applicable for calculating $[Ca^{2+}]_{mt}$, and fluorescence intensity at a given wavelength $F\lambda$ must be used instead (we use $\lambda_{Ex} = 550$ nm for mito-R-GECO1). As stated above, F (or F_{550}) depends on dye concentration (i.e., mito-R-GECO1 expression in an individual neuron), optical length path and illumination intensity, which substantially complicates the calibration procedure. Therefore, for single wavelength indicators, $[Ca^{2+}]_{cyt}$ ($[Ca^{2+}]_{mt}$) changes are commonly expressed as:

$$" F / F_0 = (F - F_0) / F_0 \quad (2)$$

where F is current fluorescence intensity of mito-R-GECO1 ($\lambda_{Ex} = 550$ nm) and F_0 is fluorescence intensity at rest or in the beginning of recording [87]. F is always corrected for background that is found by measuring fluorescence ($\lambda_{Ex} = 550$ nm) in a non-transfected neuron.

If $[Ca^{2+}]_{mt}$ quantification in Ca^{2+} concentration terms is needed, one can use the following formula:

$$[Ca^{2+}]_{mt} = ([Ca^{2+}]_{mt,\mathrm{rest}} + K_d f / f_{\max}) / (1 - f / f_{\max}) \quad (3)$$

where $[Ca^{2+}]_{mt,rest}$ is $[Ca^{2+}]_{mt}$ in a resting cell, K_d is Ca^{2+} dissociation coefficient of mito-R-GECO1, $f = \Delta F/F_0$ (see formula 2), and f_{max} is the maximal value of $\Delta F/F_0$ [87, 88]. The latter can be determined by applying strong stimulation, such as 90 mM KCl, or by treating cells with 10 μM ionomycin at the end of an experiment. K_d for R-GECO1 (and for mito-R-GECO1) was reported to be 480 nM [63]. $[Ca^{2+}]_{mt,rest}$ is estimated to be ~100 nM in neurons [89, 90], which is similar to $[Ca^{2+}]_{cyt}$ levels at rest. However, an X-ray microanalysis of total Ca concentration in mitochondria suggests that $[Ca^{2+}]_{mt,rest}$ is substantially lower than 100 nM [91, 92]. $[Ca^{2+}]_{mt,rest}$ can be measured using mitochondria-targeted dual wavelength genetic Ca^{2+} indicators, such as mito-GEM-GECO1 (Addgene Cat.#32461; [63]). For that, the ratiometric formula (1) and the corresponding calibration can be used as described above (Sect. 3.7, step 2). Unfortunately, mito-GEM-GECO1 is spectrally incompatible with Fura-2, which prevents simultaneous $[Ca^{2+}]_{mt}$ and $[Ca^{2+}]_{cyt}$ measurements using this pair of indicators.

Thus, $[Ca^{2+}]_{mt}$ can be quantified either as $\Delta F/F_0$ (formula 2), or using a more sophisticated formula (3). The latter case requires separate determination of $[Ca^{2+}]_{mt,rest}$ that is typically

≤100 nM. Note, however, that upon stimulation $[Ca^{2+}]_{mt}$ levels reach 1–5 μM, which is 10–50-fold greater than $[Ca^{2+}]_{mt,rest}$. Therefore, for $[Ca^{2+}]_{mt}$ calculation, an error attributed to unprecise determination of $[Ca^{2+}]_{mt,rest}$ is expected to be <10 %.

4 Notes

1. Serums, NGF and insulin can deteriorate over time if stored at 4 °C, or when undergo multiple freeze–thaw cycles. To prevent this, we make small aliquots of the HIHS (1 ml), FBS (1 ml), NGF (20 μl of 50 μg/ml stock solution), and insulin (10 μl of 1.25 % stock solution) and store them at −20 °C until they are used.
2. We found that using a specific type of 25 mm glass coverslips (Fisher Scientific, Cat. #12-545-86) was essential for successful culturing of hippocampal, DRG and other types of neurons. These coverslips are made from preferred glass to promote cell growth and to provide an anchor for cells.
3. Using high-quality endotoxin-free plasmids is essential for successful high-efficiency transfection. We use the EndoFree Plasmid Maxi kit (Qiagen Cat.#12362).
4. In addition to the described here transfection techniques, DRG and hippocampal neurons can be transfected using lentivirus (FIV) as previously described [2, 80, 93]. Although lentivirus-based protocols provide excellent transfection efficiency (40–80 %), preparing lentivirus is costly and is difficult to use for co-transfecting cells with multiple genes. Biolistic gene transfer (or Gene gun) represents an additional technique for transfection (and co-transfection of multiple genes) in DRG and hippocampal neurons [94].
5. The spectral properties of a low-affinity mitochondrial Ca^{2+} indicator mito-LAR-GECO1.2 are very similar to those of mito-R-GECO1 (Table 1), and hence it requires essentially the same instrumentation and experimental protocol that are described for mito-R-GECO1.
6. To prevent phototoxicity, it is important to minimize cells exposure to excitation light while searching for a transfected cell.

Acknowledgments

This work was supported by NIH/NINDS grants NS087068 and NS096246. J.E.R. was supported by a predoctoral fellowship through the American Heart Association, Midwest Affiliate Grant 15PRE25310013.

References

1. Colegrove SL, Albrecht MA, Friel DD (2000) Dissection of mitochondrial Ca^{2+} uptake and release fluxes in situ after depolarization-evoked $[Ca^{2+}]_i$ elevations in sympathetic neurons. J Gen Physiol 115(3):351–369
2. Medvedeva YV, Kim MS, Usachev YM (2008) Mechanisms of prolonged presynaptic Ca2+ signaling and glutamate release induced by TRPV1 activation in rat sensory neurons. J Neurosci 28(20):5295–5311
3. Friel DD (2000) Mitochondria as regulators of stimulus-evoked calcium signals in neurons. Cell Calcium 28(5-6):307–316
4. Thayer SA, Usachev YM, Pottorf WJ (2002) Modulating Ca^{2+} clearance from neurons. Front Biosci 7:D1255–D1279
5. Nicholls DG (2005) Mitochondria and calcium signaling. Cell Calcium 38(3-4):311–317
6. Usachev YM (2015) Mitochondrial Ca2+ transport in the control of neuronal functions: molecular and cellular mechanisms. In: Hardwick JM, Gribkoff VK, Jonas EA (eds) The functions, disease-related dysfunctions, and therapeutic targeting of neuronal mitochondria. Wiley, New York, pp 103–129
7. David G, Barrett EF (2000) Stimulation-evoked increases in cytosolic $[Ca^{2+}]$ in mouse motor nerve terminals are limited by mitochondrial uptake and are temperature- dependent. J Neurosci 20(19):7290–7296
8. Billups B, Forsythe ID (2002) Presynaptic mitochondrial calcium sequestration influences transmission at mammalian central synapses. J. Neurosci. 22(14):5840–5847
9. Tang YG, Zucker RS (1997) Mitochondrial involvement in post-tetanic potentiation of synaptic transmission. Neuron 18(3): 483–491
10. Jonas E (2004) Regulation of synaptic transmission by mitochondrial ion channels. J Bioenerg Biomembr 36(4):357–361
11. Jonas EA, Buchanan J, Kaczmarek LK (1999) Prolonged activation of mitochondrial conductances during synaptic transmission. Science 286(5443):1347–1350
12. Kim MS, Usachev YM (2009) Mitochondrial Ca2+ cycling facilitates activation of the transcription factor NFAT in sensory neurons. J Neurosci 29(39):12101–12114
13. Wheeler DG, Groth RD, Ma H, Barrett CF, Owen SF, Safa P, Tsien RW (2012) Ca(V)1 and Ca(V)2 channels engage distinct modes of Ca(2+) signaling to control CREB-dependent gene expression. Cell 149(5): 1112–1124
14. Hajnoczky G, Robb GL, Seitz MB, Thomas AP (1995) Decoding of cytosolic calcium oscillations in the mitochondria. Cell 82(3): 415–424
15. Denton RM (2009) Regulation of mitochondrial dehydrogenases by calcium ions. Biochim Biophys Acta 1787(11):1309–1316
16. Glancy B, Balaban RS (2012) Role of mitochondrial Ca2+ in the regulation of cellular energetics. Biochemistry 51(14):2959–2973
17. Reynolds IJ (1999) Mitochondrial membrane potential and the permeability transition in excitotoxicity. Ann NY Acad Sci 893:33–41
18. Murphy AN, Fiskum G, Beal MF (1999) Mitochondria in neurodegeneration: bioenergetic function in cell life and death. J Cereb Blood Flow Metab 19(3):231–245
19. Nicholls DG, Budd SL (2000) Mitochondria and neuronal survival. Physiol Rev 80(1): 315–360
20. Panov AV, Gutekunst CA, Leavitt BR, Hayden MR, Burke JR, Strittmatter WJ, Greenamyre JT (2002) Early mitochondrial calcium defects in Huntington's disease are a direct effect of polyglutamines. Nat Neurosci 5(8):731–736
21. Bezprozvanny I (2009) Calcium signaling and neurodegenerative diseases. Trends Mol Med 15(3):89–100
22. Moskowitz MA, Lo EH, Iadecola C (2010) The science of stroke: mechanisms in search of treatments. Neuron 67(2):181–198
23. Bernardi P (1999) Mitochondrial transport of cations: channels, exchangers, and permeability transition. Physiol Rev 79(4):1127–1155
24. Rizzuto R, De Stefani D, Raffaello A, Mammucari C (2012) Mitochondria as sensors and regulators of calcium signalling. Nat Rev Mol Cell Biol 13(9):566–578
25. Kirichok Y, Krapivinsky G, Clapham DE (2004) The mitochondrial calcium uniporter is a highly selective ion channel. Nature 427(6972):360–364
26. Baughman JM, Perocchi F, Girgis HS, Plovanich M, Belcher-Timme CA, Sancak Y, Bao XR, Strittmatter L, Goldberger O, Bogorad RL, Koteliansky V, Mootha VK (2011) Integrative genomics identifies MCU as an essential component of the mitochondrial calcium uniporter. Nature 476(7360): 341–345
27. De Stefani D, Raffaello A, Teardo E, Szabo I, Rizzuto R (2011) A forty-kilodalton protein of the inner membrane is the mitochondrial calcium uniporter. Nature 476(7360):336–340

28. Oxenoid K, Dong Y, Cao C, Cui T, Sancak Y, Markhard AL, Grabarek Z, Kong L, Liu Z, Ouyang B, Cong Y, Mootha VK, Chou JJ (2016) Architecture of the mitochondrial calcium uniporter. Nature 533(7602):269–273
29. Raffaello A, De Stefani D, Sabbadin D, Teardo E, Merli G, Picard A, Checchetto V, Moro S, Szabo I, Rizzuto R (2013) The mitochondrial calcium uniporter is a multimer that can include a dominant-negative pore-forming subunit. EMBO J 32(17): 2362–2376
30. Sancak Y, Markhard AL, Kitami T, Kovacs-Bogdan E, Kamer KJ, Udeshi ND, Carr SA, Chaudhuri D, Clapham DE, Li AA, Calvo SE, Goldberger O, Mootha VK (2013) EMRE is an essential component of the mitochondrial calcium uniporter complex. Science 342(6164):1379–1382
31. Perocchi F, Gohil VM, Girgis HS, Bao XR, McCombs JE, Palmer AE, Mootha VK (2010) MICU1 encodes a mitochondrial EF hand protein required for Ca(2+) uptake. Nature 467(7313):291–296
32. Plovanich M, Bogorad RL, Sancak Y, Kamer KJ, Strittmatter L, Li AA, Girgis HS, Kuchimanchi S, De Groot J, Speciner L, Taneja N, Oshea J, Koteliansky V, Mootha VK (2013) MICU2, a paralog of MICU1, resides within the mitochondrial uniporter complex to regulate calcium handling. PLoS One 8(2):e55785
33. Mallilankaraman K, Cardenas C, Doonan PJ, Chandramoorthy HC, Irrinki KM, Golenar T, Csordas G, Madireddi P, Yang J, Muller M, Miller R, Kolesar JE, Molgo J, Kaufman B, Hajnoczky G, Foskett JK, Madesh M (2012) MCUR1 is an essential component of mitochondrial Ca2+ uptake that regulates cellular metabolism. Nat Cell Biol 14(12):1336–1343
34. Tomar D, Dong Z, Shanmughapriya S, Koch DA, Thomas T, Hoffman NE, Timbalia SA, Goldman SJ, Breves SL, Corbally DP, Nemani N, Fairweather JP, Cutri AR, Zhang X, Song J, Jana F, Huang J, Barrero C, Rabinowitz JE, Luongo TS, Schumacher SM, Rockman ME, Dietrich A, Merali S, Caplan J, Stathopulos P, Ahima RS, Cheung JY, Houser SR, Koch WJ, Patel V, Gohil VM, Elrod JW, Rajan S, Madesh M (2016) MCUR1 is a scaffold factor for the MCU complex function and promotes mitochondrial bioenergetics. Cell Rep 15(8):1673–1685
35. Kamer KJ, Mootha VK (2014) MICU1 and MICU2 play nonredundant roles in the regulation of the mitochondrial calcium uniporter. EMBO Rep 15(3):299–307
36. Patron M, Checchetto V, Raffaello A, Teardo E, Vecellio Reane D, Mantoan M, Granatiero V, Szabo I, De Stefani D, Rizzuto R (2014) MICU1 and MICU2 finely tune the mitochondrial Ca2+ uniporter by exerting opposite effects on MCU activity. Mol Cell 53(5):726–737
37. Tsai MF, Phillips CB, Ranaghan M, Tsai CW, Wu Y, Willliams C, Miller C (2016) Dual functions of a small regulatory subunit in the mitochondrial calcium uniporter complex. Elife 5
38. Vais H, Mallilankaraman K, Mak DO, Hoff H, Payne R, Tanis JE, Foskett JK (2016) EMRE is a matrix Ca(2+) sensor that governs gatekeeping of the mitochondrial Ca(2+) uniporter. Cell Rep 14(3):403–410
39. Paupe V, Prudent J, Dassa EP, Rendon OZ, Shoubridge EA (2015) CCDC90A (MCUR1) is a cytochrome c oxidase assembly factor and not a regulator of the mitochondrial calcium uniporter. Cell Metab 21(1):109–116
40. Chaudhuri D, Artiga DJ, Abiria SA, Clapham DE (2016) Mitochondrial calcium uniporter regulator 1 (MCUR1) regulates the calcium threshold for the mitochondrial permeability transition. Proc Natl Acad Sci U S A 113(13):E1872–E1880
41. Pivovarova NB, Andrews SB (2010) Calcium-dependent mitochondrial function and dysfunction in neurons. FEBS J 277(18): 3622–3636
42. Palty R, Silverman WF, Hershfinkel M, Caporale T, Sensi SL, Parnis J, Nolte C, Fishman D, Shoshan-Barmatz V, Herrmann S, Khananshvili D, Sekler I (2010) NCLX is an essential component of mitochondrial Na+/Ca2+ exchange. Proc Natl Acad Sci U S A 107(1):436–441
43. Jiang D, Zhao L, Clapham DE (2009) Genome-wide RNAi screen identifies Letm1 as a mitochondrial Ca2+/H+ antiporter. Science 326(5949):144–147
44. Jiang D, Zhao L, Clish CB, Clapham DE (2013) Letm1, the mitochondrial Ca2+/H+ antiporter, is essential for normal glucose metabolism and alters brain function in Wolf-Hirschhorn syndrome. Proc Natl Acad Sci U S A 110(24):E2249–E2254
45. Doonan PJ, Chandramoorthy HC, Hoffman NE, Zhang X, Cardenas C, Shanmughapriya S, Rajan S, Vallem S, Chen X, Foskett JK, Cheung JY, Houser SR, Madesh M (2014) LETM1-dependent mitochondrial Ca2+ flux modulates cellular bioenergetics and proliferation. FASEB J 28(11):4936–4949
46. Nowikovsky K, Froschauer EM, Zsurka G, Samaj J, Reipert S, Kolisek M, Wiesenberger G, Schweyen RJ (2004) The LETM1/YOL027 gene family encodes a factor of the mitochondrial K+ homeostasis with a potential

role in the Wolf-Hirschhorn syndrome. J Biol Chem 279(29):30307–30315
47. Nowikovsky K, Pozzan T, Rizzuto R, Scorrano L, Bernardi P (2012) Perspectives on: SGP symposium on mitochondrial physiology and medicine: the pathophysiology of LETM1. J Gen Physiol 139(6):445–454
48. David G, Talbot J, Barrett EF (2003) Quantitative estimate of mitochondrial [Ca2+] in stimulated motor nerve terminals. Cell Calcium 33(3):197–206
49. Fonteriz RI, de la Fuente S, Moreno A, Lobaton CD, Montero M, Alvarez J (2010) Monitoring mitochondrial [Ca(2+)] dynamics with rhod-2, ratiometric pericam and aequorin. Cell Calcium 48(1):61–69
50. Davidson SM, Duchen MR (2012) Imaging mitochondrial calcium signalling with fluorescent probes and single or two photon confocal microscopy. Methods Mol Biol 810: 219–234
51. Trollinger DR, Cascio WE, Lemasters JJ (2000) Mitochondrial calcium transients in adult rabbit cardiac myocytes: inhibition by ruthenium red and artifacts caused by lysosomal loading of Ca2+–indicating fluorophores. Biophys J 79(1):39–50
52. Rizzuto R, Simpson AWM, Brini M, Pozzan T (1992) Rapid changes of mitochondrial Ca^{2+} revealed by specifically targeted recombinant aequorin. Nature 358:325–327
53. Brini M (2008) Calcium-sensitive photoproteins. Methods 46(3):160–166
54. Grienberger C, Konnerth A (2012) Imaging calcium in neurons. Neuron 73(5):862–885
55. Miyawaki A, Llopis J, Heim R, McCaffery JM, Adams JA, Ikura M, Tsien RY (1997) Fluorescent indicators for Ca2+ based on green fluorescent proteins and calmodulin. Nature 388(6645):882–887
56. Baird GS, Zacharias DA, Tsien RY (1999) Circular permutation and receptor insertion within green fluorescent proteins. Proc Natl Acad Sci U S A 96(20):11241–11246
57. Romoser VA, Hinkle PM, Persechini A (1997) Detection in living cells of Ca2+–dependent changes in the fluorescence emission of an indicator composed of two green fluorescent protein variants linked by a calmodulin-binding sequence—a new class of fluorescent indicators. J Biol Chem 272(20): 13270–13,274
58. Nagai T, Sawano A, Park ES, Miyawaki A (2001) Circularly permuted green fluorescent proteins engineered to sense Ca2+. Proc Natl Acad Sci U S A 98(6):3197–3202
59. Nakai J, Ohkura M, Imoto K (2001) A high signal-to-noise Ca2+ probe composed of a single green fluorescent protein. Nat Biotechnol 19(2):137–141
60. Palmer AE, Tsien RY (2006) Measuring calcium signaling using genetically targetable fluorescent indicators. Nat Protoc 1(3): 1057–1065
61. Perez Koldenkova V, Nagai T (2013) Genetically encoded Ca(2+) indicators: properties and evaluation. Biochim Biophys Acta 1833(7):1787–1797
62. Palmer AE, Giacomello M, Kortemme T, Hires SA, Lev-Ram V, Baker D, Tsien RY (2006) Ca2+ indicators based on computationally redesigned calmodulin-peptide pairs. Chem Biol 13(5):521–530
63. Zhao Y, Araki S, Wu J, Teramoto T, Chang YF, Nakano M, Abdelfattah AS, Fujiwara M, Ishihara T, Nagai T, Campbell RE (2011) An expanded palette of genetically encoded Ca(2)(+) indicators. Science 333(6051): 1888–1891
64. Chen TW, Wardill TJ, Sun Y, Pulver SR, Renninger SL, Baohan A, Schreiter ER, Kerr RA, Orger MB, Jayaraman V, Looger LL, Svoboda K, Kim DS (2013) Ultrasensitive fluorescent proteins for imaging neuronal activity. Nature 499(7458):295–300
65. Wu J, Prole DL, Shen Y, Lin Z, Gnanasekaran A, Liu Y, Chen L, Zhou H, Chen SR, Usachev YM, Taylor CW, Campbell RE (2014) Red fluorescent genetically encoded Ca2+ indicators for use in mitochondria and endoplasmic reticulum. Biochem J 464:13–22
66. Akerboom J, Carreras Calderon N, Tian L, Wabnig S, Prigge M, Tolo J, Gordus A, Orger MB, Severi KE, Macklin JJ, Patel R, Pulver SR, Wardill TJ, Fischer E, Schuler C, Chen TW, Sarkisyan KS, Marvin JS, Bargmann CI, Kim DS, Kugler S, Lagnado L, Hegemann P, Gottschalk A, Schreiter ER, Looger LL (2013) Genetically encoded calcium indicators for multi-color neural activity imaging and combination with optogenetics. Front Mol Neurosci. 6:2
67. Suzuki J, Kanemaru K, Ishii K, Ohkura M, Okubo Y, Iino M (2014) Imaging intraorganellar Ca2+ at subcellular resolution using CEPIA. Nat Commun 5:4153
68. Inoue M, Takeuchi A, Horigane S, Ohkura M, Gengyo-Ando K, Fujii H, Kamijo S, Takemoto-Kimura S, Kano M, Nakai J, Kitamura K, Bito H (2015) Rational design of a high-affinity, fast, red calcium indicator R-CaMP2. Nat Methods 12(1):64–70

69. Berlin S, Carroll EC, Newman ZL, Okada HO, Quinn CM, Kallman B, Rockwell NC, Martin SS, Lagarias JC, Isacoff EY (2015) Photoactivatable genetically encoded calcium indicators for targeted neuronal imaging. Nat Methods 12(9):852–858
70. Rose T, Goltstein PM, Portugues R, Griesbeck O (2014) Putting a finishing touch on GECIs. Front Mol Neurosci 7:88
71. Pendin D, Greotti E, Filadi R, Pozzan T (2015) Spying on organelle Ca(2)(+) in living cells: the mitochondrial point of view. J Endocrinol Invest 38(1):39–45
72. Grynkiewicz G, Poenie M, Tsien RY (1985) A new generation of Ca^{2+} indicators with greatly improved fluorescence properties. J Biol Chem 260(6):3440–3450
73. Cobbold PH, Rink TJ (1987) Fluorescence and bioluminescence measurement of cytoplasmic free calcium. Biochem J 248:313–328
74. Shuttleworth TJ, Thompson JL (1991) Effect of temperature on receptor-activated changes in [Ca2+]i and their determination using fluorescent probes. J Biol Chem 266(3):1410–1414
75. Yuste R, Lanni F, Konnerth A (2000) Imagins neurons: a laboratory manual. Cold Spring Harbor Laboratory Press, Cold Spring Harbor, NY
76. Hamel EJ, Grewe BF, Parker JG, Schnitzer MJ (2015) Cellular level brain imaging in behaving mammals: an engineering approach. Neuron 86(1):140–159
77. Ji N, Freeman J, Smith SL (2016) Technologies for imaging neural activity in large volumes. Nat Neurosci 19(9):1154–1164
78. Thayer SA, Sturek M, Miller RJ (1988) Measurement of neuronal Ca^{2+} transients using simultaneous microfluorimetry and electrophysiology. Pflugers Arch 412:216–223
79. Hille B (ed) (2001) Ion channels of excitable membranes, 3rd edn. Sinauer Associates, Sunderland, MA, p 814
80. Shutov LP, Kim MS, Houlihan PR, Medvedeva YV, Usachev YM (2013) Mitochondria and plasma membrane Ca2+–ATPase control presynaptic Ca2+ clearance in capsaicin-sensitive rat sensory neurons. J Physiol 591(Pt 10):2443–2462
81. Thayer SA, Miller RJ (1990) Regulation of the free intracellular calcium concentration in rat dorsal root ganglion neurones in vitro. J Physiol 425:85–115
82. Usachev YM, Thayer SA (1997) All-or-none Ca^{2+} release from intracellular stores triggered by Ca^{2+} influx through voltage-gated Ca^{2+} channels in rat sensory neurons. J Neurosci 17(19):7404–7414
83. Lu Y, Zhang M, Lim IA, Hall DD, Allen M, Medvedeva Y, McKnight GS, Usachev YM, Hell JW (2008) AKAP150-anchored PKA activity is important for LTD during its induction phase. J Physiol 586(Pt 17):4155–4164
84. Medvedeva YV, Kim MS, Schnizler K, Usachev YM (2009) Functional tetrodotoxin-resistant Na(+) channels are expressed presynaptically in rat dorsal root ganglia neurons. Neuroscience 159(2):559–569
85. Garaschuk O, Schneggenburger R, Schirra C, Tempia F, Konnerth A (1996) Fractional Ca2+ currents through somatic and dendritic glutamate receptor channels of rat hippocampal CA1 pyramidal neurones. J Physiol 491(Pt 3):757–772
86. Kauffman RF, Taylor RW, Pfeiffer DR (1980) Cation transport and specificity of ionomycin. Comparison with ionophore A23187 in rat liver mitochondria. J Biol Chem 255(7):2735–2739
87. Helmchen F (2000) Calibration of fluorescent calcium indicators. In: Yuste R, Lanni F, Konnerth A (eds) Imaging neurons. Cold Spring Harbor Laboratory Press, Cold Spring Harbor, NY, pp 32.1–32.9
88. Lev-Ram V, Miyakawa H, Lasser-Ross N, Ross WN (1992) Calcium transients in cerebellar Purkinje neurons evoked by intracellular stimulation. J Neurophysiol 68(4):1167–1177
89. Babcock DF, Hille B (1998) Mitochondrial oversight of cellular Ca2+ signaling. Curr Opin Neurobiol 8(3):398–404
90. David G (1999) Mitochondrial clearance of cytosolic Ca^{2+} in stimulated lizard motor nerve terminals proceeds without progressive elevation of mitochondrial matrix [Ca^{2+}]. J Neurosci 19(17):7495–7506
91. Pivovarova NB, Hongpaisan J, Andrews SB, Friel DD (1999) Depolarization-induced mitochondrial Ca accumulation in sympathetic neurons: spatial and temporal characteristics. J Neurosci 19(15):6372–6384
92. Stanika RI, Villanueva I, Kazanina G, Andrews SB, Pivovarova NB (2012) Comparative impact of voltage-gated calcium channels and NMDA receptors on mitochondria-mediated neuronal injury. J Neurosci 32(19):6642–6650
93. Merrill RA, Dagda RK, Dickey AS, Cribbs JT, Green SH, Usachev YM, Strack S (2011) Mechanism of neuroprotective mitochondrial remodeling by PKA/AKAP1. PLoS biology 9(4):e1000612

94. Usachev YM, Khammanivong A, Campbell C, Thayer SA (2000) Particle-mediated gene transfer to rat neurons in primary culture. Pflugers Arch 439(6):730–738
95. Filippin L, Magalhaes PJ, Di Benedetto G, Colella M, Pozzan T (2003) Stable interactions between mitochondria and endoplasmic reticulum allow rapid accumulation of calcium in a subpopulation of mitochondria. J Biol Chem 278(40):39224–39234
96. Kettlewell S, Cabrero P, Nicklin SA, Dow JA, Davies S, Smith GL (2009) Changes of intra-mitochondrial Ca2+ in adult ventricular cardiomyocytes examined using a novel fluorescent Ca2+ indicator targeted to mitochondria. J Mol Cell Cardiol 46(6):891–901
97. Palmer AE, Jin C, Reed JC, Tsien RY (2004) Bcl-2-mediated alterations in endoplasmic reticulum Ca2+ analyzed with an improved genetically encoded fluorescent sensor. Proc Natl Acad Sci U S A 101(50):17404–17409
98. Jean-Quartier C, Bondarenko AI, Alam MR, Trenker M, Waldeck-Weiermair M, Malli R, Graier WF (2012) Studying mitochondrial Ca(2+) uptake—a revisit. Mol Cell Endocrinol 353(1-2):114–127
99. Qiu J, Tan YW, Hagenston AM, Martel MA, Kneisel N, Skehel PA, Wyllie DJ, Bading H, Hardingham GE (2013) Mitochondrial calcium uniporter Mcu controls excitotoxicity and is transcriptionally repressed by neuroprotective nuclear calcium signals. Nat Commun 4:2034
100. Tallini YN, Ohkura M, Choi BR, Ji G, Imoto K, Doran R, Lee J, Plan P, Wilson J, Xin HB, Sanbe A, Gulick J, Mathai J, Robbins J, Salama G, Nakai J, Kotlikoff MI (2006) Imaging cellular signals in the heart in vivo: cardiac expression of the high-signal Ca2+ indicator GCaMP2. Proc Natl Acad Sci U S A 103(12):4753–4758
101. Marland JR, Hasel P, Bonnycastle K, Cousin MA (2016) Mitochondrial calcium uptake modulates synaptic vesicle endocytosis in central nerve terminals. J Biol Chem 291(5): 2080–2086
102. Akerboom J, Chen TW, Wardill TJ, Tian L, Marvin JS, Mutlu S, Calderon NC, Esposti F, Borghuis BG, Sun XR, Gordus A, Orger MB, Portugues R, Engert F, Macklin JJ, Filosa A, Aggarwal A, Kerr RA, Takagi R, Kracun S, Shigetomi E, Khakh BS, Baier H, Lagnado L, Wang SS, Bargmann CI, Kimmel BE, Jayaraman V, Svoboda K, Kim DS, Schreiter ER, Looger LL (2012) Optimization of a GCaMP calcium indicator for neural activity imaging. J Neurosci 32(40):13819–13840
103. Li H, Wang X, Zhang N, Gottipati MK, Parpura V, Ding S (2014) Imaging of mitochondrial Ca2+ dynamics in astrocytes using cell-specific mitochondria-targeted GCaMP5G/6 s: mitochondrial Ca2+ uptake and cytosolic Ca2+ availability via the endoplasmic reticulum store. Cell Calcium 56(6): 457–466

Chapter 9

Live Imaging of Mitochondrial ROS Production and Dynamic Redox Balance in Neurons

Karolina Can, Sebastian Kügler, and Michael Müller

Abstract

Mitochondria are the most prominent cellular source of reactive oxygen species. As a by-product of cellular respiration, superoxide constantly escapes from the electron transport chain and is converted into other reactive oxygen and nitrogen species, which then may mediate downstream redox changes also in neighboring compartments and organelles. Such mitochondria-derived redox signals crucially contribute to the modulation of normal cell function but may as well cause random oxidative damage to various cellular constituents and evoke aberrant signaling. The resulting redox stress is considered to contribute to the onset and progression of various neuropathologies. Hence, there is a tremendous interest in mapping subcellular ROS levels as well as redox changes and to understand their spatiotemporal dynamics.

It is only since the development of genetically encoded fluorescent redox sensors that such analysis has become possible in a reliable manner. These advanced optical sensors overcome the severe disadvantages of oxidation-sensitive synthetic fluorescent dyes. For the first time they allow to monitor both reducing as well as oxidizing changes on the subcellular level, to decipher their detailed dynamics, and to quantify their very extent. In this chapter we summarize the properties of protein-based optical redox indicators and explain their superiority to synthetic dyes. Furthermore, we address the challenges of a proper and efficient delivery of the sensor-coding DNA, with a special emphasis on viral transduction and vector design. Finally, we give a detailed description of redox live-imaging applications in different neuronal preparations and point out to potential pitfalls.

Key words Redox signaling, Oxidative stress, Reactive oxygen species (ROS), Redox imaging, Optical redox probes, 2-photon microscopy, Ratiometric imaging, Viral vectors, Viral transduction

1 Introduction

Mitochondria are often referred to as cellular powerplants, as they provide the majority of cellular ATP [1]. Yet, at the same time, they constitute the major source of superoxide ($^{\cdot}O_2^-$) and other reactive oxygen species (ROS). It is estimated that 2–5 % of electrons constantly escape from the electron transport chain and react with O_2 to produce $^{\cdot}O_2^-$ at respiratory complexes I and III [2–4]. The $^{\cdot}O_2^-$ is then converted into the less reactive hydrogen peroxide

Stefan Strack and Yuriy M. Usachev (eds.), *Techniques to Investigate Mitochondrial Function in Neurons*, Neuromethods, vol. 123, DOI 10.1007/978-1-4939-6890-9_9, © Springer Science+Business Media LLC 2017

(H_2O_2) [5], or it may interact with nitric oxide (NO) to form reactive nitrogen species (RNS) such as peroxynitrite ($ONOO^-$) [6].

It is now widely accepted that ROS and RNS as well as the resulting downstream changes in redox balance are crucial components of physiological cellular signaling [7–9]. However, accumulation of reactive oxidants—as a result of excessive mitochondrial ROS production or insufficient oxidant scavenging—may cause uncontrolled cell damage. Various neuropathological and/or neurodegenerative conditions are closely associated with disturbed cellular redox balance and oxidative stress. Prominent examples include Parkinson's disease [10], Huntington's disease [11], Leigh disease [12], Alzheimer's disease [13], amyotrophic lateral sclerosis [14], epilepsy [15], Rett syndrome [16, 17], and ischemia–reperfusion injury [18].

Oxidative stress or an imbalance of oxidizing and reducing agents may cause direct and indirect damage to a variety of cellular targets. This does not only include oxidation of macromolecules such as proteins, membrane lipids, and nucleic acids [18], but may as well extend to the nitrosylation of tyrosine and cysteine sulfhydryls, thereby severely disturbing tyrosine phosphorylation-mediated signaling [19] and sulfhydryl-based redox sensing [6, 20].

To decipher the detailed role of such reactive compounds and the resulting changes in cellular redox balance in living tissue in real time, optical probes are the method of choice. Unfortunately, feasible optical sensors were not available for a long time, and less ideal synthetic fluorescent dyes had to be dealt with. About 14 years ago the development of the first genetically encoded optical ROS/redox sensor [21] slowly began to bridge this gap. Meanwhile, a selection of advanced redox probes is available (see: [22, 23]). They overcome almost all of the earlier shortcomings and can be targeted to various desired cellular locations to monitor ROS/redox conditions directly on the level of for example mitochondria. Hence, they allow not only to analyze mitochondrial ROS formation in these organelles but also to decipher the functional consequences in other cell compartments.

2 Materials and Methods

2.1 Choosing an Ideal Optical ROS/Redox Sensor

Since $^{\cdot}O_2^-$ may be released into the mitochondrial matrix and also into the intermembrane space [4], from where it may easily escape into the cytosol, analyses of mitochondria-related ROS production should include both, the mitochondrial and the cytosolic compartments. Accordingly, both reliable cytosolic and reliable mitochondria-based redox sensors are essential, to fully understand the downstream effects of ROS on cellular physiology.

Since more than 25 years, a variety of synthetic oxidation-sensitive fluorescent dyes have been established (Table 1). Widely

Table 1
Frequently used oxidation-sensitive synthetic fluorescent dyes and genetically encoded redox sensors

	Sensor	Cellular localization / Targeting	Selectivity	Excitation (nm)	Emission (nm)
Synthetic dyes	Amplex Red	Cytosol	Oxidation, H_2O_2	570	585
	DCF and derivatives	Cytosol	Oxidation	505	530
	Dihydroethidium	Cytosol	Oxidation	500	590
	Dihydrorhodamine	Cytosol, mitochondria	Oxidation	500	535
	MitoSOX Red	Cytosol, mitochondria	Oxidation, superoxide	500	580
	MitoTracker Orange	Cytosol, mitochondria	Oxidation	550	575
	MitoTracker Red	Cytosol, mitochondria	Oxidation	580	600
	RedoxSensor Red	Cytosol, mitochondria, lysosomes	Oxidation	540	600
Genetically encoded sensors	HyPer	Cytosolic Any other subcellular target can be specified by proper tagging sequences	Oxidation, reduction	490/420 ratiometric	515
	roGFP1		Oxidation, reduction	395/470 ratiometric	510
	rxYFP		Oxidation, reduction	510	525
	CFP-HSP33-YFP		Oxidation, reduction	Donor 435 Accpt 515	490, 530 FRET ratio
	CFP-RL5-YFP		Oxidation, reduction	Donor 435 Accpt 515	490, 530 FRET ratio
	Redoxfluor (Cerulean-Yap1-Citrine)		Oxidation, reduction	Donor 430 Accpt 515	500, 530 FRET ratio

Listed excitation/emission wavelengths are approximate starting points and need to be optimized for the very type of experiment

used are dichlorodihydrofluorescein or its numerous derivatives [24], dihydroethidium (hydroethidine) [25], and Amplex Red [26]. These compounds freely enter cells, distribute within the cytosol, and upon oxidation by various types of ROS become fluorescent. Labeling can easily be achieved by bulk loading of the cells and tissue of interest, using micromolar dye concentrations (diluted in physiological saline or culture medium) and incubation times of typically 15–30 min. Other synthetic dyes were developed to directly assess ROS formation in mitochondria. Examples include dihydrorhodamine [27], chemically reduced variants of MitoTracker compounds (MitoTracker Red CM-H_2XROS, MitoTracker Orange CM-H_2TMROS) [28], and MitoSOXRed, a cationic derivative of dihydroethidium [29]. These dyes accumulate in the mitochondrial matrix and upon oxidation become fluorescent, causing intense mitochondrial staining. Also, RedoxSensor

Red CC-1 Stain becomes fluorescent upon oxidation and then labels mitochondria and lysosomes, but its exact distribution among these organelles depends on the current cellular redox conditions [30].

All of these synthetic dyes respond, however, only to oxidation but not reduction. Also, they are oxidized irreversibly and then remain fluorescent. Accordingly, equilibrium experiments or absolute quantification of ROS levels are not possible. Furthermore, these dyes are quite sensitive to oxidation by ambient O_2 (auto-oxidation) or illumination (photo-oxidation), and their cell retention is far from ideal. Additional problems may arise when the oxidized dyes secondarily enter other cell compartments. Oxidized dihydroethidium for example binds to DNA and stains the nucleus, whereas oxidized dihydrorhodamine accumulates in mitochondria. Therefore, the very location of fluorescence is misleading as it does not necessarily represent the original subcellular site of oxidation.

A reliable functional analysis of redox signaling demands optical probes that distinguish among oxidized and reduced states, respond reversibly and accurately, and report redox balance down to the subcellular level. Unfortunately, none of the synthetic dyes fulfills these criteria.

In comparison, genetically encoded optical redox indicators are more reliable (see [22, 23]). They derive from fluorescent proteins (GFP, YFP), which were rendered redox-sensitive by targeted inclusion of reactive cysteines or fusion to nonfluorescent redox-sensitive linker proteins (Table 1). Redox sensors consisting of modified fluorescent proteins are rxYFP [21] and roGFP [31]. Probes in which a fluorescent protein is coupled to redox-sensitive linker include HyPer [32] as well as the FRET (fluorescence resonance energy transfer, also known as Förster resonance energy transfer) constructs CFP-HSP33-YFP [33], CFP-RL5-YFP [34], and Redoxfluor (Cerulean-Yap1-Citrine; [35]).

Thus, proper optical tools are now available for functional analyses of subcellular redox conditions. These sensors show high photostability and superior cell retention, withstand tissue fixation, reversibly respond to oxidation as well as reduction, and upon calibration yield semiquantitative measures of redox dynamics [22, 31, 36, 37]. The selection of the ideal redox sensor depends on the specific experimental conditions and questions addressed. Yet it also depends on the detailed response properties of the different sensors, which so far have been only partly available. Therefore, characterizing the detailed sensor responses and confirming their reliability in various preparations, cellular compartments and experimental conditions is still an ongoing challenge. Among the fascinating features of these redox sensors is that they are not restricted to the cytosol, but can be targeted to various compartments and organelles such as nucleus, mitochondria, endoplasmic reticulum, or plasma membrane [31, 32, 38, 39]. It is this genetic targeting opportunity which allows to define local specializations of redox balance and to decipher downstream signaling events as well as their particular features in the different subcellular compartments. These redox sensors can even be linked to other proteins and enzymes to address more specific aspects of redox signaling. Fusion of roGFP to glutaredoxin-1 for example specifically reports glutathione redox potentials [40].

2.2 Targeted Delivery of Genetically Encoded Fluorescent Sensors

As mentioned above, genetically encoded redox sensors allow for qualitative and quantitative readout of subcellular (patho-)physiological states with high temporal and spatial resolution. While this is excellent news, it remains to be discussed how these sensor proteins are brought into the cell or into its compartments.

In the ideal case, the investigator determines a physiological situation and a tissue or cell type of interest, and orders an available transgenic mouse or rat expressing the sensor protein of choice. Given the recent advances in transgenic rodent technologies and the ever-growing repositories of genetically engineered animals this scenario is by no means inconceivable, but realistically spoken seems to be roughly one or two decades away for the majority of novel experimental approaches. Alternatively, if having a good core facility at hands or sufficient funds available to order at commercial sources you may be able to generate your own transgenic animals within roughly a year (realistically estimated). Yet this process needs to be repeated with every iteration of the sensor under study, i.e., with other subcellular locations, other target cells, novel developments in sensor technology, and so forth.

So what may be the alternative to save time, money, and efforts? While DNA transfection techniques work very well for many cultured cell lines, they are quite inefficient for several preparations important in the context of these pages, namely cultured primary brain cells, acute slices and organotypic cultures of different brain nuclei. Also, the cellular expression pattern obtained is usually rather random and unspecific. This is why one may consider an alternative that Mother Nature has created with amazing elegance, i.e., the use of recombinant viruses as gene transfer vehicles. The "raison d'être" of a virus is to deliver genetic material to its host/target cell with unprecedented efficacy. Engineered recombinant versions of lentiviruses (LV), adeno-associated viruses (AAV), adenoviruses (AD), herpes-simplex viruses (HSV), or Semliki Forest viruses (SFV) available today do simply this, without being capable to reproduce themselves or cause diseases [41, 42]. For sure, there is not much to say against the argument that establishing production of recombinant viruses from scratch is at least partially as painstaking as establishing production of transgenic mice. Therefore, we concentrate here on viruses which can be made easily (LV, SFV) or which can be obtained quite quickly and at comparably low cost from several university vector core facilities or commercial sources (LV, AAV).

2.2.1 LV Vectors

Production of recombinant LV vectors is a straightforward process, because viral particles can be harvested from the cell culture supernatant as described in detail elsewhere [43]. Most laboratories capable of pursuing recombinant DNA technologies and with some experience in cell culture practice will be able to generate their own infective particles with reasonable efficacy, although reproducible production of high-titer stocks needs some training, or may be outsourced to commercial facilities. The advantage of easy production is somewhat counterbalanced by the fact that the vast majority of LV vectors generated today are still based on the HIV, which in most countries necessitates their handling in biosafety level (BSL) 2 laboratories. Recombinant LV can be pseudotyped with the standard VSV-G glycoprotein, directing the virus predominately to neurons, or with the Mokola glycoprotein, directing the virus predominately to astrocytes [44]. If exclusive neuronal or astrocytic sensor expression is of importance, then appropriate promoters and introduction of cell type-specific miR binding sequences may be helpful [45]. A good starting point for expression of sensors in cultured neurons/organotypic cultures, but also in vivo, i.e., for 2-photon imaging, would be a VSV-G pseudotyped vector driving expression from the ubiquitin promoter, as found in the "FUGW" series of LV genome plasmids.

2.2.2 AAV Vectors

Gene transfer tools based on recombinant adeno-associated viruses are our preferred working horses. Almost all AAV-based vectors can be handled under BSL-1, because the wild-type AAV does not cause human or animal diseases. Moreover, the simple capsid structure of this non-enveloped virus has led to the evolution of many major capsid variants (serotypes) and roughly hundred minor variants (sub-serotypes) [46]. Since the very constituents of the capsid determine binding, uptake, and processing (i.e., transduction properties) of the recombinant virus, it is likely that a serotype with optimal transduction properties for the desired target cell types and tissues has already been described. In addition, the diversity of available serotypes can be used to modify the expression level of the transgene carried by the recombinant virus: for example, AAV serotypes 2 and 5 deliver relatively few vector genomes, i.e., copies of the transgene to be expressed, into individual brain cells, while serotype 6, 8, and 9 are more "efficient" and result in much higher levels of transgene expression. On top of that, diffusion through tissue is quite dependent on the serotype used, due to differential binding to cell surface receptors. Thus, AAV-2 or -6 will stay in close proximity to their injection tract into brain tissue in vivo, or into a slice culture in vitro, while AAV-5 or -9 will diffuse over much larger areas [47].

In brain tissue samples and cultured neurons most AAV serotypes with the exception of AAV-2 are able to infect multiple cell types. Therefore, the choice of cell type-specific promoters is important. As a good starting point we would recommend to use the human synapsin 1 gene promoter for exclusive neuronal expression, and the full length GFAP promoter for exclusive astrocytic expression [48]. Exactly this strategy we also followed for the design of the roGFP1-expressing AAV-6 vectors (see below).

There are a few drawbacks to consider if planning to use AAV vectors: the maximum genome capacity (transgene plus all transcriptional control elements) must not exceed 5 kb, which should not be a problem for the vast majority of genetically encoded fluorescent sensors. Production of high titer stocks is more tedious than with LV vectors [49], and if only one or just a few vectors are needed, it might be worthwhile to have them produced by vector core facilities or commercial sources.

2.2.3 SFV Vectors

Gene transfer with AAV and LV vectors will result in transgene expression for the live-time of the respective specimen, be it cultured brain cells, organotypic cultures or the rodent or nonhuman primate brain in vivo. Depending on the paradigm and applied amount of viral vector, expression of the transgene may need a few days to reach a level sufficient for analysis, which should suit most applications. In case that “overnight expression” is needed, Semliki Forest-based vectors might be considered [50, 51]. These can be produced relatively easily in small-scale cell culture (6 well plates) and as pure RNA viruses will not need any processing steps after infecting their target cells, but will immediately start to produce their transgene. However, as SFV replicate their RNA genomes within their host cells they can result in cytotoxicity after just a few days, leaving only a window of ~1 day for analysis, which for certain applications may be too narrow [52].

2.2.4 General Aspects of Viral Vector-Mediated Transgenesis

The major advantage of using recombinant viruses as vehicles for gene transfer of genetically encoded sensors into the living brain, organotypic cultures, or cultured brain cells is the enormous flexibility of these tools. Transcriptional control elements include promoters or expression enhancers like the woodchuck hepatitis virus posttranscriptional control element (WPRE) are easily exchanged, as are vector sero-/pseudotypes, the titer applied, the brain region of interest or the sensor under investigation. Given the rapid development of novel sensors it is possible to test various constructs with different subcellular targeting in due course and in a variety of preparations. Viral vectors can be faithfully combined with classical mouse genetics, e.g., by using floxed transgenes in Cre-expressing mouse lines, or *vice versa*, applying Cre expressing viruses into mice carrying floxed transgenes. Their production is fast and economical as compared to generation of mouse lines. In order to become really valuable and reproducible tools, a certain effort in establishing production routines is necessary, which, however, can be bypassed by ordering recombinant virus from external sources (companies and University Vector Core Facilities) for just a few thousand Euro per vector. Even the common obstacle of having to use stereotaxic injections to deliver the viral vector into the brain, as a source for intravital 2-photon imaging or generation of acute tissue slices, turns into an advantage, namely, when considering that this allows to direct transgene expression exactly to the neuronal populations of interest. Thus, while viral vector mediated transgenesis will never completely displace transgenic animal technology, it can substantially speed up evaluation of both, genetically encoded fluorescent sensors and (patho)physiological processes, be they related to diseases or to normal cellular homeostasis.

2.3 ROS/Redox Imaging

At the core of our mitochondrial research is the optical redox sensor roGFP1 [31], which contains a thiol-disulfide switch and responds reversibly to oxidation/reduction with characteristic changes in fluorescence emission (Fig. 1). It exhibits two distinct absorption peaks (395 and 470 nm), both of which are independently and oppositely modulated by the current relative redox state of the cell (Fig. 1c). Thus, ratiometric excitation is possible, ensuring a broader dynamic range and more stable recordings [31]. The roGFP1 sensor is fully integrated into cellular thiol homeostasis—behaving as other cell endogenous proteins with disulfide redox switches—and quantitatively reports thiol redox status. We confirmed that roGFP1 responds reliably in mouse and rat hippocampal cell and slice cultures and is not noticeably affected by intracellular pH or Cl^- changes [16, 36]. Of the different roGFP variants we selected roGFP1, because it shows a pronounced absorption at the 395 nm peak [31], where light sources become less intense and light transmission of the objectives is declining. Furthermore, its 470 nm absorption peak responds intensely to redox modulation which ensures a wide dynamic range even with single wavelength excitation, e.g., during 2-photon imaging. Furthermore, roGFP1 is also feasible for fluorescence lifetime imaging microscopy (FLIM) [53], and we found it to be clearly superior to other genetically encoded redox sensors such as for example HyPer (Table 2, [36, 54]).

2.3.1 Transfection/Transduction of Hippocampal Cell- and Slice Cultures

Dissociated cell and organotypic slice cultures are standard preparations to perform studies on the cellular level, especially in combination with genetically encoded optical sensors. We prepare primary dissociated cell cultures as well as organotypic slice cultures from the hippocampus of neonatal mice (postnatal day P2-P5) following standard procedures [55, 56] with some modifications. For both types of cultures only one animal is used at a time. Typically, this yields dissociated cells for ~12 coverslips or ~16 organotypic slices. To deliver cytosolic roGFP1-coding DNA, we initially used transfection which results in a random unspecific expression in neurons and glia.

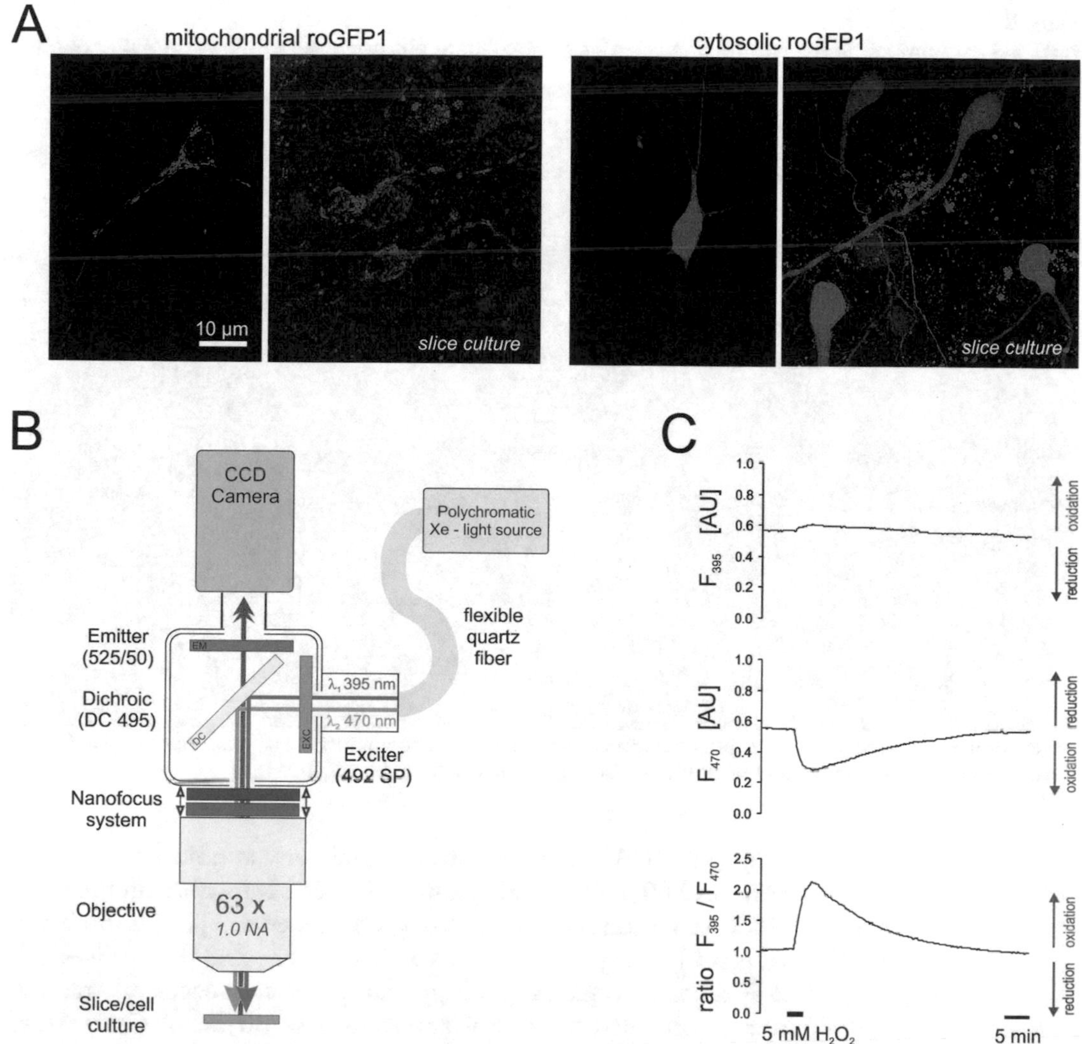

Fig. 1 Compartment-specific expression of the ratiometric redox sensor roGFP1. (**a**) Transduced hippocampal neurons expressing mitochondrial or cytosolic roGFP1. By means of viral vectors (AAV-6), we expressed roGFP1 in neurons of dissociated cell cultures and organotypic slices. Scale bar represents 10 μm, images were taken with a 2-photon laser scanning microscope, excitation wavelength was 890 nm. (**b**) Principle of dynamic live cell redox imaging. Excitation-ratiometric imaging demands alternating exposure to two defined wavelengths. This can be achieved by using a switchable polychromatic xenon light source. Indicated wavelengths and optical filters are those needed for ratiometric epifluorescence imaging of roGFP1. (**c**) The ratiometric principle of roGFP1 is based on two excitation peaks (395 nm, 475 nm) which respond oppositely to redox challenge. Fluorescence of cytosolic roGFP1 is plotted in arbitrary units (AU). Calculating the ratio F_{395}/F_{470} yields a reliable and stable measure of redox balance, which increases upon oxidation by for example H_2O_2. Traces were recorded using a CCD camera imaging system as depicted in panel (**b**)

Transfection can be performed in both cell cultures and organotypic slices cultures using Lipofectamine 2000 (Invitrogen). In dissociated cell cultures, roGFP1 coding plasmids are transfected at day in vitro (DIV) 1–4. The transfection solution (Opti-MEM, Invitrogen) is complemented with 1 % Lipofectamine and

Table 2
Detailed comparison of the response properties of the genetically encoded roGFP1 and HyPer redox sensors

Sensor	roGFP1	HyPer
Specificity	General redox sensor	H_2O_2 specific
Redox sensitivity	Oxidation AND reduction	Oxidation only
Structural composition	SH-groups in fluorophore	SH groups in linker protein
Excitation	Excitation ratiometric 1-Photon (395/470 nm) 2-Photon (740/910 nm)	Excitation ratiometric 1-Photon (490/420 nm) 2-Photon (920/760 nm)
Cellular localization	Cytosol, mitochondria	Cytosol, mitochondria
pH sensitivity	pH sensitivity negligible	pH sensitivity problematic
Cl^- sensitivity	Negligible	Negligible
H_2O_2 sensitivity	EC_{50} = 97 μM H_2O_2	EC_{50} = 104 μM H_2O_2
FLIM	~2.3 ns lifetime	~1.5 ns lifetime
Mode of DNA delivery	Lipofection, electroporation, transduction	Lipofection

Listed features are based on our in-depth assessment in rat and mouse hippocampal cell and slice cultures [36, 53, 54]. In these trials, reasonable responses of HyPer to reducing stimuli could not be obtained [54]

1 μg/ml DNA, and should not contain any antibiotics. Each well receives 200 μl of this solution, and after 1 h incubation the transfection solution is replaced with growth medium [36]. Lipofection results in an expression of roGFP1 in only ~2–3 % of cells within 2–3 days. In organotypic hippocampal slices successful transfection is obtained by topical application of 30–50 μl Opti-MEM-based transfection solution supplemented with 0.4 % Lipofectamine and 1.1 μg DNA on each slice. Sufficient roGFP1 expression is obtained within 2–3 days, but in slices the transfection efficiency is even lower than in dissociated cultures (<<1 % of cells) [16, 36].

More recently, we have established viral constructs (AAV-6) encoding cytosolic and mitochondrial roGFP1 under the control of the synapsin-1 promoter, to optimize gene transfer and to ensure that roGFP1 is expressed specifically in neurons [57]. Dissociated hippocampal cell cultures are transduced on DIV 2 with 2.5 μl of the vector (1:50 dilution in PBS; undiluted titer 1.3×10^8 particles/μl) added directly into 800 μl growth medium and are cultured for additional 6–7 days to ensure proper roGFP1 expression. Organotypic slices are transduced on DIV 3–4 by applying topically 2.5 μl of the vector (1:50 dilution). Sufficient roGFP1 expression (~15 % of neurons) is obtained within 5–6 days post-transduction. Figure 1a shows sample images of transduced cell and slice cultures expressing cytosolic and mitochondrial roGFP1.

A drawback of transfection and transduction procedures is that, due to the time required for sufficient expression, they are usually limited to cultured preparations, which in turn can be obtained from the neonatal brain only. Adult tissue can only be studied when viral vectors are directly injected into the brain region of interest of each single animal a couple of days before the experiments. To overcome this limitation, and to circumvent additional surgical procedures, transgenic mice would have to be generated which stably express optical redox sensors in the desired cell type and the specific subcellular compartment (see Conclusions and Outlook).

2.3.2 Imaging Equipment

Successful live cell imaging demands a collection of reliable accessories, i.e., fluorescence microscope, proper objectives, a stable light source, and a sensitive detector; all of which have to be synchronized by control software and a computer for data storage. The choice of a microscope depends on the type of preparation under investigation. An upright microscope is mandatory if acute brain slices are to be analyzed, but even for cell cultures it may offer more flexibility than an inverted microscope.

To take full advantage of the favorable properties of roGFP redox sensors, excitation ratiometric analyses should be performed. This requires alternating excitation at the two defined absorption peaks, i.e., 395 and 470 nm for roGFP1 while recording emission at 500–550 nm. Accordingly, a switchable polychromatic light source or a broad band (mercury, tungsten-halogen, or LED) light source with an automated filter wheel is needed to rapidly alternate excitation wavelengths. Also, proper excitation, dichroic, and emission filters are crucial to optimize fluorophore excitation, separation of emission, and detection efficiency (Fig. 1b). For 395 nm excitation an objective with a sufficient near-UV transmission has to be chosen. Furthermore, we mount the objective to a remote-controlled Pifoc nanofocusing system (Physik Instrumente), as this enables refocusing during the experiments without touching the microscope.

ROS formation and the associated redox changes are rather slow signals. Therefore, moderate sampling rates should be applied, which also helps preventing over-illumination of the samples under investigation. Usually, we apply frame rates of 0.1–0.2 Hz (5–10 s/image pair). Yet when highly localized redox changes are to be analyzed, i.e., in individual mitochondria or refined dendritic/axonal segments, frame rates in the order of 1–2 Hz may be required. Excitation exposure time is another critical issue for sample viability. For CCD camera imaging, exposure times of 15–50 ms should be sufficient. Unless high resolution imaging is desired, binning (signal combination of adjacent CCD pixels) of 2 × 2 or 4 × 4 pixels should be applied to increase detection sensitivity, reduce detector noise, and to shorten exposure times. Under these conditions, stable roGFP1 recordings of >1 h can be performed easily (Figs. 1c, 2a).

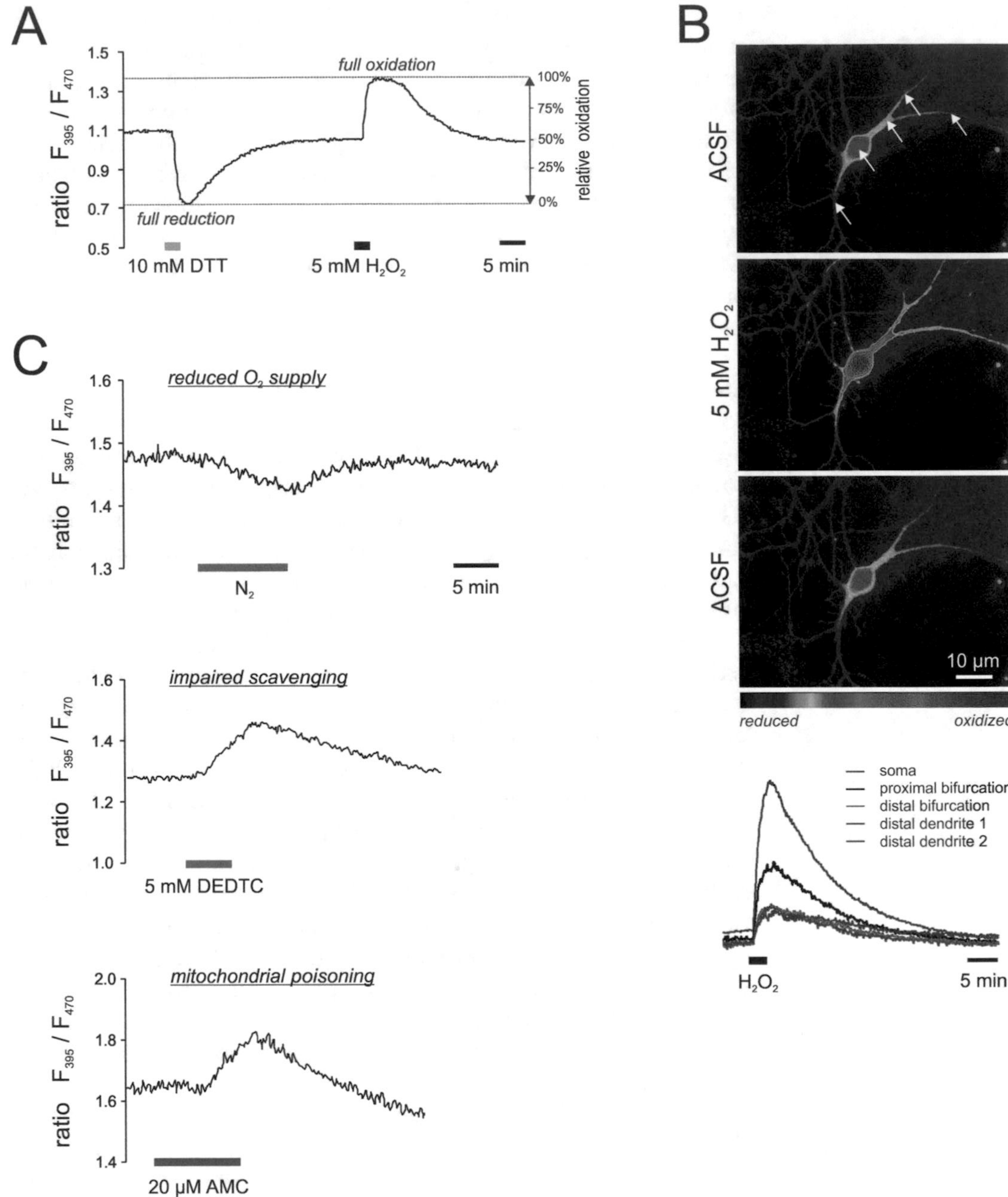

Fig. 2 Calibration of roGFP1 enables subcellular semiquantitative analysis. (**a**) Exposing roGFP1 to saturating oxidizing and reducing stimuli defines its ratiometric response range for the current experimental conditions. Displayed is a calibration of mitochondrial roGFP1 in a mouse hippocampal cell culture. (**b**) Hippocampal neuron expressing cytosolic roGFP1. Note the regionally different redox-responses to oxidant challenge in the five distinct regions of interest analyzed (indicated by arrows in the *upper panel*). Shown are CCD camera images. (**c**) Changes in mitochondrial redox balance upon modulation of cell endogenous ROS production by O_2 withdrawal, inhibition of superoxide dismutase by DEDTC, or blocking mitochondrial complex III by antimycin A (AMC). Time scale is identical for all three traces displayed in this panel

If laser imaging is desired with roGFP redox sensors, ratiometric excitation becomes more challenging, as two separate laser lines are required. For 2-photon imaging, we recently demonstrated that ratiometric excitation of roGFP1 and HyPer redox sensors is also possible (Table 2, [53, 54]). Yet wavelengths have to be switched manually between the individual exposures, which is only feasible if only a few ratiometric image pairs are needed. For continuous dynamic recordings, this is hardly applicable. Such analyses would rather require the rare luxury of two separate 2-photon laser systems, each set to one of the desired wavelengths and coupled into the scanning system by the proper beam-combining optics [53].

2.3.3 Response Range Calibration and Semiquantitative Analysis

The roGFP sensors exhibit two distinct excitation peaks responding oppositely to redox modulation (Fig. 1c) [31, 38]. This excitation ratiometric feature enables quantitative analyses, as the fluorescence intensity ratio (F_{395}/F_{470}) does not depend on sensor concentration [58]. Similar excitation ratiometric properties also apply to HyPer [32, 54], whereas the FRET-based redox sensors (Table 1) are ratiometric by emission, reporting the ratio of acceptor and donor fluorescence.

Accordingly, one has to calibrate the response range of any given redox-sensor for the specific conditions of the experiment. As already shown in the first description of roGFP, determining the responses to full oxidation and reduction is sufficient to perform semiquantitative assays and to estimate the relative degree of oxidation and the corresponding redox potential [22, 31]. Once calibrated, the absolute roGFP ratio can then be used to compare various treatment conditions and/or redox balance in different cellular compartments, cell types or even genotypes, as we have recently shown in a mouse model of Rett syndrome [16, 17, 57].

For response range calibration of roGFP1 we apply saturating doses of H_2O_2 (5 mM, 3 min) and DTT (10 mM, 3 min) to induce full oxidation and reduction, respectively (Fig. 2a). Exposure to H_2O_2 massively increases the roGFP1 fluorescence ratio (F_{395}/F_{470}), whereas DTT markedly decreases it. Once calibrated, the very same experimental setup, identical excitation parameters, and hardware settings have to be used. It is important to understand that the determined response range applies only to the specific recording conditions. Separate calibrations are required for each type of sensor, subcellular targeting, preparation (cell culture, slice culture, acute slice) or objective used, possibly even for different tissues or brain regions, as optical parameters (scattering, autofluorescence) may differ markedly among specimen.

So far, we have calibrated cytosolic and mitochondrial roGFP1 especially in neurons of hippocampal cell and slice cultures. This now allows to compare cytosolic and mitochondrial compartments in dissociated cells and more complex organotypic slices.

Indeed, the relative baseline oxidation of roGFP1 differs among the compartments. For example, based on calibrations in organotypic slices, oxidation of neuronal roGFP1 at resting baseline conditions amounts to ~35 % in the cytosol and ~50 % in mitochondria.

As cytosolic roGFP1 distributes homogenously, its fluorescence ratio reports redox balance within the cytosolic sub-compartments such as soma, dendrites, and axon, i.e., those environments housing various intracellular signaling cascades and organelles. Clear differences become obvious on the sub-cellular level upon, e.g., acute oxidant stimulation (Fig. 2b). Along this line of thoughts, mitochondrial roGFP1 may serve to reveal redox differences among individual mitochondria, depending on their cellular location, activity status, membrane potential, and morphological appearance. Indeed, we recently confirmed such differences on the mitochondrial level, using the mitochondria-targeted HyPer sensor [54].

The roGFP1 redox sensor is sufficiently sensitive to report reliably subtle redox changes that may arise in response to altered endogenous ROS formation in mitochondria or the cytosol. Severe anoxia, for example, elicits a reducing shift, whereas inhibition of the scavenging enzyme superoxide dismutase or blocking of mitochondrial complex III mediate an oxidizing shift of the mitochondrial roGFP1 fluorescence ratio (Fig. 2c).

3 Notes

1. Genetically encoded redox sensors derive from fluorescent proteins, which—depending on the very fluorophore—may respond also to environmental changes in pH and halide content. Therefore, a thorough testing of such undesired interference is essential. Whereas roGFP1 is rather insensitive to these ambient variables, other sensors such HyPer may show a pronounced pH sensitivity [54]. In this case parallel pH control experiments are mandatory to prevent misinterpretation of data.
2. It is also important to understand that most redox probes lack specificity (Table 1). Sensors as roGFP resemble general redox indicators, reporting thiol redox balance. Even for those probes assumed to be specific to H_2O_2, such as HyPer [32], recovery of the reduced form of the sensor nevertheless relies on glutathione-dependent disulfide reduction [22]. Even if more specific redox sensors were available, another source of bias would be the different reactivity of the particular ROS/RNS. Both $^{\cdot}O_2^-$ and $OH^{\cdot}$ show extremely short lifetimes, whereas H_2O_2 and $ONOO^-$ are less reactive and hence are the primary detectable oxidants.

3. For redox imaging at the tissue level, a potential interference of redox sensor fluorescence with autofluorescence has to be tested. NADH/FAD autofluorescence markedly changes with metabolic activity, O_2 tension and mitochondrial function. Though being less intense than the pronounced fluorescence of synthetic dyes or genetically encoded redox sensors, some overlap may still occur. This especially applies when shorter wavelengths (<450 nm) are used for excitation.
4. Antioxidants are already included in most culturing media (in concentrations and compositions not further specified by the vendor) to optimize viability of cultured cells and tissues. Yet such antioxidants, e.g., in B27 supplement, may affect baseline redox balance and the responses to acute redox challenge, as we confirmed for mouse organotypic hippocampal slice cultures [16]. Therefore, control experiments with antioxidant-free media should be included in each experimental series.
5. Any movement or swelling of the tissue under investigation during imaging experiments constitutes a problem to be dealt with, especially when strong stimuli such as uncouplers, mitochondrial poisoning and anoxia/ischemia are applied. While out-of-focus swelling may be corrected easily by refocusing during the experiments (e.g., by using a nanofocus system), lateral drifts are more difficult to compensate for, as they will affect the defined regions of interest. In case of brightly labeled cells, tracking algorithms may help to correct the drift and to realign the individual image planes before analysis. For small subcellular structures, this certainly will be more challenging. In the case of mitochondria, further complications can be expected to arise from their motility and pronounced degree of subcellular trafficking [59].
6. Finally, it has to be kept in mind that any kind of transfection procedure may challenge cellular viability. In part, this also applies to viral transduction. Optimized viral titers are crucial to obtain the desired cell-specific expression of a given sensor.

4 Conclusions and Outlook

For ROS/redox imaging in mitochondria, the cytosol or any other subcellular compartment, genetically encoded redox sensors are by far superior to synthetic fluorescent dyes. Fortunately for all of us, the general and even commercial availability of these sensors has markedly improved in recent years. Their use requires, however, transfection or transduction followed by a sufficient expression time of several days, which often limits an application to cultured preparations and hence, neonatal stages. Therefore, synthetic oxidation-sensitive dyes—despite their severe disadvantages—are still quite commonly used to assess ROS/redox conditions in acute and/or adult tissue preparations, such as brain slices. With the constant advances in transgenic animal development, quite a selection of redox-sensor expressing transgenic mouse lines should become available soon. Indeed, first mice expressing mitochondrial roGFP2 redox sensors in monoaminergic neurons and erythrocytes or glutaredoxin 1-coupled roGFP2 in neuronal mitochondria were already generated to elucidate oxidant stress in Parkinson's disease [60], red blood cell redox status [61], and mitochondrial glutathione redox potentials in individual organelles [62], respectively. Other models express cytosolic or mitochondrial roGFP1 in keratinocytes to study oxidative stress in the skin [63]. We just recently extended this list by generating mice expressing roGFP1 in cytosol and mitochondrial matrix of excitatory neurons [53]. Availability of such redox-sensor mice can be expected to further revolutionize ROS/redox imaging in mitochondria and other subcellular compartments. Ultimately, it will allow to successfully extend reliable redox imaging to adult and more complex preparations and to also address developmental aspects of ROS formation and redox alterations.

Acknowledgments

We thank Belinda Kempkes for excellent technical assistance, and we are grateful to Professor S. James Remington, Institute of Molecular Biology, University of Oregon, Eugene OR USA, for making available to us the plasmids expressing roGFP1 redox-sensitive proteins. Our research was funded by the Cluster of Excellence and Research Center for Nanoscale Microscopy and Molecular Physiology of the Brain (CNMPB) and the International Rett Syndrome Foundation (IRSF, grant #2817) as well as by the University Medical Center Göttingen and the State of Lower Saxony (large scale equipment grant INST 1525/14-1 FUGG).

References

1. Senior AE (1988) ATP synthesis by oxidative phosphorylation. Physiol Rev 68:177–231
2. Boveris A, Chance B (1973) The mitochondrial generation of hydrogen peroxide. General properties and effect of hyperbaric oxygen. Biochem J 134:707–716
3. Dröge W (2002) Free radicals in the physiological control of cell function. Physiol Rev 82:47–95
4. Brand MD (2010) The sites and topology of mitochondrial superoxide production. Exp Gerontol 45:466–472
5. Cadenas E, Davies KJ (2000) Mitochondrial free radical generation, oxidative stress, and aging. Free Radic Biol Med 29:222–230
6. Lipton SA, Nicotera P (1998) Calcium, free radicals and excitotoxins in neuronal apoptosis. Cell Calcium 23:165–171
7. Kamsler A, Segal M (2003) Hydrogen peroxide modulation of synaptic plasticity. J Neurosci 23:269–276
8. Atkins CM, Sweatt JD (1999) Reactive oxygen species mediate activity-dependent neuron-glia signaling in output fibers of the hippocampus. J Neurosci 19:7241–7248
9. Finkel T (2011) Signal transduction by reactive oxygen species. J Cell Biol 194:7–15
10. Schulz JB, Beal MF (1994) Mitochondrial dysfunction in movement disorders. Curr Opin Neurol 7:333–339

11. Cooper JM, Schapira AH (1997) Mitochondrial dysfunction in neurodegeneration. J Bioenerg Biomembr 29:175–183
12. Dahl HH (1998) Getting to the nucleus of mitochondrial disorders: identification of respiratory chain-enzyme genes causing Leigh syndrome. Am J Hum Genet 63:1594–1597
13. Behl C, Moosmann B (2002) Oxidative nerve cell death in Alzheimer's disease and stroke: antioxidants as neuroprotective compounds. Biol Chem 383:521–536
14. Kong J, Xu Z (1998) Massive mitochondrial degeneration in motor neurons triggers the onset of amyotrophic lateral sclerosis in mice expressing a mutant SOD1. J Neurosci 18: 3241–3250
15. Kovács R, Schuchmann S, Gabriel S, Kann O, Kardos J, Heinemann U (2002) Free radical-mediated cell damage after experimental status epilepticus in hippocampal slice culture. J Neurophsiol 88:2909–2918
16. Großer E, Hirt U, Janc OA, Menzfeld C, Fischer M, Kempkes B, Vogelgesang S, Manzke TU, Opitz L, Salinas-Riester G, Müller M (2012) Oxidative burden and mitochondrial dysfunction in a mouse model of Rett syndrome. Neurobiol Dis 48:102–114
17. Müller M, Can K (2014) Aberrant redox homoeostasis and mitochondrial dysfunction in Rett syndrome. Biochem Soc Trans 42: 959–964
18. Chan PH (1996) Role of oxidants in ischemic brain damage. Stroke 27:1124–1129
19. Martin BL, Wu D, Jakes S, Graves DJ (1990) Chemical influences on the specificity of tyrosine phosphorylation. J Biol Chem 265: 7108–7111
20. Lipton SA, Choi YB, Takahashi H, Zhang D, Li W, Godzik A, Bankston LA (2002) Cysteine regulation of protein function--as exemplified by NMDA-receptor modulation. Trends Neurosci 25:474–480
21. Ostergaard H, Henriksen A, Hansen FG, Winther JR (2001) Shedding light on disulfide bond formation: engineering a redox switch in green fluorescent protein. EMBO J 20:5853–5862
22. Meyer AJ, Dick TP (2010) Fluorescent protein-based redox probes. Antioxid Redox Signal 13:621–650
23. Björnberg O, Ostergaard H, Winther JR (2006) Measuring intracellular redox conditions using GFP-based sensors. Antioxid Redox Signal 8:354–361
24. LeBel CP, Ischiropoulos H Bondy SC (1992) Evaluation of the probe 2′,7′-dichlorofluorescin as an indicator of reactive oxygen species formation and oxidative stress. Chem Res Toxicol 5:227–231
25. Gallop PM, Paz MA, Henson E, Latt SA (1984) Dynamic approaches to the delivery of reporter reagents into living cells. BioTechniques 2:32–36
26. Mohanty JG, Jaffe JS, Schulman ES, Raible DG (1997) A highly sensitive fluorescent microassay of H_2O_2 release from activated human leukocytes using a dihydroxyphenoxazine derivative. J Immunol Methods 202:133–141
27. Dugan LL, Sensi SL, Canzoniero LM, Handran SD, Rothman SM, Lin TS, Goldberg MP, Choi DW (1995) Mitochondrial production of reactive oxygen species in cortical neurons following exposure to N-methyl-D-aspartate. J Neurosci 15:6377–6388
28. Esposti MD, Hatzinisiriou I, McLennan H, Ralph S (1999) Bcl-2 and mitochondrial oxygen radicals. New approaches with reactive oxygen species-sensitive probes. J Biol Chem 274:29831–29837
29. Robinson KM, Janes MS, Pehar M, Monette JS, Ross MF, Hagen TM, Murphy MP, Beckman JS (2006) Selective fluorescent imaging of superoxide in vivo using ethidium-based probes. Proc Natl Acad Sci U S A 103: 15038–15043
30. Chen CS, Gee KR (2000) Redox-dependent trafficking of 2,3,4,5, 6-pentafluorodihydrotetr amethylrosamine, a novel fluorogenic indicator of cellular oxidative activity. Free Radic Biol Med 28:1266–1278
31. Hanson GT, Aggeler R, Oglesbee D, Cannon M, Capaldi RA, Tsien RY, Remington SJ (2004) Investigating mitochondrial redox potential with redox-sensitive green fluorescent protein indicators. J Biol Chem 279: 13044–13053
32. Belousov VV, Fradkov AF, Lukyanov KA, Staroverov DB, Shakhbazov KS, Terskikh AV, Lukyanov S (2006) Genetically encoded fluorescent indicator for intracellular hydrogen peroxide. Nat Methods 3:281–286
33. Guzy RD, Hoyos B, Robin E, Chen H, Liu L, Mansfield KD, Simon MC, Hammerling U, Schumacker PT (2005) Mitochondrial complex III is required for hypoxia-induced ROS production and cellular oxygen sensing. Cell Metab 1:401–408
34. Kolossov VL, Spring BQ, Sokolowski A, Conour JE, Clegg RM, Kenis PJ, Gaskins HR (2008) Engineering redox-sensitive linkers for genetically encoded FRET-based biosensors. Exp Biol Med 233:238–248
35. Yano T, Oku M, Akeyama N, Itoyama A, Yurimoto H, Kuge S, Fujiki Y, Sakai Y (2010)

A novel fluorescent sensor protein for visualization of redox states in the cytoplasm and in peroxisomes. Mol Cell Biol 30:3758–3766

36. Funke F, Gerich FJ, Müller M (2011) Dynamic, semi-quantitative imaging of intracellular ROS levels and redox status in rat hippocampal neurons. Neuroimage 54:2590–2602
37. Foster KA, Galeffi F, Gerich FJ, Turner DA, Müller M (2006) Optical and pharmacological tools to investigate the role of mitochondria during oxidative stress and neurodegeneration. Prog Neurobiol 79:136–171
38. Dooley CT, Dore TM, Hanson GT, Jackson WC, Remington SJ, Tsien RY (2004) Imaging dynamic redox changes in mammalian cells with green fluorescent protein indicators. J Biol Chem 279:22284–22293
39. Enyedi B, Varnai P, Geiszt M (2010) Redox state of the endoplasmic reticulum is controlled by Ero1L-alpha and intraluminal calcium. Antioxid Redox Signal 13:721–729
40. Gutscher M, Pauleau AL, Marty L, Brach T, Wabnitz GH, Samstag Y, Meyer AJ, Dick TP (2008) Real-time imaging of the intracellular glutathione redox potential. Nat Methods 5:553–559
41. Lentz TB, Gray SJ, Samulski RJ (2012) Viral vectors for gene delivery to the central nervous system. Neurobiol Dis 48:179–188
42. Machida CA (ed) (2003) Viral vectors for gene therapy, Methods in molecular medicine, vol 76. Humana Press, Totowa, NJ
43. Zhang F, Gradinaru V, Adamantidis AR, Durand R, Airan RD, de Lecea L, Deisseroth K (2010) Optogenetic interrogation of neural circuits: technology for probing mammalian brain structures. Nat Protoc 5:439–456
44. Delzor A, Escartin C, Deglon N (2013) Lentiviral vectors: a powerful tool to target astrocytes in vivo. Curr Drug Targets 14:1336–1346
45. Delzor A, Dufour N, Petit F, Guillermier M, Houitte D, Auregan G, Brouillet E, Hantraye P, Deglon N (2012) Restricted transgene expression in the brain with cell-type specific neuronal promoters. Hum Gene Ther Methods 23:242–254
46. Wu Z, Asokan A, Samulski RJ (2006) Adeno-associated virus serotypes: vector toolkit for human gene therapy. Mol Ther 14:316–327
47. Shevtsova Z, Malik JM, Michel U, Bähr M, Kügler S (2005) Promoters and serotypes: targeting of adeno-associated virus vectors for gene transfer in the rat central nervous system in vitro and in vivo. Exp Physiol 90:53–59
48. Drinkut A, Tereshchenko Y, Schulz JB, Bähr M, Kügler S (2012) Efficient gene therapy for Parkinson's disease using astrocytes as hosts for localized neurotrophic factor delivery. Mol Ther 20:534–543
49. Ayuso E, Mingozzi F, Bosch F (2010) Production, purification and characterization of adeno-associated vectors. Curr Gene Ther 10:423–436
50. Ehrengruber MU (2002) Alphaviral vectors for gene transfer into neurons. Mol Neurobiol 26:183–201
51. Shevtsova Z, Malik JM, Michel U, Schöll U, Bähr M, Kügler S (2006) Evaluation of epitope tags for protein detection after *in vivo* CNS gene transfer. Eur J Neurosci 23:1961–1969
52. Lingor P, Schöll U, Bähr M, Kügler S (2005) Functional applications of novel Semliki Forest virus vectors are limited by vector toxicity in cultures of primary neurons in vitro and in the substantia nigra in vivo. Exp Brain Res 161: 335–342
53. Wagener KC, Kolbrink B, Dietrich K, Kizina KM, Terwitte LS, Kempkes B, Bao G, Müller M (2016) Redox-indicator mice stably expressing genetically-encoded neuronal roGFP: versatile tools to decipher subcellular redox dynamics in neuropathophysiology. Antioxid Redox Signal 25:41–58
54. Weller J, Kizina KM, Can K, Bao G, Müller M (2014) Response properties of the genetically encoded optical H_2O_2 sensor HyPer. Free Radic Biol Med 76C:227–241
55. Stoppini L, Buchs PA, Muller D (1991) A simple method for organotypic cultures of nervous tissue. J Neurosci Methods 37:173–182
56. Malgaroli A, Tsien RW (1992) Glutamate-induced long-term potentiation of the frequency of miniature synaptic currents in cultured hippocampal neurons. Nature 357: 134–139
57. Can K, Toloe J, Kügler S, Müller M (2014) Aberrant redox homeostasis in Rett syndrome affects cytosol and mitochondria. Soc Neurosci Abstr 515.17
58. Grynkiewicz G, Poenie M, Tsien RY (1985) A new generation of Ca^{2+} indicators with greatly improved fluorescence properties. J Biol Chem 260:3440–3450
59. Müller M, Mironov SL, Ivannikov MV, Schmidt J, Richter DW (2005) Mitochondrial organization and motility probed by two-photon microscopy in cultured mouse brainstem neurons. Exp Cell Res 303:114–127
60. Guzman JN, Sanchez-Padilla J, Wokosin D, Kondapalli J, Ilijic E, Schumacker PT, Surmeier DJ (2010) Oxidant stress evoked by pacemaking in dopaminergic neurons is attenuated by DJ-1. Nature 468:696–700

61. Xu X, von Lohneysen K, Soldau K, Noack D, Vu A, Friedman JS (2011) A novel approach for in vivo measurement of mouse red cell redox status. Blood 118:3694–3697

62. Breckwoldt MO, Pfister FM, Bradley PM, Marinkovic P, Williams PR, Brill MS, Plomer B, Schmalz A, St Clair DK, Naumann R, Griesbeck O, Schwarzländer M, Godinho L, Bareyre FM, Dick TP, Kerschensteiner M, Misgeld T (2014) Multiparametric optical analysis of mitochondrial redox signals during neuronal physiology and pathology in vivo. Nat Med 20:555–560

63. Wolf AM, Nishimaki K, Kamimura N, Ohta S (2014) Real-time monitoring of oxidative stress in live mouse skin. J Invest Dermatol 134:1701–1709

Chapter 10

Monitoring of Permeability Transition Pore Openings in Isolated Individual Brain Mitochondria

Nickolay Brustovetsky and Tatiana Brustovetsky

Abstract

Mitochondrial permeability transition is a process marked by a significant increase in the permeability of the inner mitochondrial membrane that can be initiated by massive Ca^{2+} influx into mitochondria with or without oxidative stress. The increase in membrane permeability is due to induction of a gigantic proteinaceous pore in the inner mitochondrial membrane called the permeability transition pore (PTP). The detrimental role of PTP induction in neuronal mitochondria in various neuropathologies is well documented. There are different methods to study PTP in isolated mitochondria and in live cells. In the present chapter, we describe a method to monitor PTP induction in individual isolated brain mitochondria. Although this method is not often used, it could be an excellent addition to the arsenal of methodologies utilized in studies of PTP. In addition, we present here detailed description of a method used by us to isolate rat or mouse brain nonsynaptic and synaptic mitochondria.

Key words Permeability transition pore, Mitochondria, Neuron, Calcium, Mitochondrial membrane potential, Fluorescent microscopy

1 Introduction

Mitochondrial permeability transition is a process leading to induction of a gigantic proteinaceous pore in the inner mitochondrial membrane permeable for solutes with molecular weight up to 1.5 kDa and termed the mitochondrial permeability transition pore (PTP) [1]. Rapid and excessive Ca^{2+} accumulation in mitochondria and oxidative stress are the main inducers of the PTP [2]. The PTP can be induced and studied in isolated mitochondria, including brain synaptic and nonsynaptic mitochondria. The PTP induction is manifested in mitochondrial depolarization and swelling of organelles [1]. Mitochondrial depolarization inhibits Ca^{2+} uptake by mitochondria [3]. In addition, previously accumulated Ca^{2+} can be released from mitochondria with activated PTP. Mitochondrial swelling can lead to the rupture of the outer mitochondrial membrane and release of mitochondrial apoptogenic proteins such as

Stefan Strack and Yuriy M. Usachev (eds.), *Techniques to Investigate Mitochondrial Function in Neurons*, Neuromethods, vol. 123, DOI 10.1007/978-1-4939-6890-9_10, © Springer Science+Business Media LLC 2017

cytochrome c, Smac/DIABLO, and apoptosis-inducing factor (AIF). Mitochondrial depolarization and swelling can be monitored in the suspension of isolated brain mitochondria by following, for example, the transmembrane distribution of the lipophilic cation tetraphenylphosphonium (TPP^+) with TPP^+-sensitive electrode and by measuring light scattering of mitochondrial suspension, respectively [4]. Accumulation of TPP^+ in mitochondria indicates an increase in mitochondrial membrane potential, whereas release of TPP^+ from mitochondria indicates mitochondrial depolarization. The decrease in light scattering of a mitochondrial suspension reflects mitochondrial swelling. The main advantage of this approach is its relative simplicity and the opportunity to record an averaged response from the entire population of isolated mitochondria. The disadvantage is the lack of information from individual mitochondria and a need to use relatively high concentrations of $CaCl_2$ (25–100 μM) to induce PTP. The method introduced by Huser and Blatter in 1998 [5] and used with isolated brain mitochondria by Ian Reynolds' group [6–8] and by us [9, 10] allows for monitoring of mitochondrial membrane potential in individual isolated mitochondria and observation of cyclosporin A-sensitive PTP induction with as low as 0.5 μM of free Ca^{2+}. In this chapter, we describe the methods for isolating and Percoll-gradient purifying of nonsynaptic and synaptic mitochondria and for monitoring PTP induction in individual isolated brain mitochondria.

2 Materials

The following solutions have to be prepared before the experiment.

1. *Isolation Buffer 1*: 225 mM mannitol, 75 mM sucrose, 10 mM Hepes, pH 7.4 adjusted with KOH. Prepare 200 ml of this solution and then divide into two portions, 100 ml each. Add 0.1% BSA free from fatty acids (MP Biomedicals, Cat # 152401) and 1 mM EGTA (2 ml of 50 mM EGTA stock solution) to the first 100 ml portion described above.
2. *Isolation Buffer 2*: Add 0.1 mM EGTA (0.2 ml of 50 mM EGTA stock solution) to the second 100 ml portion described above.
3. *Isolation Buffer 3*: 395 mM sucrose, 0.1 mM EGTA, 10 mM Hepes, pH 7.4.
4. *Percoll Buffer*: 320 mM sucrose, 1 mM EGTA, 10 mM Hepes, pH 7.4.
5. *26% Percoll in Percoll Buffer*: Mix 5.2 ml Percoll (Sigma, Cat # GE17-0891-01) and 14.8 ml Percoll Buffer (for nonsynaptic mitochondria).

6. *24% Percoll in Percoll Buffer*: Mix 4.8 ml Percoll (Sigma, Cat # GE17-0891-01) and 15.2 ml Percoll Buffer (for synaptic mitochondria).
7. *40% Percoll in Percoll Buffer*: Mix 8 ml Percoll (Sigma, Cat # GE17-0891-01) and 12 ml Percoll Buffer.
8. *Standard incubation medium*: 125 mM KCl, 10 mM Hepes, pH 7.4 (KOH), 0.5 mM $MgCl_2$, 3 mM KH_2PO_4, 10 μM EGTA. In the experiments with membrane potential recordings in individual mitochondria, the standard incubation medium was supplemented with 0.5 mM ATP. Oxidative substrates are added into the standard incubation medium (final concentrations): either malate (1 mM) plus pyruvate (3 mM) or glutamate (3 mM) plus succinate (3 mM). Glutamate is used with succinate to remove oxaloacetate in transaminase reaction to prevent oxaloacetate accumulation and oxaloacetate-mediated inhibition of succinate dehydrogenase [12]. All stock solutions of substrates have pH adjusted to 7.4. Keep all solutions in a refrigerator at 2–4 °C; stock solutions of substrates in the freezer at least at −20 °C.

3 Methods

3.1 Isolation and Percoll-Gradient Purification of Rat Brain Nonsynaptic and Synaptic Mitochondria

This method is based on the procedure published by Neil Sims in 1990 [11] with some modifications introduced by us [4] (illustrated in Fig. 2).

1. One hour before the procedure put all instruments and solutions in the cold room, turn on centrifuges, set centrifuge temperature 2–4 °C.
2. Bring an animal to cold room vicinity, decapitate, quickly dissect out brain and place into a 90 mm petri dish filled with 15 ml of ice-cold Isolation Buffer 1, rinse well, transfer into a 35 mm petri dish filled with 2 ml of Isolation Buffer 1.
3. Mince the brain with small 110 mm scissors while in the 35 mm petri dish, then pour into a 15 ml glass Dounce homogenizer (Fisher Scientific, Cat # 06-435B) make ten gentle strokes with a pestle A ("loose," pestle clearance 0.114 ± 0.025 mm), then 30 strokes with a pestle B ("tight," pestle clearance: 0.05±0.025 mm).
4. Place brain homogenate into a 50 ml polycarbonate centrifuge tube (Nalgene), rinse Dounce homogenizer once with Isolation Buffer 1 and pour rinse into a centrifuge tube, then add more Isolation Buffer 1 to the centrifuge tube to bring the final volume to about 35 ml, mix well by pipetting with a plastic single-use pipette.

5. **1st centrifugation**: 10 min at 2400 rpm in the Beckman Centrifuge Avanti J-26XP, rotor JA-25.50 (650 × *g* for 10 min). Temperature 2–4 °C.
6. Transfer supernatant into a fresh centrifuge tube.
7. **2nd centrifugation:** 10 min at 12,500 rpm (18,850 × *g* for 10 min) in the Beckman Centrifuge Avanti J-26XP, rotor JA-25.50. Temperature 2–4 °C.
8. Discard supernatant, gently resuspend the pellet in 10 ml of Isolation Buffer 2, then add more Isolation Buffer 2, the final volume should be about 35 ml, mix well by pipetting with a plastic single-use pipette.
9. **3rd centrifugation**: 10 min at 12,500 rpm (18,850 × *g* for 10 min) Beckman Centrifuge Avanti J-26XP, rotor JA-25.50. Temperature 2–4 °C. During this centrifugation, prepare discontinuous Percoll gradient in the Beckman Ultra-Clear centrifuge tubes (13.2 ml, Beckman Coulter, Part # 34405914). You will need two tubes for one rat brain or 2–3 mouse brains. First, add 3.5 ml of 26% Percoll solution into the tube, then layer on the bottom of the tube, beneath 26% Percoll solution, 3 ml of 40% Percoll using 5 ml syringe with a needle and thin (inner Ø ~1–1.5 mm) PVC tube.
10. Discard supernatant, resuspend the pellet in 5 ml of Isolation Buffer 3. Layer the suspension onto the top of Percoll gradient in Beckman Ultra-Clear centrifuge tubes.
11. **4th centrifugation**: 28 min at 15,500 rpm (41,100 × *g* for 28 min) Beckman Ultracentrifuge Optima L100K, bucket rotor SW41Ti, temperature 2–4 °C. After the centrifugation, there will be five layers (Fig. 1): (a) clear (*top*); (b) thick white-yellowish, this are *synaptosomes*; (c) slightly cloudy thick layer; (d) thin turbid layer, this are *nonsynaptic mitochondria*; (e) clear (bottom).
12. Aspirate the first layer very accurately with a vacuum or water pump. Collect the *second layer (synaptosomes)* with a plastic single-use pipette and place into an ice-cold 10 ml glass beaker for preparing synaptic mitochondria. Aspirate the third layer very accurately, leaving in the tube about 5 mm height of this layer.
13. Collect the *fourth, mitochondrial layer* with a glass Pasteur pipette, 140 mm size, place into a fresh empty Beckman Ultra-Clear centrifuge tube, add Isolation Buffer 3, final volume ~12 ml.
14. **5th centrifugation**: 20 min at 15,500 rpm (41,100 × *g* for 20 min) in Beckman Ultracentrifuge Optima L100K, bucket rotor SW41Ti, temperature 2–4 °C.

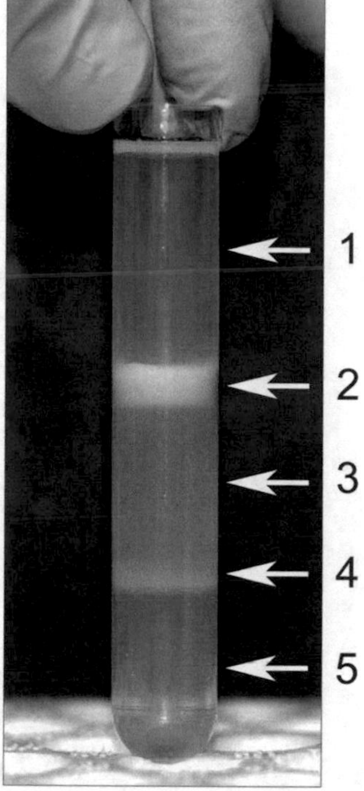

Fig. 1 A centrifuge tube with discontinuous Percoll gradient and synaptosomal and mitochondrial fractions after centrifugation in the ultracentrifuge. *1*—clear (*top*); *2*—thick *white-yellowish*, this are synaptosomes; *3*—slightly cloudy thick layer; *4*—thin turbid layer, this are mitochondria; *5*—clear (*bottom*)

15. Aspirate supernatant very accurately, as much as possible. Resuspend the pellet with Isolation Buffer 3 in the same centrifuge tube, final volume ~12 ml.
16. **6th centrifugation**: 20 min at 15,500 rpm (41,100 × *g* for 20 min) in Beckman Ultracentrifuge Optima L100K, bucket rotor SW41Ti, temperature 2–4 °C.
17. Aspirate supernatant very accurately, add 250–300 μl of Isolation Buffer 3 and resuspend the pellet.
18. Transfer mitochondrial suspension with glass Pasteur pipette into a small 2 ml Dounce homogenizer, add Isolation Buffer 3, the final volume should be about 0.5 ml. This is a stock suspension of nonsynaptic mitochondria. Protein concentration in stock mitochondrial suspension should be about 3.5–4 mg/ml. Store on ice and use in the experiment within the next 5–6 hours.

3.2 Isolation of Brain Synaptic Mitochondria

Synaptic mitochondria can be isolated from synaptosomes by the nitrogen cavitation method using a nitrogen cell disruption bomb, model 4639 (Parr Instrument Company, Moline, IL, USA), cooled on ice as described by Brown et al. [13] with some modifications.

1. Transfer the synaptosomes obtained during preparation of nonsynaptic mitochondria (Sect. 3.1, step 11, 4th centrifugation) into an ice-cold 10 ml glass beaker and place it into the nitrogen bomb on ice under nitrogen with 1100 psi for 13 min.
2. Then, layer the synaptosomes on top of the discontinuous Percoll gradient (24%/40%) and centrifuge for 28 min at 15,500 rpm (41,100 × *g* for 28 min) in Beckman Ultracentrifuge Optima L100K, bucket rotor SW41Ti, temperature 2–4 °C (**4′ centrifugation**).
3. The next centrifugations can be performed together with nonsynaptic mitochondria. Transfer the mitochondrial fraction from the interface between Percoll layers into a fresh tube, add Isolation Buffer 3 and centrifuge for 20 min at 15,500 rpm (**5th centrifugation**, 41,100 × *g* for 20 min) in Beckman Ultracentrifuge Optima L100K, bucket rotor SW41Ti, temperature 2–4 °C.
4. Aspirate supernatant very gently, as much as possible. Resuspend the pellet with Isolation Buffer 3 in the same centrifuge tube, final volume ~12 ml.
5. Centrifuge for 20 min at 15,500 rpm (**6th centrifugation**, 41,100 × *g* for 20 min) in Beckman Ultracentrifuge Optima L100K, bucket rotor SW41Ti, temperature 2–4 °C.
6. Resuspend the pellet in 0.1 ml of Isolation Buffer 3 and keep on ice. Protein concentration in stock mitochondrial suspension should be about 3.5–4 mg/ml. Store on ice and use in the experiment within the next 5–6 h.

The graphic scheme shown in Fig. 2 illustrates the described isolation and purification procedures. Figure 3 shows representative recordings made with isolated synaptic mitochondria. Mitochondrial respiration was followed with Clark-type oxygen electrode and mitochondrial membrane potential was monitored by following tetraphenylphosphonium (TPP^+) distribution across the inner membrane with TPP^+-sensitive electrode.

3.3 Monitoring of PTP Opening in Individual Mitochondria Loaded with Rhodamine 123

An induction of PTP in response to Ca^{2+} or oxidative stress can be monitored by following changes in membrane potential of individual isolated brain mitochondria assessed with fluorescence of Rhodamine 123 (Rh123) in non-quenching mode [9, 10].

1. Dilute mitochondrial stock suspension to 0.7–0.8 mg protein/ml using Isolation Buffer 3. You can use commercially available perfusion chambers as described previously [6] or, alternatively, you can build a single-use perfusion chamber using a glass-bottomed 35 mm petri dish and a glass bridge made out of the coverslip. Cover mitochondria with a 4 × 12 mm glass bridge (distance to the bottom is about 1 mm,

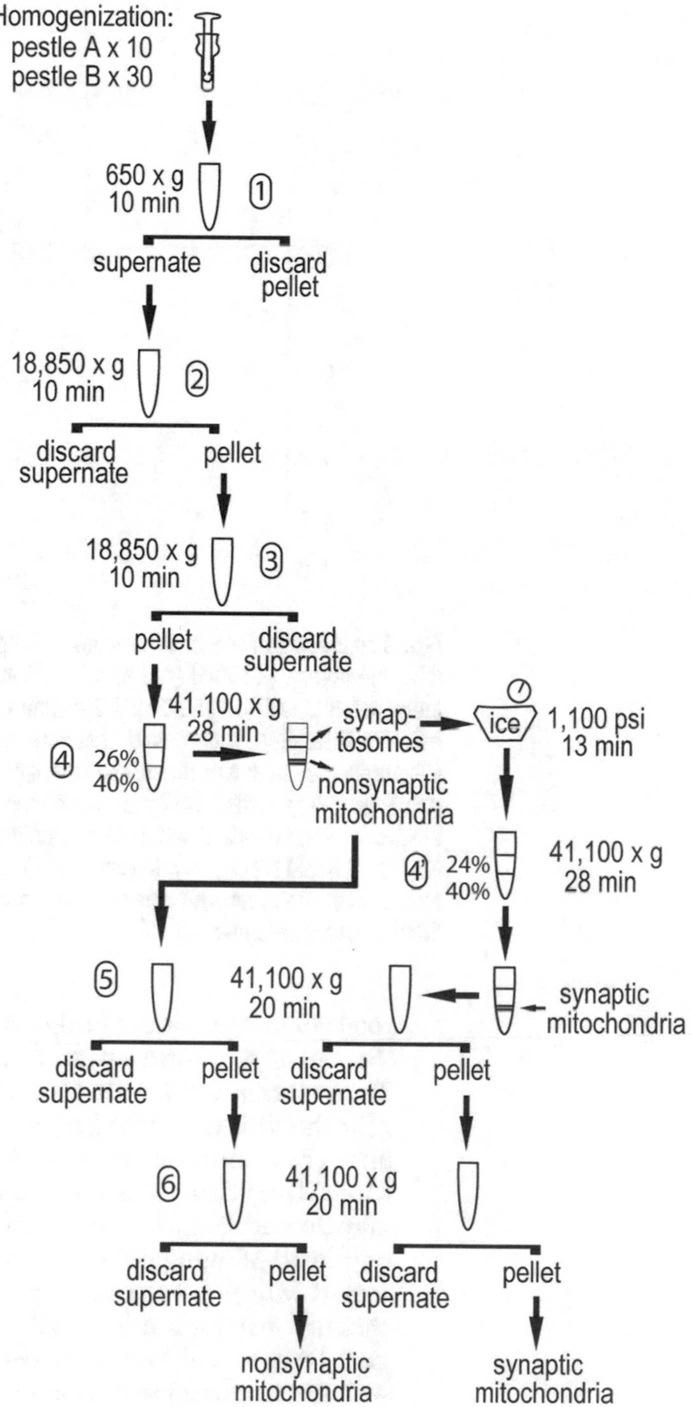

Fig. 2 The scheme illustrating isolation and purification of brain mitochondria. The explanation is in the text

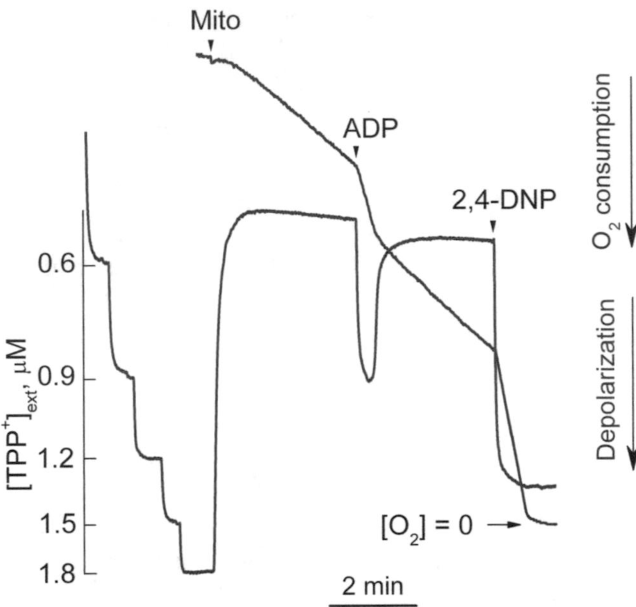

Fig. 3 Representative measurements of respiration (*blue trace*) and mitochondrial membrane potential (*red trace*) in isolated synaptic mitochondria. Where indicated, 300 μM ADP and 60 μM 2,4-dinitrophenolm (2,4-DNP) were applied to mitochondria. Respiration was measure with Clark-type oxygen electrode. Membrane potential was monitored by following distribution of tetraphenylphosphonium (TPP^+) across the inner mitochondrial membrane with TPP^+-sensitive electrode. The standard incubation medium contained 125 mM KCl, 3 mM KH_2PO_4, 0.5 mM $MgCl_2$, 10 mM Hepes, pH 7.4, 10 μM EGTA, 0.1% BSA (free from fatty acids). The measurements were performed under continuous stirring at 37 °C in a 0.4 ml chamber

total volume under the bridge is about 50 mm^3) made out of the coverslip to form a miniature perfusion chamber (Fig. 4). The coverslip can be cut with a diamond tipped glass cutter. Mitochondria under the bridge are perfused with the standard incubation medium using a ValveBank 8 perfusion system (AutoMate Scientific, San Francisco, CA). Place 5 μl of diluted mitochondrial stock suspension in the middle of the glass-bottomed 35 mm petri dish (preliminary, the glass bottom is coated with poly-L-lysine (Sigma, Cat # P-2636)), let mitochondria stay for 1 min to adhere to the poly-L-lysine coated glass bottom and then start perfusion at the rate 1 ml/min with the standard incubation medium supplemented with 0.2 μM Rh123. To induce PTP, the standard incubation medium is supplemented with 10 μM $CaCl_2$. The calculated free Ca^{2+} concentration in this solution is 0.5 μM (calculated with Webmaxc Standard by C. Patton, http://web.stanford.edu/~cpatton/webmaxcS.htm).

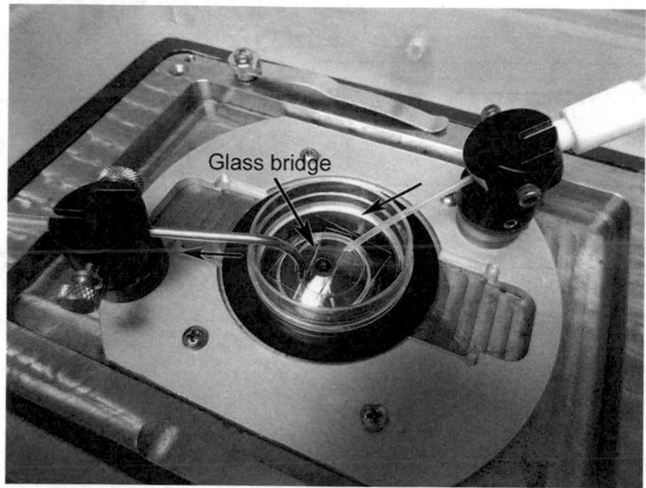

Fig. 4 A 35 mm petri dish with a 4 × 12 mm bridge made out of glass coverslip to form a perfusion chamber. The distance between coverslip and the glass bottom is about 1 mm, total volume under the bridge is about 50 mm^3. *Arrows* show the direction of perfusion. Note, the pockets outside the glass bridge are dry

2. We use a Nikon Eclipse TE2000-S inverted microscope (100% output at the right optical port) equipped with objective Nikon CFI Plan Apo VC 100× 1.4 NA and SimplePCI 6.1 software (Compix, Sewickley, PA; now belongs to Hamamatsu) to visualize individual mitochondria loaded with Rh123 in non-quenching mode (polarized mitochondria produce bright fluorescence, depolarized mitochondria—dimmed or no fluorescence).
3. Mitochondria are illuminated at 480 ± 20 nm using a Lambda LS lighting system (Sutter Instrument, Novato, CA) with a 175 W xenon lamp. If necessary, the light beam can be attenuated by neutral density filters to 10% to diminish photobleaching. Fluorescence is collected through a 505 nm dichroic mirror and a 535 ± 25 nm emission filter by a back-illuminated EM-CCD Hamamatsu C9100-12 camera (Hamamatsu, Bridgewater, NJ). Images are taken at room temperature or at 37 °C every 1–15 s throughout the experiment.
4. In post-experiment analysis performed with SimplePCI 6.1 software, individual mitochondria are marked as Regions of Interest (ROI), from which fluorescence signals are recorded (Fig. 5a). (This research was originally published in Journal of Biological Chemistry: Shalbuyeva et al. Lithium desensitizes brain mitochondria to calcium, antagonizes permeability transition, and diminishes cytochrome c release. J. Biol. Chem., 2007; 282:18,057–18,068. ©The American Society

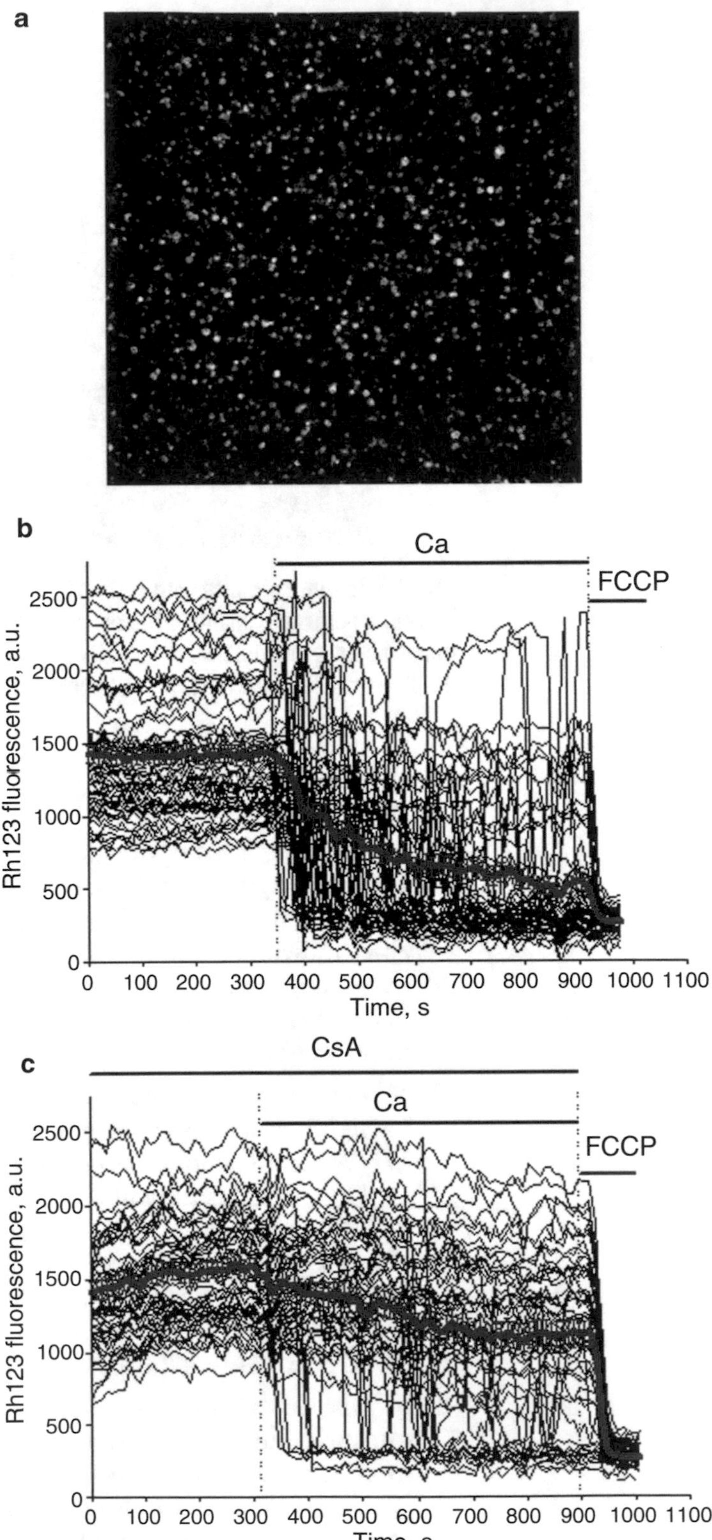

Fig. 5 Detection of Ca^{2+}-induced PTP openings in individual isolated brain mitochondria perfused with Rhodamine 123 (Rh123). The effect of cyclosporin A (CsA). The standard KCl-based incubation medium was supplemented with 1 mM ATP.

for Biochemistry and Molecular Biology.) After background subtraction, a plot of R123 fluorescence over time is constructed. Figure 5b, c, shows the effect of Ca^{2+} and cyclosporin A (CsA) on membrane potential in individual isolated brain mitochondria. PTP opening is detected as mitochondrial depolarizations (a loss of Rh123 fluorescence), which could be short- and long-lasting. Prevention of these depolarizations with cyclosporin A (CsA) attributes them to PTP opening. Note, not all mitochondria appeared to be protected by CsA (Fig. 5c).

4 Notes

4.1 Isolation of Mitochondria

1. Check pH in all buffers, including Percoll solutions, before each experiment while the solutions are at 4 °C.
2. Strokes with pestle B must be energetic and strong. Weak, too gentle strokes or insufficient number of strokes may result in a lower yield of mitochondria at the end of the isolation procedure. If mitochondrial yield is low, try to increase number of strokes with pestle B or increase the amount of starting material for mitochondrial isolation.

4.2 Monitoring of PTP Opening in Individual Mitochondria

1. Before recording, the space under the glass bridge (perfusion camera) should be perfused for 1–2 min to remove loosely attached mitochondria.
2. Before recording, use bright field image to focus on mitochondria. Then, switch to fluorescence mode.
3. Make sure that perfusion flow is smooth without pulsation.
4. Microscope must be installed on the anti-vibration table (we use TMC anti-vibration table).
5. Keep exposure to excitation light to minimum to avoid photobleaching.

Fig. 5 (continued) BSA was omitted. In (**a**), a fluorescent image of individual mitochondria attached to glass-bottomed petri dish and perfused with 0.2 μM Rh123. In (**b**) and (**c**), *thin black traces* are fluorescent signals from individual mitochondria, *red thick traces* are averaged fluorescent signals ± SEM. In (**b**) and (**c**), where indicated, the perfusion solution contained 10 μM Ca^{2+}. According to Webmaxc Standard software (by C. Patton, http://web.stanford.edu/~cpatton/webmaxcS.htm), free Ca^{2+} concentration was 0.5 μM. In (**c**), the perfusion solution was supplemented with 1 μM cyclosporin A (CsA). At the end of experiments, 1 μM FCCP was added to the perfusion solution to completely depolarize mitochondria. *a.u.* arbitrary units

Acknowledgment

The authors are grateful to Jessica Pellman and James Hamilton (Department of Pharmacology and Toxicology, Indiana University School of Medicine) for help with preparing Fig. 2. This work was supported by Indiana Spinal Cord and Brain Injury Research Fund grant and NIH/NINDS grant R01 NS078008 to N.B.

References

1. Bernardi P, Krauskopf A, Basso E et al (2006) The mitochondrial permeability transition from in vitro artifact to disease target. FEBS J 273: 2077–2099
2. Bernardi P, Rasola A (2007) Calcium and cell death: the mitochondrial connection. Subcell Biochem 45:481–506
3. Bernardi P (1999) Mitochondrial transport of cations: channels, exchangers, and permeability transition. Physiol Rev 79:1127–1155
4. Brustovetsky N, Brustovetsky T, Jemmerson R et al (2002) Calcium-induced cytochrome c release from CNS mitochondria is associated with the permeability transition and rupture of the outer membrane. J Neurochem 80:207–218
5. Huser J, Rechenmacher CE, Blatter LA (1998) Imaging the permeability pore transition in single mitochondria. Biophys J 74: 2129–2137
6. Vergun O, Votyakova TV, Reynolds IJ (2003) Spontaneous changes in mitochondrial membrane potential in single isolated brain mitochondria. Biophys J 85:3358–3366
7. Vergun O, Reynolds IJ (2004) Fluctuations in mitochondrial membrane potential in single isolated brain mitochondria: modulation by adenine nucleotides and Ca2+. Biophys J 87: 3585–3593
8. Vergun O, Reynolds IJ (2005) Distinct characteristics of Ca(2+)-induced depolarization of isolated brain and liver mitochondria. Biochim Biophys Acta 1709:127–137
9. Shalbuyeva N, Brustovetsky T, Brustovetsky N (2007) Lithium desensitizes brain mitochondria to calcium, antagonizes permeability transition, and diminishes cytochrome C release. J Biol Chem 282:18057–18068
10. Brustovetsky T, Brittain MK, Sheets PL et al (2011) KB-R7943, an inhibitor of the reverse Na+/Ca2+ exchanger, blocks N-methyl-D-aspartate receptor and inhibits mitochondrial complex I. Br J Pharmacol 162:255–270
11. Sims NR (1990) Rapid isolation of metabolically active mitochondria from rat brain and subregions using Percoll density gradient centrifugation. J Neurochem 55:698–707
12. Wojtczak L, Wojtczak AB, Ernster L (1969) The inhibition of succinate dehydrogenase by oxalacetate. Biochim Biophys Acta 191:10–21
13. Brown MR, Sullivan PG, Dorenbos KA et al (2004) Nitrogen disruption of synaptoneurosomes: an alternative method to isolate brain mitochondria. J Neurosci Methods 137:299–303

Chapter 11

Examination of Mitochondrial Ion Conductance by Patch Clamp in Intact Neurons and Mitochondrial Membrane Preparations

Elizabeth A. Jonas and Nelli Mnatsakanyan

Abstract

Mitochondrial ion channels are involved in numerous cellular processes. Membrane pores and transporters regulate the influx and efflux of H^+, calcium, sodium, potassium, zinc and determine the membrane compartmentalization of numerous cytosolic metabolites. The permeability of the inner membrane to the various ions determines the membrane potential of the inner membrane. The recently described mitochondrial calcium uniporter (MCU) plays a major role in inner membrane calcium uptake into the matrix. The permeability of the outer membrane, controlled in part by VDAC and by BCL-2 family proteins, regulates the trafficking of metabolites into and out of mitochondria and the release of important signaling molecules that determine the onset of programmed cell death. Very recently discovered is that the membrane portion of the F_1F_OATP synthase contains a highly regulated ion channel that participates in formation of the death channel known as the mitochondrial permeability transition pore (mPTP). mPTP opening in response to pro-death stimuli compromises the osmotic barrier of the inner membrane. The ion channels of the mitochondrial inner and outer membranes may come together in a complex of proteins during programmed cell death, particularly during neuronal ischemia, where elevated levels of calcium and zinc activate inner membrane ion channel conductances. The variety of possible molecular participants within the mPTP ion channel complex may be matched only by the variety of different types of programmed cell death. Perhaps equally interesting, however, is the potential role of the ion channel activity of the F_O of the ATP synthase as a metabolic regulator, since a relative decrease in channel activity enhances inner mitochondrial membrane coupling, producing salubrious effects on cell metabolism including the ability of synaptic mitochondria to enhance the efficiency of ATP production and to handle calcium loads during synaptic plasticity.

Key words ATP synthase, BCL-2 family proteins, Mitochondria, Mitochondrial permeability transition pore (mPTP), Mitoplast, Neurons, Patch clamp, Synaptic transmission, VDAC

1 Introduction

The processes of mitochondrial localization, morphologic changes, docking, and ion channel activity at synaptic sites demonstrate plasticity. Properties of mitochondrial plasticity are particularly relevant in cells requiring high energy production such as heart

Stefan Strack and Yuriy M. Usachev (eds.), *Techniques to Investigate Mitochondrial Function in Neurons*, Neuromethods, vol. 123, DOI 10.1007/978-1-4939-6890-9_11, © Springer Science+Business Media LLC 2017

and brain, but also during proliferation or growth of cancer tissue, developing organs, or synapses. High frequency synaptic activity produces synaptic plasticity that underlies memory formation and learning, during which mitochondria affect intracellular calcium homeostasis and provide the energy needed for synaptic vesicle recycling and for the continued operation of membrane ion pumps. Recent discoveries have altered our ideas about the role of mitochondria in the synapse. In response to neuronal activity these special mitochondria alter the ability of the neuron to release or receive neurotransmitter, to extend neurites and to form contacts with partner neurons. The mutability of synaptic mitochondria is correlated with and may indeed be causally implicated in onset of synaptic plasticity, neuronal development, and prevention of neurodegeneration.

1.1 Overview of Mitochondrial Ion Conductances

1.1.1 Cytosolic to Mitochondrial Ca^{2+} Cycling Regulates ATP production and release and Rerelease of Ca^{2+}

The regulated targeting of mitochondria to sites of high energy demand in neurons suggests that the mechanisms of ATP production and release by mitochondria could very well be regulated during changes in excitability or long-term changes in the frequency of synaptic events. Although there may be many second messengers, the main messenger is Ca^{2+}. After Ca^{2+} enters the cytosol it is buffered by uptake into the endoplasmic reticulum (ER) and by uptake via the mitochondrial Ca^{2+} uniporter ion channel (MCU) [1, 2] which helps regulate mitochondrial matrix Ca^{2+} to stimulate production and release of ATP.

Following stimulation of ATP production in the matrix, ATP is released from mitochondria via the adenine nucleotide transporter in the inner membrane and VDAC in the outer membrane. VDAC has three isoforms in humans, VDAC1, VDAC2, and VDAC3. Its regulation is important for normal cell function and cell death. VDAC forms a channel with the diameter of ~2.5 nm at the mitochondrial outer membrane. VDAC is permeable to molecules up to 5 kDa in its high conductance state but is insufficient to pass folded proteins like cytochrome c [3]. All three isoforms of VDAC contribute to the maintenance of mitochondrial membrane potential ($\Delta\Psi$), but the greatest decrease in $\Delta\Psi$ occurs after knockdown of VDAC3, which is the least abundant isoform of VDAC in the cell. The knockdown of VDAC3, but not other isoforms significantly decreases cellular levels of ATP, ADP, NAD(P)H, and mitochondrial redox state [4]. The closure of VDAC leads eventually to mitochondrial outer membrane permeabilization and apoptosis. Tubulin negatively modulates mitochondrial metabolism by closing VDAC1/2, but not VDAC3, contributing to the Warburg phenomenon. The tubulin–VDAC interaction is reversed by erastin, which antagonizes the Warburg metabolism and restores oxidative mitochondrial metabolism [4].

1.1.2 Recording Ca^{2+} Release from Mitochondria: Contribution of Ca^{2+} Rerelease After Uptake to Short-Term Synaptic Plasticity

Ca^{2+} rerelease after uptake into the matrix determines short-term synaptic plasticity in neuronal synapses [5] by providing for residual Ca^{2+} in presynaptic endings during subsequent stimuli. Residual Ca^{2+} enhances synaptic vesicle fusion, providing for short-term facilitation of neurotransmitter release [6, 7]. Ca^{2+} rerelease is performed in part by Ca^{2+}-sensitive ligand gated mitochondrial channels which are widely conserved from invertebrates to mammals; they open in response to elevated Ca^{2+} within the ER and mitochondrial matrix. We first recorded such calcium-induced mitochondrial ion channel activity during synaptic transmission by performing electrophysiological recordings on mitochondrial membranes within the living presynaptic terminal of squid stellate ganglion [8]. Through the use of a double-barreled patch pipette, recordings were made at rest, during and after intense synaptic stimulation. We found that in resting presynaptic terminals, the conductance of mitochondrial membranes is low [9]. In contrast, during high frequency electrical stimulation of the presynaptic nerve, a large increase in mitochondrial membrane ion channel activity occurs [9]. The delay in onset of the mitochondrial activity, the persistence of the mitochondrial activity for approximately 1 min after stimulation and its Ca^{2+} dependence suggest that a channel and/or exchanger rereleases Ca^{2+} from mitochondria to regulate short-term plasticity [9–11]. Indeed, short-term plasticity was prevented by inhibiting activity of contributing mitochondrial channels. Identifying the channel or channels that contribute to activity of mitochondria during synaptic activity in the living presynaptic terminal continues to contribute to the work of our laboratory. Described in the subsequent sections are some of the findings that we have made defining the contributions of different channels to this activity.

1.1.3 Contribution of Bcl-2 Family Mitochondrial Ion Channel Activity to Synaptic Plasticity and Hypoxic-Ischemic Cell Death

In order to identify the outer and inner mitochondrial membrane channels contributing to normal synaptic physiology we employed pharmacological and biological studies. One of the main regulators of mitochondrial permeability or leakiness are the proteins of the Bcl-2 family that regulate programmed cell death (apoptosis) in vertebrate cells in response to intracellular or extracellular pro-death signals [12]. The final common pathway for programmed cell death in many systems is mitochondrial outer membrane permeabilization (MOMP) [13–16]. Pro-apoptotic Bcl-2 family members such as Bax regulate MOMP by inducing the formation of large outer membrane pores comprising activated oligomerized proteins, aided by other pro-apoptotic moieties [13, 15, 17]. In their canonical role, the anti-apoptotic Bcl-2 family proteins such as Bcl-x_L protect cells against MOMP by interacting with, and inhibiting the pore forming properties of, the pro-apoptotic family members [16, 18]. In addition, direct interaction of Bcl-2 family proteins with VDAC has been demonstrated by biochemical assays and NMR spectroscopy [19–22]. Bax increases the conductance of reconstituted VDAC and this increase is blocked by Bcl-x_L [19]. It has been reported recently that the BH4 domain of Bcl-x_L, but not that of Bcl-2, selectively targets VDAC1 and inhibits apoptosis by decreasing VDAC1-mediated Ca^{2+} uptake into the mitochondria [23].

Previous work has shown that Bcl-2 family proteins alone conduct ions when reconstituted into artificial lipid bilayers [3, 24, 25]. The three-dimensional structure of Bcl-x_L consists of two central hydrophobic helices surrounded by five amphipathic helices [26]. The structure is similar to that of pore-forming bacterial toxins [27]. In lipid vesicles or planar lipid bilayers, the induction of ion channel activity by Bcl-x_L is related to its known ability to target to, and insert into, lipid membranes [24, 28]. Bcl-x_L-induced channel activity induces metabolite exchange across mitochondrial membranes during apoptotic stimuli such as growth factor deprivation [29–31]. It appears to do this by maintaining VDAC in its open configuration, thereby enhancing ATP and phosphocreatine release from mitochondria [29], while also preventing the release of cytochrome *c* by binding to and sequestering pro-apoptotic molecules [18, 32]. Therefore, both the channel activity of Bcl-x_L and its ability to interact with other proteins may comprise the anti-apoptotic functions of Bcl-x_L.

We explored the ion channel activity of Bcl-x_L in the context of its role in synaptic plasticity. In the squid synapse, full-length recombinant human Bcl-x_L protein produces channel activity with multiple conductances when applied by patch pipette to mitochondrial patches within the living presynaptic terminal [28], consistent with activity recorded on mitochondria during normal synaptic transmission. In contrast, patch clamp recordings of endogenous mitochondrial channels in hypoxic synapses reveal much larger activity than that induced by full length Bcl-x_L protein. This hypoxia-induced activity is similar to that of recordings made with recombinant activated Bax or with the N-terminal proteolytic cleavage product of Bcl-x_L, ΔN-Bcl-x_L, a killer protein similar to Bax which induces cell death and cytochrome *c* release [33–36]. Both hypoxia and injection of ΔN-Bcl-x_L into the presynaptic terminal cause a decrease in synaptic responses, suggesting that large conductance mitochondrial activity may be correlated with synaptic depression. In support of a role for ΔN-Bcl-x_L in forming the hypoxia-induced activity, the appearance of the large conductance mitochondrial ion channel activity and synaptic depression are both prevented by pretreatment of the synapse with a pan-caspase/calpain inhibitor (that prevents the cleavage of Bcl-x_L) [34, 36].

1.1.4 ABT-737 Inhibits Both Pro- and Anti-apoptotic Forms of Bcl-x_L

ABT-737 is a mimetic of the BH3-only protein BAD that binds to Bcl-x_L with high affinity within a pocket of the three-dimensional structure that usually binds BH3-only proteins. In intact tumor cells and in cancer cell lines, ABT-737 effectively induces cell death possibly via its ability, as a BAD mimetic, to displace from Bcl-x_L pre-bound pro-apoptotic proteins [37]. In the squid synapse, ABT-737 prevents the large conductance mitochondrial channel activity associated with hypoxia or with exposure to ΔN BCL-x_L, but also prevents the mitochondrial activity associated with normal synaptic plasticity [38], suggesting that both full length Bcl-x_L and pro-apoptotic ΔN-Bcl-x_L contribute to synaptic plasticity by regulating mitochondrial ion channels.

In mammalian brain after global ischemia, neurons from the CA1 region of the hippocampus are vulnerable to delayed death with characteristics of programmed cell death, in particular the release of pro-apoptotic molecules such as cytochrome c from mitochondria into the cytosol and the activation of downstream caspases [39]. Just as in hypoxic squid neurons, mitochondria isolated from the hippocampus of ischemic rats demonstrate large conductance ion channel activity, inhibited by a specific antibody against Bcl-x_L [40] and by ABT-737 [41] which also inhibits delayed ischemic death of the CA1 neurons [41]. Therefore, the channel activity of cleaved Bcl-x_L is necessary for neuronal death in the CA1 after ischemic injury. Post-ischemic mammalian brain mitochondria also have a decreased frequency of large conductance channel activity when exposed to the VDAC inhibitor NADH, suggesting that ischemia-induced channel activity requires a complex formed of VDAC and ΔN-Bcl-x_L [36, 40, 42–44].

1.1.5 Outer Membrane Channel Activity May Be Regulated by Divalent-Induced Channel Activity of the Inner Membrane

Recordings of outer membranes reconstituted into liposomes or of intact mitochondria within living cells reveal that the permeability of the outer membrane is tightly controlled [9, 45–48]. Thus, outer membrane channel regulation could occur by activation at the cytosolic or intermembrane face by ligands, by second messengers, or by contact with inner membrane proteins such as other ion channels or transporters [44]. In recordings of preparations of isolated inner membranes ("mitoplasts"), that lack outer membrane components [49, 50], introduction of Ca^{2+} or Zn^{2+} to the matrix side of the patch or via the bath perfusion [51] activates a nS channel in the inner membrane. This inner membrane channel is blocked by CsA [52] suggesting that the channel shares features with the mitochondrial permeability transition pore (mPTP). In ischemic brain, Zn^{2+} localizes to the inside of ischemic mitochondria by microscopic imaging of fluorescent zinc-specific indicators. Large conductance activity present in the outer membranes of post-ischemic mitochondria is exquisitely sensitive to the specific Zn^{2+} chelator TPEN [40]. The findings suggest that Zn^{2+} activates a divalent-sensitive channel on the inner membrane that in turn activates a VDAC/ΔN Bcl-xL complex on the outer membrane.

1.1.6 A Mitochondrial Channel Complex Spanning Two Membranes Opens in Response to Matrix Ca^{2+} or Zn^{2+} Overload: mPTP?

We have shown that in ischemic mitochondria, an increase in mitochondrial outer membrane permeability may be triggered by an acute inner membrane depolarization [53], particularly after cytosolic and mitochondrial Ca^{2+} or Zn^{2+} overload. Although Ca^{2+} uptake and rerelease from mitochondria is a normal physiological event in cells, accumulation of Ca^{2+} in the matrix can have detrimental effects, including diminution in energy production by the ATP synthase [54]. Ca^{2+} overload can also produce an uncoupling process or rapid inner membrane permeabilization described historically [55–57] as permeability transition (PT).

PT, although initially reversible, can become irreversible [55, 58–63]. Irreversible PT leads to catastrophic mitochondrial changes including structural breakdown of the mitochondrial matrix accompanied by outer mitochondrial membrane rupture (MOMP) and cell death. Pathological PT is often associated with necrotic cell death such as is found in ischemia or injury whereas MOMP occurring without inner membrane permeabilization in the presence of sufficient amounts of ATP may regulate developmental and genetically predetermined death [64, 65]. Intermembrane space pro-apoptotic factors such as cytochrome *c* and Smac/DIABLO are released during both forms of cell death [53].

PT has been extensively studied in ischemic injury in brain, heart and other organs as well as in neurodegenerative conditions [66]. In addition to calcium influx into the matrix, PT is also induced by ROS, elevated inorganic phosphate and intracellular acidification [67, 68]. In contrast, it is inhibited by ATP/ADP and Mg^{2+} [58, 69]. The pharmacological agent most efficient in inhibiting PT is cyclosporine A (CsA), an immunosuppressant drug which binds to cyclophilin D (CypD) and inhibits the channel activity associated with PT. Inhibitors of ANT can either attenuate (bongkrekic acid) or enhance (atractyloside) PT opening [52, 56, 70]. Recent reports have also confirmed increased activity of PT by polyphosphates, chains of 10s–100s of repeating phosphates linked by ATP-like high energy bonds [71–74]. The actions of Ca^{2+} may also require polyhydroxybutyrate (PHB), which enters mitochondria and enhances the ability of Ca^{2+} to induce PT [75].

1.2 The Quest for the PT Pore

1.2.1 Electrophysiologic Properties of the Mitochondrial PT Pore (mPTP)

Thus, PT is an important event that performs both physiological and pathophysiological functions. PT may begin as the opening of a Ca^{2+} sensitive ion channel in the inner mitochondrial membrane similar to the ion channel activity initiated by mitochondrial Ca^{2+} influx occurring during physiological mitochondrial calcium cycling (such as during synaptic transmission). Such a calcium release channel is heavily regulated; therefore, it is assumed that only after prolonged opening does pathological PT (with MOMP) occur [76]. The conversion of a physiological Ca^{2+} extrusion mechanism into a pathological channel opening is perhaps correlated with energy failure as a result of arrest of ATP synthesizing activity and slowing of energy dependent Ca^{2+} extrusion mechanisms. Despite these hypothetical models, the factors that regulate the transition from physiological to pathophysiological events are not completely understood. Therefore, identification of the molecular structure of the pore will allow for a greater understanding of PT modulation during health and disease.

Description of the biophysical properties of the pore that opens in the inner membrane during PT (the mPTP) provided the earliest indication that PT was initiated by the opening of an ion channel. The first patch clamp recordings of mitochondrial inner membrane were published in 1987. This early report highlighted a ~100 pS channel recorded by patch-clamping giant mouse liver mitochondria produced by cuprizone application [77]. In the late 1980s, a putative mPTP was recorded by patch-clamping mitochondrial inner membrane or mitoplast preparations [50]. The multiconductance activity with peak single channel conductances of 1.3 nS was found either in whole organelle mode or in single channel recordings in the organelle-attached configuration [50].

Also in 1989, Kinnally et al. recorded a similar mitochondrial multiconductance channel (MMC) in mouse liver mitoplasts [49]. This channel changed over time, with low activity at the onset of the recording followed by progressively higher activity at later times during the recording. Channel activity had varied voltage dependence and frequently displayed multiple conductances ranging from 10–1000 pS. It was weakly cation-selective. These early studies established expected criteria for activity of mPTP.

Shortly after the first recordings of the putative mPTP were performed, similar inner membrane activity was found to be inhibited by CsA in a manner consistent with the expression of the CsA binding site on the matrix side of the inner membrane [52]. The large conductance channel was also sensitive to Mg^{2+}, Mn^{2+}, Ba^{2+}, and Sr^{2+} in that order, which inhibited the activity in a competitive manner with Ca^{2+}, the main activator of the channel [67].

1.2.2 Characterization of a Molecular Complex Regulating the Pore

The recent identification of the molecular structure matching the biophysical properties of mPTP was aided by findings that Bcl-x_L enhances metabolic efficiency (decreases uncoupling) by binding to the β-subunit of the ATP synthase; that CypD, known for many years to regulate PT, binds to the stator arm of ATP synthase on the oligomycin-sensitivity conferring protein (OSCP) subunit; that closure of the mPTP at the onset of respiration in mammalian heart is related to decreasing activity of CypD; and that ATP synthase assembles into a very large complex with other proteins that regulate the mPTP.

1.2.3 Bcl-x_L Regulates Metabolic Efficiency by Binding to the β-Subunit of the ATP Synthase

Inefficiency of metabolism is correlated with cell death under conditions of neurodegeneration or acute cellular injury such as occurs during PT [78–80]. In contrast, it has been found recently that Bcl-2 family proteins regulate this efficiency by binding directly to the ATP synthase [81–85]. Bcl-2 family proteins may form part of a large protein complex that regulates mPTP.

Interaction of Bcl-x_L with the β-subunit of the ATP synthase maximizes the efficiency of ATP production [82, 83] by enhancing enzyme activity of the F_1 of the ATP synthase and by decreasing a leak conductance in the mitochondrial inner membrane. We measured this by performing patch clamp recordings of submitochondrial vesicles (SMVs) enriched in F_1F_O ATP synthase [82].

Closure of the leak within the inner membrane in the presence of Bcl-x_L aids actively firing neurons to increase neurotransmitter release [86, 87], consistent with a correlation between the increase in metabolic efficiency and the long-term higher efficacy of synaptic transmission found in Bcl-x_L expressing neurons. In contrast, opening of the Bcl-x_L-regulated leak decreases metabolic efficiency and predisposes neurons to death, providing a clue that the Bcl-x_L-regulated inner membrane leak could be mPTP [83]. These data suggest that Bcl-x_L regulates inner membrane coupling and cell death via direct effects on F_1F_O ATP synthase.

1.2.4 CypD Binding to OSCP Is Associated with mPTP opening

Regulation of the mPTP may occur via the interaction of CyPD and other molecules with F_1F_O ATP synthase and its binding partners. Recent work on the binding of CypD to ATP synthase indicates that this occurs exclusively with OSCP [68]. Purified ATP synthase dimers produce current consistent with mPTP in artificial lipid membranes suggesting that the pore is located within the membrane-embedded portion of the ATP synthase complex [68].

1.2.5 PT Activity Regulates Cardiac Development

A third line of evidence that helped unravel the identity of the mPTP was a series of studies of mitochondrial function during cardiac development. In the heart, the mPTP is open in myocytes at early embryonic stages, and this opening is not associated with any form of cell death. However, by mid-embryonic development, the mPTP is closed [88], coinciding with activation of complex I of the electron transport chain, assembly of electron transport chain supercomplexes called respirasomes, and activation of oxidative phosphorylation [89]. These changes cause a fall in mitochondrial-derived ROS that signals the myocyte to undergo further differentiation [88]. Furthermore, pharmacologically inhibiting mPTP or genetically deleting CypD enhances myocyte differentiation, while opening mPTP inhibits differentiation [88, 90–93].

1.2.6 PT Regulatory Molecules Do Not Form the Pore of mPTP

The F_1F_OATP synthase interacts with a large number of proteins many of which have been candidates for mPTP. ANT was an early candidate to form the mPTP since atractyloside and bongkrekic acid, which inhibit ANT, affect the mPTP [56] and ANT interacts with CypD [94]. VDAC was also an early candidate to form the mPTP due to its high conductance and its association with ANT in immunoprecipitation experiments [95]. A complex of ANT,

VDAC, hexokinase, and mitochondrial creatine kinase (mtCK) forms high conductance pores when reconstituted into membranes [96, 97]. Finally, the PiC is a more recent candidate to form the mPTP [98].

However, genetic deletion of ANT1 and 2 and of the PiC demonstrated that these proteins were not essential to mPTP formation, although these studies still support their regulatory roles [99–102]. Furthermore, deletion of VDAC does not affect pore formation [103]. Therefore, these candidate molecules clearly regulate the mPTP, but evidence suggests that they form large macromolecular structures with F_1F_OATP synthase in the inner mitochondrial membrane [96, 97, 104] and thereby modulate the opening of the mPTP.

1.2.7 The c-Subunit of F_1F_O ATP Synthase Comprises the PT Pore

These various reports agree that F_1F_OATP synthase is a major factor in the formation of the mPTP. Recent evidence suggests that the F_O or membrane portion of F_1F_O ATP synthase in fact forms the pore [66, 105–109]. Mammalian F_1F_O ATP synthase is a ~600 kDa complex of 15 subunits. The membrane portion, or F_O, contains a ring of eight hydrophobic c-subunits and subunits a, b, e, f, g, and A6L. A stalk composed of the δ, ε, and γ subunits connects the c-subunit ring to the catalytic F_1 component made of a hexamer of alternating α and β subunits, where ATP synthesis and hydrolysis occur. Finally, a stator containing the b, d, F6, and OSCP subunits connects the lateral portion of F_O to the top of the F_1. Movement of protons between the c-subunit and the a-subunit causes rotation of the c-subunit ring, the energy of which is transferred to F_1 to synthesize ATP [110–114].

CypD and Bcl-x_L regulate ATP synthesis by interacting with the stator and β subunit of ATPase, respectively, but these interacting proteins are not embedded in the inner mitochondrial membrane. The channel or pore forming subunits, however, must be embedded in the membrane. The structure of purified bacterial c-subunit rings suggests that the center of the mammalian c-subunit ring could form an ion conducting channel that would allow for uncoupling if the stalk partially or completely dissociated from it [115].

1.2.8 The c-Subunit of ATP Synthase Creates the High Conductance mPTP Pore: Proteoliposome Recordings of Purified c-Subunit

Our recent experiments have directly tested the hypothesis that the main membrane embedded portion of mammalian F_1F_OATP synthase, i.e., the c-subunit ring, forms the pore of the mPTP [107, 116]. Electrophysiological recordings of the purified mitochondrial c-subunit in liposomes yield a multiconductance, voltage dependent channel with prominent subconductance states [106]. Peak single channel conductances of ~1.5–2 nS are similar to activity described previously for the mitochondrial multiple conductance channel (MCC) [49] as are current-voltage (*I-V*) curves [49, 50, 106].

1.2.9 F_1 Regulates Biophysical Characteristics of the Purified c-Subunit in Proteoliposomes

Purified c-subunit protein reconstituted into liposomes lacks extrinsic moieties that are important for mPTP regulation including components of the F_1 and stator complex. Ca^{2+} may affect directly interaction of CypD with OSCP [96, 97]. Therefore, although the new models of the pore must account for all known mPTP inducers and inhibitors, these molecules will not be present when purified c-subunit is prepared; in addition, some regulators such as Ca^{2+} may not bind directly to the c-subunit but may instead bind to sites in the F_1 or to other molecules such as ANT and CypD that undergo structural rearrangements to open and close the pore.

1.2.10 F1 Components Regulate Inner Membrane Channel Activity

In order to determine the location of the regulators, we performed recordings using purified mitochondrial and F_1F_OATP synthase (SMV) preparations. The absence of an effect of a modulator indicated that the ligand or binding site for that modulator had been removed by the purification process. The studies demonstrated that the inner membrane leak channel is regulated by the overlying F_1 and peripheral regulatory proteins [106]. In inner membrane preparations lacking the outer membrane such as SMVs [104], Ca^{2+} activates the leak channel while CsA and ATP/ADP inhibit it, suggesting that the Ca^{2+} and CsA sensitive sites are present at the inner membrane. In contrast, removal of the F_1 and other peripheral membrane proteins by urea treatment of the inner membrane or removal of CypD by purification of ATP synthase monomers abrogates regulation of the c-subunit channel by CsA and Ca^{2+} and greatly diminishes sensitivity to ATP/ADP. These studies suggest that the CypD/calcium binding site is contained within or associated with the F_1 portion of the ATP synthase and are consistent with reports identifying the binding site of CypD on OSCP [117, 118].

Channel activity of the purified c-subunit is inhibited by the purified F_1, suggesting a structural rearrangement whereby the stalk of the ATP synthase inhibits opening of the c-subunit channel, aided by ADP/ATP/Bcl-x_L binding to the β-subunit and opposed by CypD/Ca^{2+} interaction with OSCP. Our current model suggests that the channel of the mPTP forms within the c-subunit ring itself upon reversible CypD and Ca^{2+}-dependent movement of the stalk away from the c-subunit [106].

That the loss of protein/protein interaction between F_1 and F_O is likely to be reversible has been shown upon chelation of Ca^{2+} in mitoplasts [96], intact mitochondria [119], intact neurons [9] as well as in reconstituted dimers of F-ATP synthase [120], suggesting that the F_1 and the c-subunit can recombine to close the mPTP, reforming intact F_1F_O ATP synthase and reinitiating enzymatic function [121]. However, under certain conditions, this separation may become irreversible, forming pathophysiological PT (with MOMP).

F_1 also has binding sites that accommodate the effects of Ca^{2+}, Mg^{2+}, adenine nucleotides and P_i; and through CypD (un)binding those of H^+, CsA and possibly of oxidants [122]. Therefore, in summary, the new model of mPTP describes either direct or indirect interaction with all known inducers, inhibitors, and modulators of pore function.

1.2.11 Structural Location of the Pore Within the c-Subunit Ring

The exact location of the pore within the c-subunit is becoming increasingly understood. In experiments using fluorescence to label the c-subunit, we demonstrate that the c-subunit ring expands when it conducts ions, making it likely that the pore is formed by the c-subunit ring [106]. In addition, mutagenesis to increase the diameter of the c-subunit ring highlights an increased conductance due to ring expansion [106].

Although it has been suggested that phospholipids occupy the central cavity of the c-subunit ring [123–125] other evidence provides for formation of a proteolipid or proteophospholipid channel structure within the central lipid region [71, 75, 107, 126, 127]. Data suggest a working model whereby the c-subunit pore forms within the proteolipid milieu upon activation of mPTP, whereupon the ring expands and F_1 shifts; the pore is closed by a decrease in diameter of the ring and inactivated by binding of the F_1 components to the ring. The details of these changes and their regulation remain a work in progress.

2 Methods and Materials

2.1 Protocol for Performing Intracellular Membrane Patch Clamp Recordings in Living Invertebrate Neurons and Synapses

For intracellular membrane patch clamp recordings, a conventional patch pipette [128] is placed inside of an outer microelectrode that ensheaths it and allows the patch clamp electrode to be inserted cleanly into the interior of a cell (Figs. 1, 2 and 3). The inner patch pipette is made of 0.9 mm borosilicate glass (World Precision Instruments), and is pulled on a horizontal puller (see equipment Table 1). The diameter of the inner pipette tip is ~0.2–0.4 μm, and it has a long shaft. Its resistance with mammalian intracellular solution is ~50–120 MΩ. The outer electrode can be pulled of softer glass (1.5 mm capillary tubing, Drummond Scientific) or of borosilicate glass similar to the inner pipette but with an outer diameter of 1.8 mm. It has a short, more balloon-like shank with a tip diameter of ~1 μm.

Shortly after pulling, the inner patch pipette is filled with intracellular solution and mounted in a standard patch pipette holder with a side port for applying suction. The tip of the inner electrode is then inserted into the shank of the outer electrode. The outer electrode contains a small amount of intracellular solution at its tip, and has an abbreviated shaft that is surrounded by two rubber

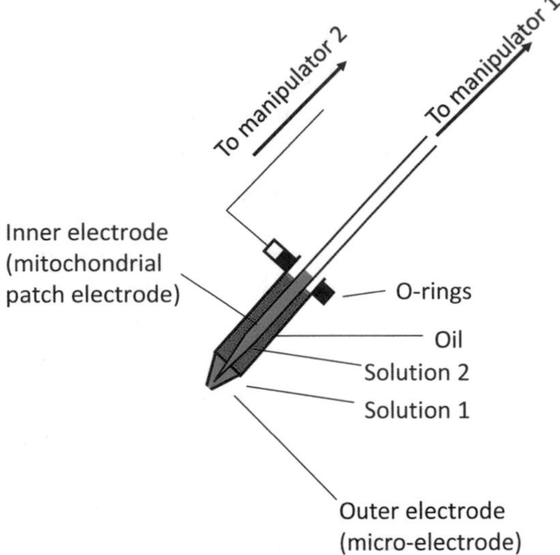

Fig. 1 Double patch clamp setup for recording intracellular membranes within invertebrate cells. The mitochondrial patch electrode is contained in the outer electrode. The outer electrode is used to obtain an intracellular recording after which the outer electrode is withdrawn, exposing the inner electrode tip. The inner electrode forms a giga-ohm seal on an intracellular membrane (for details see text, Sect. 2.1)

1. Maneuver manipulator 1 toward cell. This moves all elements except headstage and manipulator 2.
2. Using manipulator 1, maneuver electrode complex to achieve whole cell patch.
3. Using manipulator 2, maneuver headstage attached to wire to push inner electrode through outer electrode.

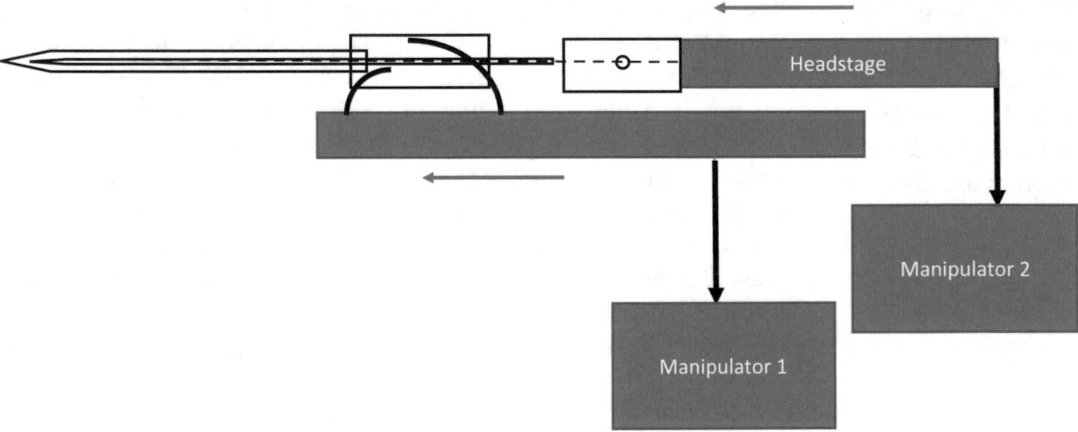

Fig. 2 Double patch clamp setup for recording intracellular membranes within mammalian cells. The mitochondrial patch electrode is contained in the outer electrode. The outer electrode is used to obtain a whole cell patch on the plasma membrane after which the inner electrode is positioned inside the cell to form a giga-ohm seal on an intracellular membrane (for details see text, Sect. 2.2)

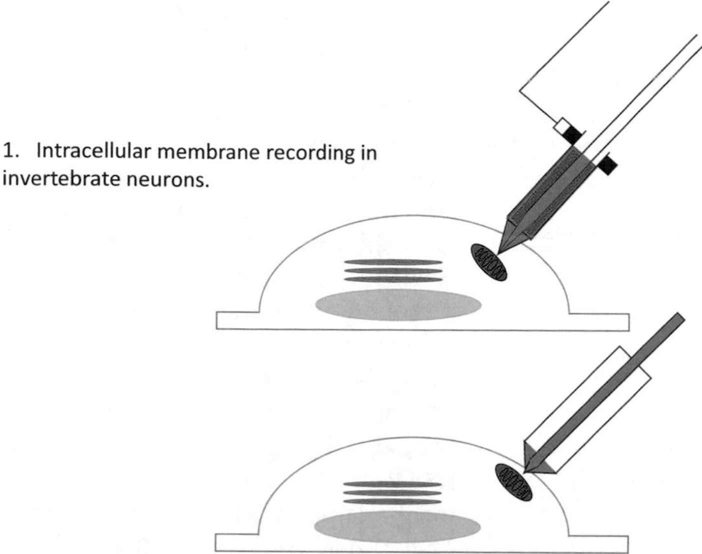

Fig. 3 Comparison of two different recording techniques to form giga-ohm seals on intracellular membranes within living cells. For details, see Figs. 1 and 2 and text, Sects. 2.1 and 2.2

Table 1
Equipment

Name	Company	Catalogue number
Potter-Elvehjem Tissue Grinder with PTFE Pestle	Krackeler Scientific, Inc.	1-7725 T-5
Eppendorf Centrifuge 5424	Eppendorf	5424 000.410
4639 Cell Disruption Vessel	Parr Instrument Company	4639
Ficoll	Sigma-Aldrich	F5415
Polycarbonate centrifuge tubes	Beckman Coulter	P20314
SW-50.1 rotor	Beckman Coulter	
L8-70 M Ultracentrifuge	Beckman Coulter	
Digitonin	Sigma-Aldrich	D5628
Lubrol PX (C12E9)	Calbiochem	205,534
Axopatch 200B	Axon Instruments	
Digidata 1440A	Molecular Device	
pClamp10.0	Molecular Device	
Manipulator	Sutter Instrument	
Borosilicate glass capillary	World Precision Instruments	1,308,325
Flaming/Brown Micropipette Puller Model P-87	Sutter Instrument	

O-rings (Fig. 1). Heavy oil, such as Zeiss microscope oil, is placed at the back of the outer electrode and allowed to fill the space between the two electrodes.

The oil reduces stray electrode capacitance and allows the relative motion of the electrodes to proceed smoothly. The two electrodes are mounted on the headstage of the patch clamp amplifier, which is in turn mounted on a mechanical manipulator (Narishige). A second, hydraulic manipulator moves the outer electrode independently by means of a stiff U-shaped copper wire surrounding the rubber O-rings (Fig. 1). The two electrodes are moved in unison toward the cell. While the electrodes are in the bath, the resistance of the electrode combination is calculated. The tips of the electrodes are visualized using a low power objective, and the tip of the outer electrode is moved into a position just in front of that of the inner electrode, using the hydraulic manipulator. When the electrode tips are resting within ~2 μm of each other, the double electrode unit is manipulated past the plasma membrane of the cell and deep into the cytoplasm. At this point, an intracellular whole cell recording can be obtained. Using the hydraulic manipulator, the outer electrode is then rapidly withdrawn from the cell. Subsequent application of suction reduces the amplitude of evoked macroscopic currents and yields a giga-ohm seal on an internal membrane in ~30–50% of attempts. Single channel recordings can be made using the organelle-attached configuration, or the patch can be excised either into the cytoplasm of the cell or into the cell-free environment of the bath. Recordings can be obtained with the same rate of success as when using standard on-cell patch clamp techniques. Moreover, the additional equipment needed for patch clamping on internal membranes is relatively inexpensive, and can be assembled in a few hours.

2.2 Protocol for Performing Intracellular Membrane Patch Clamp Recordings in Living Mammalian Neurons

To perform the experiment in mammalian cell bodies, a double patch technique was designed (Figs. 2 and 3). This technique requires that a stiff wire within the inner electrode is under the control of a manipulator. The outer electrode is placed in the electrode holder and held in place by a clamp attached to a metal rod. To patch the cell, the outer electrode and inner electrode are maneuvered in unison toward the cell and a giga-ohm seal is made on the cell plasma membrane with the outer electrode by applying a brief pulse of suction to the electrode holder side port. After achieving a whole cell recording, the inner electrode is maneuvered through the outer pipette tip by means of a separate manipulator that pushes a wire placed inside the inner electrode. After the tip is resting inside the cell, suction will cause a seal to form on an intracellular membrane.

Table 2
Isolation buffer

	Final concentration
Sucrose	250 mM
Hepes	20 mM
EDTA	1 mM
BSA	0.5%

2.3 Protocol for Differential Centrifugation and Isolation of Brain Mitochondria (Adapted from [129, 130])

1. Sacrifice the rat using methods approved by the Institutional Animal Care and Use Committee (IACUC).
2. Cut the head of the animal by decapitation, cut the skin and expose the skull.
3. Open the skull gently by cutting with a scissor or rongeur. Remove the brain.
4. Mince finely the brain without cerebellum in Isolation Buffer (see Table 2) and transfer it to a 5 ml glass/teflon homogenizer (see equipment list).
5. Homogenize tissue gently ten times (no bubbles), approximately 5 min.
6. Centrifuge sample at 1500 × *g* for 10 min at 4 °C in a benchtop centrifuge.
7. Save the supernatant (mitochondrial and synaptosomes) and discard the pellet (nuclear material and cell debris). Centrifuge at 16000 × *g* for 10 min at 4 °C in a benchtop centrifuge.
8. To separate synaptosomal mitochondria from somatic mitochondria, perform Percoll gradients. Percoll gradients: Layer mitochondria and synaptosomal fraction obtained from step 7 onto discontinuous Percoll gradient with the bottom layer containing 40% Percoll solution in isolation buffer, followed by a 24% Percoll solution, and finally the sample on a 15% Percoll solution. Centrifuge the density gradients in a Sorvall RC-5C plus superspeed refrigerated centrifuge (Asheville, NC) in a fixed angle SE-12 rotor at 30,400 × *g* for 10 min. Usage of two Percoll density gradients for cortical regions from each animal improves resolution of mitochondria on the Percoll density gradient (the band in between the 40% and 24% is the mitochondrial layer; the synaptosomal layer is less dense). Wash samples in 10–15 ml of Isolation buffer by centrifugation at 16,700 × *g* for 15 min. Resuspend the pellets in 1 ml Isolation buffer and centrifuge at 16,700 × *g* for 15 min and subsequently at 11,000 × *g* for 10 min. The resultant pellet can be resuspended in 1 ml of isolation buffer without EGTA and microcentrifuged at 10,000 × *g* for 10 min. The final

mitochondrial pellet resuspended in isolation buffer without EGTA should yield a protein concentration of ~10 mg protein/ml. All mitochondrial and synaptosomal preparations must be kept on ice.

9. Discard supernatant and resuspend pellet in 500 μl of Isolation Buffer. Disrupt synaptosomes with a cell disruption vessel (see equipment list). Apply a pressure of 1200 psi for 10 min, followed by rapid decompression.
10. Layer the mixture onto Ficoll gradients (see Table 3), place in SW-50.1 rotor and centrifuge at 126,500 × *g* for 20 min at 4 °C in an ultracentrifuge (see equipment list). The pellet is the purified mitochondria, the layer between the different densities of Ficoll is the still undisrupted synaptosomes.
11. Wash pellet by centrifuging in Isolation Buffer at 16,000 × *g* for 10 min at 4 °C in a benchtop centrifuge.

2.4 Protocol for Submitochondrial Vesicles (SMV) Isolation (Adapted from [131])

1. Resuspend mitochondria in 200 μl Isolation Buffer combined with an equal volume of 0.1% digitonin and allow to sit on ice for 15 min.
2. Add more Isolation Buffer and centrifuge at 16,000 × *g* for 10 min at 4 °C in a benchtop centrifuge. Do this twice.
3. Resuspend pellet in 200 μl Isolation Buffer and add 2 μl of 10% Lubrol PX (C12E9). Mix and allow to sit on ice for 15 min.
4. Layer mixture onto Isolation Buffer, place in SW-50.1 rotor and centrifuge at 182,000 × *g* at 4 °C for 1 h.
5. Wash final pellet by centrifuging in a desktop centrifuge in Isolation Buffer at 16,000 × *g* for 10 min at 4 °C.
6. Submitochondrial vesicles can be stored in −80 freezer in 10% DMSO for later usage in 500 μl Iso w/EGTA.

Table 3
Ficoll stock

Ficoll	22 ml 20%
Sucrose	12 ml 1 M
EDTA	18.75 μl 0,1 M
Tris-HCl	375 μl 1 M (ph 7.4)
Top layer 7.5% of ficoll	
Ficoll stock	10 ml
Isolation buffer	6 ml
Bottom layer 10% of ficoll	
Ficoll stock	15.5 ml
Isolation buffer	2.5 ml

2.5 Protocol for Mitoplast Preparation

To prepare "mitoplasts" or inner membrane preparations for recording, place whole mitochondria (see Sect. 2.3) in hypo-osmotic solution containing 30 mM KCl, 10 mM HEPES at pH 7.3. After 30 min, spin mitochondria at 10,000 × *g* on a desktop centrifuge for 2 min. Resuspend in normal recording buffer.

2.6 Electrophysiological Recording from Whole Mitochondria, Mitoplasts or SMVs

1. A typical electrophysiology rig includes an amplifier, a PC computer equipped with a Digidata 1440A analog-to-digital converter interface in conjunction with pClamp10.0 software, manipulators, a microscope, a vibration isolation table, Faraday Cage.
2. Borosilicate glass capillary tubes are inserted into a Flaming/Brown Micropipette Puller Model P-87 or 97. A pipette-puller program is optimized to generate pipettes with resistances between 80 and 100 MΩ.
3. Mitochondria, mitoplasts or SMVs are placed in a physiological intracellular solution (pharmacological agents are added to the bath at the appropriate time during the recording) (Table 4). Patch clamp pipettes (50–120 MΩ) are filled with the same solution (no ATP). Recordings are made by forming a giga-ohm seal onto SMVs at room temperature. Vesicles are visualized by phase-contrast microscopy with a Nikon or Zeiss inverted microscope. Patch electrode is maneuvered to bottom of dish. Under visual control, tip is placed on top of a mitochondrion and suction is applied. After seal formation, electrode is lifted from bottom of dish. This yields an on-organelle recording. With increases in patch gain, single channel events may be discerned.
4. The membrane potential is maintained at voltages ranging from −100 to + 100 mV for periods of 10 s. Recordings are filtered at 5 kHz using the amplifier circuitry. Level of stray electrical or non-specific noise of less than 1pA is desirable for successful recording.

Table 4
Internal solution (recording buffer)

KCl	120 mM
NaCl	8 mM
EGTA	0.5 mM
Hepes	10 mM
pH	7.3

5. Data are analyzed with Clampfit 10.0 software for example to determine frequency and amplitude of single channel events. Voltage dependence is determined by plotting a current voltage relationship.

2.7 Protocol for Making Proteoliposomes for Recording ATP Synthase Monomers or Purified c-Subunit Protein Incorporated into Liposomes

1. Dry glass slides. Wash in 99.5% chloroform 2–3 times. Dry in between and after for 2 min.
2. Do not touch weighing paper with hands. Weigh phosphatidylcholine (azolectin) from Sigma. Place 50 mg in 1 ml chloroform (use filtered tips to measure chloroform) in glass scintillation vial.
3. Dissolve in scintillation vial. Solution will be pale yellow or clear.
4. Place 50 μl of the azolectin–chloroform solution on a dry, clean slide (treated with chloroform as above). Dry for 2–4 min.
5. Drop protein onto slide. Spread over lipids carefully with the side of a pipette tip. Dry this mixture for 0.5 h or longer. If concentration of stock of protein is as high as 2 mg/ml protein, put 5 μl protein in buffer onto 50 μl dried lipids on the slide. Since it is generally less, adjust accordingly.
6. Place 50 μl rehydration buffer (Rehydration buffer recipe (50 ml):
   - KCl: 932 mg.
   - Hepes: 59 mg.
   - EDTA: 1.46 mg.

(in 50 ml water) onto slide. Collect and redry this in the recording chamber (can be glass or plastic). Then add recording buffer just before recording.

3 Notes

1. Prepare all materials in advance: Materials for dissection- ice, cold Isolation buffers, dissection instruments, body bag, guillotine, rats, Petri dish, two sets of labeled tubes, 10 ml glass homogenizers
2. To obtain synaptosomal mitochondria, perform Percoll gradients
3. To disrupt synaptosomal membranes (use one of the following two methods):
   (a) Place samples in tubes into nitrogen bomb at 1000 psi for 5′ or up to 1200 psi for 10′.
   (b) Or use 0.01% digitonin for 10′ on ice (use 10× stock of 0.1% digitonin in Isolation buffer store at 4 °C). After digitonin incubation, top off in Iso w/EGTA and spin samples for 10′ at 10,000 rcf at 4 °C to remove all of the digitonin.

4. Always perform BCA Protein Assay (Pierce).
5. To check health of mitochondria, use a Clark electrode to measure respiratory control ratio which is defined as the ratio of oxygen uptake during ATP production over oxygen uptake at rest (after all ADP is used up). This ratio should be 5 or higher. If it is lower than 5, damage to mitochondria during preparation has occurred. This measure is inaccurate when performed on mitoplasts or SMVs and should be performed just after fresh mitochondria are prepared.
6. Although we record in the whole organelle-attached configuration typically, in some cases the procedure may result in excision of the patch. We observe this rarely in mitochondrial recordings but if it occurs, the individual channel openings become square and lack series resistance artifacts that are observed in on-organelle recordings. These series resistance artifacts arise as a result of the change in resistance of the membrane as channels open under the pipette tip. The patched membrane exists in series with the membrane that forms the remainder of the unpatched mitochondrion (Fig. 4). The unpatched membrane in mitochondrial recordings is a high resistance membrane unlike the unpatched portion of the plasma membrane which is low resistance compared to that of an on-cell patch. Since the two mitochondrial membranes are in series, when a channel opens in the unpatched membrane and the resistance drops on that membrane, the actual voltage change across the patched membrane drops (Fig. 4). Therefore,

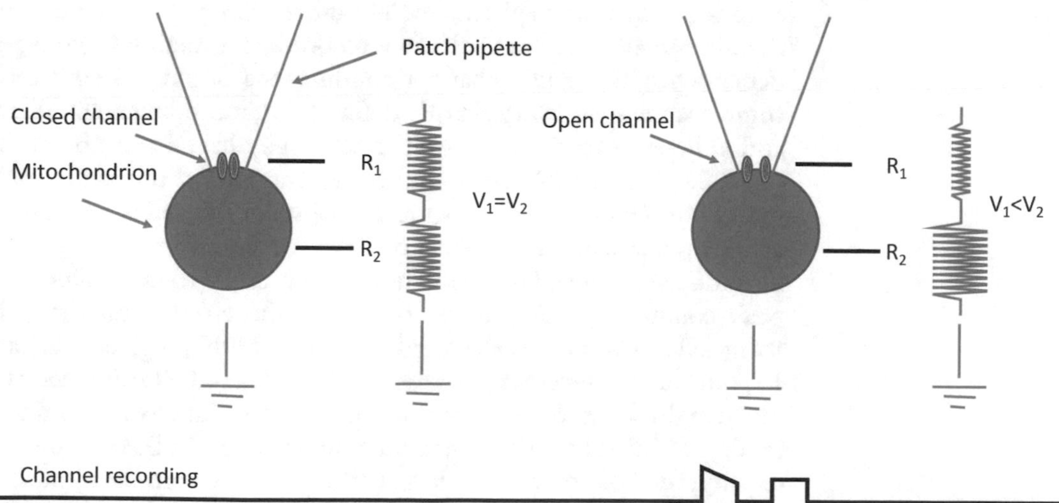

Fig. 4 The patched membrane exists in series with the membrane that forms the remainder of the unpatched mitochondrion. On the *left* is shown a recording of a closed channel. On the *right*, the channel opens, reducing the resistance of the patch. Therefore, the voltage across the patched membrane drops and, as shown in the channel recording, the channel current amplitude decreases. An example of this is shown in Fig. 5

more of the potential difference between the holding potential and the bath will be seen by the unpatched membrane in series with the patch because it continues to have a high resistance. This reduces the current of the recorded channels (because the voltage change across the membrane is smaller). The advantage of this "artifact," however, is that when a pharmacological agent is added to the bath side of the organelle, and, for example, the conductance of the membrane exposed to the bath decreases, there will be a compensatory decrease in the amplitude of any channel activity under the patch pipette tip.

7. **Imaging of the intracellular membranes sealed to the pipette tip**. To visualize the intracellular location of the patch pipette during recordings, lipophilic fluorescent dyes such as DiOC6 and BODIPY-ceramide may be placed in the inner patch pipette solution. The cells can then be imaged during and after seal formation with video-enhanced or confocal fluorescence microscopy. In some experiments, only a discrete point of fluorescence corresponding to the tip of the pipette can be detected. In other cells, however, some staining of membranes surrounding the pipette tip can be observed. If the cell is prelabeled with a mitochondrial dye, then co-localization of the patch under the pipette tip and the organelle can be achieved.

4 Typical Results and Data Analysis

As expected, there is no typical mitochondrial, submitochondrial, or liposome recording. Mitochondrial ion channel activity may be as varied as that of the plasma membrane. Although more common than in plasma membrane channel recordings, variation in voltage dependence and single channel conductance occurs; in contrast, some channels with very sharply defined single conductance values and voltage dependence as well as selectivity have been observed [8, 9, 33, 34, 36, 132, 106]. Shown are examples of mitochondrial recordings (Figs. 5 and 6). One can observe the series resistance artifact (arrows in Fig. 5). Also one can see the difficulty assigning a conductance level. Often one has to draw an IV to determine the peak conductance. As can be observed, the channel on the left recorded at +60 mV is calculated to be about 500 pS given a linear IV, but the conductance increases to 750 pS at +100 mV (recording at right) (Fig. 5). In this case, calculation of the peak conductance was aided by a drug that shut the channel. In the process of closing, the holding current level (different from zero, leak) of the patch is observed (Fig. 6; both holding current level or "closed" and 0 pA are shown by dotted lines).

Figure 7 shows a typical recording for the purified c-subunit of a 600 pS channel with long open and closed periods. On the top

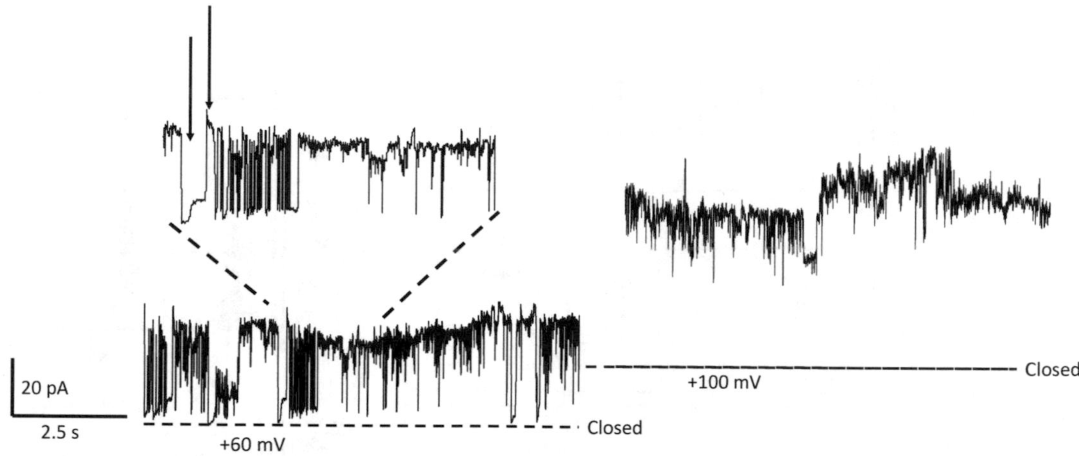

Fig. 5 Typical on-organelle whole mitochondrial recording. *Inset* shows an example of series resistance artifacts as described in Fig. 4. *Arrows* demonstrate the series resistance artifact. Also illustrated in Figs. 4 and 5 are the difficulties encountered when attempting to define the closed state of the channel. One can observe the closed position on the *right* (*closed*), that was derived from observations made on the same channel at the same holding voltage after treatment with a drug that increases the probability of channel closure (Fig. 6)

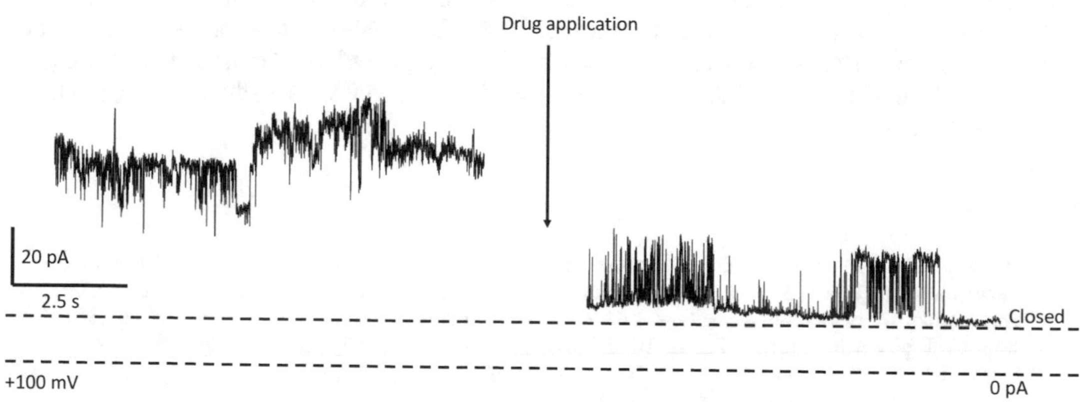

Fig. 6 Typical on-organelle whole mitochondrial recording. Drug application was used to identify the closed state

trace, one can see that the channel is not fully closed because of a subconductance state. In the bottom trace, the fully closed state is detected.

When analyzing data, we usually try to determine single channel conductance, voltage dependence of opening at different voltages between −150 and +150 mV (of the patch pipette), selectivity of the channel activity, and open probability. If single channel events cannot be discerned, we calculate a peak conductance of the activity and generate a current voltage relationship of the activity over many holding voltages.

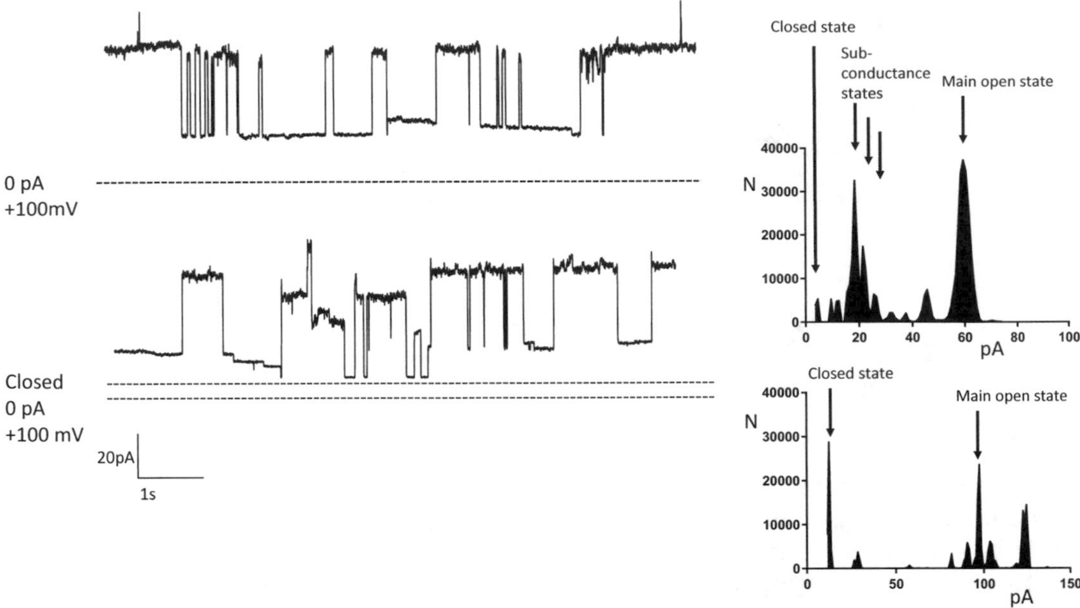

Fig. 7 Typical proteoliposome excised patch recording. Shown is the purified c-subunit of F_1F_0ATP synthase. Peak amplitude of channel activity is initially about 600 pS with prominent subconductance states but later in the recording the peak amplitude increases. Shown at *right* is an amplitude histogram of the recording shown at *left*. The main open state and several subconductance states are graphed. The second amplitude *histogram* is calculated from a later trace of the same patch clamp recording. In this trace, the peak amplitude was approximately 1.25 nS

References

1. Rizzuto R, De Stefani D, Raffaello A, Mammucari C (2012) Mitochondria as sensors and regulators of calcium signalling. Nat Rev Mol Cell Biol 13(9):566–578. doi:10.1038/nrm3412
2. Lopreiato R, Giacomello M, Carafoli E (2014) The plasma membrane calcium pump: new ways to look at an old enzyme. J Biol Chem 289(15):10261–10268. doi:10.1074/jbc.O114.555565
3. Schendel SL, Xie Z, Montal MO, Matsuyama S, Montal M, Reed JC (1997) Channel formation by antiapoptotic protein Bcl-2. Proc Natl Acad Sci U S A 94(10):5113–5118
4. Maldonado EN, Sheldon KL, DeHart DN, Patnaik J, Manevich Y, Townsend DM, Bezrukov SM, Rostovtseva TK, Lemasters JJ (2013) Voltage-dependent anion channels modulate mitochondrial metabolism in cancer cells: regulation by free tubulin and erastin. J Biol Chem 288(17):11920–11929. doi:10.1074/jbc.M112.433847
5. Zucker RS, Regehr WG (2002) Short-term synaptic plasticity. Annu Rev Physiol 64:355–405
6. Neher E, Sakaba T (2008) Multiple roles of calcium ions in the regulation of neurotransmitter release. Neuron 59(6):861–872. doi:10.1016/j.neuron.2008.08.019
7. Jonas E (2006) BCL-xL regulates synaptic plasticity. Mol Interv 6(4):208–222
8. Jonas EA, Knox RJ, Kaczmarek LK (1997) Giga-ohm seals on intracellular membranes: a technique for studying intracellular ion channels in intact cells. Neuron 19(1):7–13
9. Jonas EA, Buchanan J, Kaczmarek LK (1999) Prolonged activation of mitochondrial conductances during synaptic transmission. Science 286(5443):1347–1350
10. Tang Y, Zucker RS (1997) Mitochondrial involvement in post-tetanic potentiation of synaptic transmission. Neuron 18(3):483–491
11. Friel DD, Tsien RW (1994) An FCCP-sensitive Ca2+ store in bullfrog sympathetic neurons and its participation in stimulus-evoked changes in [Ca2+]i. J Neurosci 14(7):4007–4024
12. Kinnally KW, Martinez-Caballero S, Dejean LM (2006) Detection of the mitochondrial

apoptosis-induced channel (MAC) and its regulation by Bcl-2 family proteins. Curr Protoc Toxicol Chapter 2:Unit2.12

13. Dejean LM, Martinez-Caballero S, Guo L, Hughes C, Teijido O, Ducret T, Ichas F, Korsmeyer SJ, Antonsson B, Jonas EA, Kinnally KW (2005) Oligomeric Bax is a component of the putative cytochrome c release channel MAC, mitochondrial apoptosis-induced channel. Mol Biol Cell 16(5):2424–2432
14. Dejean LM, Martinez-Caballero S, Manon S, Kinnally KW (2006) Regulation of the mitochondrial apoptosis-induced channel, MAC, by BCL-2 family proteins. Biochim Biophys Acta 1762(2):191–201
15. Antonsson B, Montessuit S, Lauper S, Eskes R, Martinou JC (2000) Bax oligomerization is required for channel-forming activity in liposomes and to trigger cytochrome c release from mitochondria. Biochem J 345(Pt 2):271–278
16. Adams JM, Cory S (2007) The Bcl-2 apoptotic switch in cancer development and therapy. Oncogene 26(9):1324–1337
17. Kim H, Rafiuddin-Shah M, Tu HC, Jeffers JR, Zambetti GP, Hsieh JJ, Cheng EH (2006) Hierarchical regulation of mitochondrion-dependent apoptosis by BCL-2 subfamilies. [See comment]. Nat Cell Biol 8(12): 1348–1358
18. Galonek HL, Hardwick JM (2006) Upgrading the BCL-2 network. [Comment]. Nat Cell Biol 8(12):1317–1319
19. Shimizu S, Narita M, Tsujimoto Y (1999) Bcl-2 family proteins regulate the release of apoptogenic cytochrome c by the mitochondrial channel VDAC. [see comment] [Erratum appears in Nature 2000;407(6805):767]. Nature 399(6735):483–487
20. Shimizu S, Shinohara Y, Tsujimoto Y (2000) Bax and Bcl-xL independently regulate apoptotic changes of yeast mitochondria that require VDAC but not adenine nucleotide translocator. Oncogene 19(38):4309–4318
21. Tsujimoto Y, Shimizu S (2000) VDAC regulation by the Bcl-2 family of proteins. Cell Death Differ 7(12):1174–1181
22. Malia TJ, Wagner G (2007) NMR structural investigation of the mitochondrial outer membrane protein VDAC and its interaction with antiapoptotic Bcl-xL. Biochemistry 46(2):514–525
23. Monaco G, Decrock E, Arbel N, van Vliet AR, La Rovere RM, De Smedt H, Parys JB, Agostinis P, Leybaert L, Shoshan-Barmatz V, Bultynck G (2015) The BH4 domain of anti-apoptotic Bcl-XL, but not that of the related Bcl-2, limits the voltage-dependent anion channel 1 (VDAC1)-mediated transfer of pro-apoptotic Ca2+ signals to mitochondria. J Biol Chem 290(14):9150–9161. doi:10.1074/jbc.M114.622514
24. Minn AJ, Velez P, Schendel SL, Liang H, Muchmore SW, Fesik SW, Fill M, Thompson CB (1997) Bcl-x(L) forms an ion channel in synthetic lipid membranes. Nature 385(6614):353–357
25. Schendel SL, Montal M, Reed JC (1998) Bcl-2 family proteins as ion-channels. Cell Death Differ 5(5):372–380
26. Muchmore SW, Sattler M, Liang H, Meadows RP, Harlan JE, Yoon HS, Nettesheim D, Chang BS, Thompson CB, Wong SL, Ng SL, Fesik SW (1996) X-ray and NMR structure of human Bcl-xL, an inhibitor of programmed cell death. Nature 381(6580):335–341
27. Cramer WA, Heymann JB, Schendel SL, Deriy BN, Cohen FS, Elkins PA, Stauffacher CV (1995) Structure-function of the channel-forming colicins. Annu Rev Biophys Biomol Struct 24:611–641. doi:10.1146/annurev.bb.24.060195.003143
28. Jonas EA, Hoit D, Hickman JA, Brandt TA, Polster BM, Fannjiang Y, McCarthy E, Montanez MK, Hardwick JM, Kaczmarek LK (2003) Modulation of synaptic transmission by the BCL-2 family protein BCL-xL. J Neurosci 23(23):8423–8431
29. Vander Heiden MG, Li XX, Gottleib E, Hill RB, Thompson CB, Colombini M (2001) Bcl-xL promotes the open configuration of the voltage-dependent anion channel and metabolite passage through the outer mitochondrial membrane. J Biol Chem 276(22): 19414–19419
30. Vander Heiden MG, Chandel NS, Li XX, Schumacker PT, Colombini M, Thompson CB (2000) Outer mitochondrial membrane permeability can regulate coupled respiration and cell survival. Proc Natl Acad Sci U S A 97(9):4666–4671
31. Gottlieb E, Armour SM, Thompson CB (2002) Mitochondrial respiratory control is lost during growth factor deprivation. Proc Natl Acad Sci U S A 99(20):12801–12806
32. Kluck RM, Bossy-Wetzel E, Green DR, Newmeyer DD (1997) The release of cytochrome c from mitochondria: a primary site for Bcl-2 regulation of apoptosis. [See comment]. Science 275(5303):1132–1136
33. Jonas EA, Hardwick JM, Kaczmarek LK (2005) Actions of BAX on mitochondrial channel activity and on synaptic transmission. Antioxid Redox Signal 7(9–10): 1092–1100

34. Jonas EA, Hickman JA, Hardwick JM, Kaczmarek LK (2005) Exposure to hypoxia rapidly induces mitochondrial channel activity within a living synapse. J Biol Chem 280(6): 4491–4497
35. Jonas E (2004) Regulation of synaptic transmission by mitochondrial ion channels. J Bioenerg Biomembr 36(4):357–361
36. Jonas EA, Hickman JA, Chachar M, Polster BM, Brandt TA, Fannjiang Y, Ivanovska I, Basanez G, Kinnally KW, Zimmerberg J, Hardwick JM, Kaczmarek LK (2004) Proapoptotic N-truncated BCL-xL protein activates endogenous mitochondrial channels in living synaptic terminals. Proc Natl Acad Sci U S A 101(37):13590–13595
37. Oltersdorf T, Elmore SW, Shoemaker AR, Armstrong RC, Augeri DJ, Belli BA, Bruncko M, Deckwerth TL, Dinges J, Hajduk PJ, Joseph MK, Kitada S, Korsmeyer SJ, Kunzer AR, Letai A, Li C, Mitten MJ, Nettesheim DG, Ng S, Nimmer PM, O'Connor JM, Oleksijew A, Petros AM, Reed JC, Shen W, Tahir SK, Thompson CB, Tomaselli KJ, Wang B, Wendt MD, Zhang H, Fesik SW, Rosenberg SH (2005) An inhibitor of Bcl-2 family proteins induces regression of solid tumours. Nature 435(7042):677–681
38. Hickman JA, Hardwick JM, Kaczmarek LK, Jonas EA (2008) Bcl-xL inhibitor ABT-737 reveals a dual role for Bcl-xL in synaptic transmission. J Neurophysiol 99(3):1515–1522
39. Polster BM, Fiskum G (2004) Mitochondrial mechanisms of neural cell apoptosis. J Neurochem 90(6):1281–1289
40. Bonanni L, Chachar M, Jover-Mengual T, Li H, Jones A, Yokota H, Ofengeim D, Flannery RJ, Miyawaki T, Cho CH, Polster BM, Pypaert M, Hardwick JM, Sensi SL, Zukin RS, Jonas EA (2006) Zinc-dependent multi-conductance channel activity in mitochondria isolated from ischemic brain. J Neurosci 26(25):6851–6862
41. Ofengeim D, Chen YB, Miyawaki T, Li H, Sacchetti S, Flannery RJ, Alavian KN, Pontarelli F, Roelofs BA, Hickman JA, Hardwick JM, Zukin RS, Jonas EA (2012) N-terminally cleaved Bcl-xL mediates ischemia-induced neuronal death. Nat Neurosci 15(4): 574–580. doi:10.1038/nn.3054
42. Lee AC, Zizi M, Colombini M (1994) Beta-NADH decreases the permeability of the mitochondrial outer membrane to ADP by a factor of 6. J Biol Chem 269(49):30974–30980
43. Colombini M, Blachly-Dyson E, Forte M (1996) VDAC, a channel in the outer mitochondrial membrane. Ion Channels 4: 169–202
44. Jonas EA (2009) Molecular participants in mitochondrial cell death channel formation during neuronal ischemia. Exp Neurol 218(2):203–212
45. Tedeschi H, Kinnally KW (1987) Channels in the mitochondrial outer membrane: evidence from patch clamp studies. J Bioenerg Biomembr 19(4):321–327
46. Tedeschi H, Kinnally KW, Mannella CA (1989) Properties of channels in the mitochondrial outer membrane. J Bioenerg Biomembr 21(4):451–459
47. Pavlov EV, Priault M, Pietkiewicz D, Cheng EH, Antonsson B, Manon S, Korsmeyer SJ, Mannella CA, Kinnally KW (2001) A novel, high conductance channel of mitochondria linked to apoptosis in mammalian cells and Bax expression in yeast. J Cell Biochem 155(5):725–731
48. Rostovtseva TK, Bezrukov SM (2012) VDAC inhibition by tubulin and its physiological implications. Biochim Biophys Acta 1818(6): 1526–1535. doi:10.1016/j.bbamem.2011.11.004
49. Kinnally KW, Campo ML, Tedeschi H (1989) Mitochondrial channel activity studied by patch-clamping mitoplasts. J Bioenerg Biomembr 21(4):497–506
50. Petronilli V, Szabo I, Zoratti M (1989) The inner mitochondrial membrane contains ion-conducting channels similar to those found in bacteria. FEBS Lett 259(1):137–143
51. Sensi SL, Ton-That D, Sullivan PG, Jonas EA, Gee KR, Kaczmarek LK, Weiss JH (2003) Modulation of mitochondrial function by endogenous Zn2+ pools. Proc Natl Acad Sci U S A 100(10):6157–6162
52. Szabo I, Zoratti M (1991) The giant channel of the inner mitochondrial membrane is inhibited by cyclosporin A. J Biol Chem 266(6):3376–3379
53. Galluzzi L, Blomgren K, Kroemer G (2009) Mitochondrial membrane permeabilization in neuronal injury. Nat Rev Neurosci 10(7): 481–494
54. Budd SL, Nicholls DG (1996) Mitochondria, calcium regulation, and acute glutamate excitotoxicity in cultured cerebellar granule cells. J Neurochem 67(6):2282–2291
55. Haworth RA, Hunter DR (1979) The Ca2+-induced membrane transition in mitochondria. II. Nature of the Ca2+ trigger site. Arch Biochem Biophys 195(2):460–467
56. Hunter DR, Haworth RA (1979) The Ca2+-induced membrane transition in mitochondria. I The protective mechanisms. Arch Biochem Biophys 195(2):453–459

57. Hunter DR, Haworth RA (1979) The Ca2+-induced membrane transition in mitochondria. III. Transitional Ca2+ release. Arch Biochem Biophys 195(2):468–477

58. Crompton M (1999) The mitochondrial permeability transition pore and its role in cell death. Biochem J 341(Pt 2):233–249

59. Huser J, Blatter LA (1999) Fluctuations in mitochondrial membrane potential caused by repetitive gating of the permeability transition pore. Biochem J 343(Pt 2):311–317

60. Petronilli V, Miotto G, Canton M, Brini M, Colonna R, Bernardi P, Di Lisa F (1999) Transient and long-lasting openings of the mitochondrial permeability transition pore can be monitored directly in intact cells by changes in mitochondrial calcein fluorescence. Biophys J 76(2):725–734. doi:10.1016/S0006-3495(99)77239-5

61. Hausenloy D, Wynne A, Duchen M, Yellon D (2004) Transient mitochondrial permeability transition pore opening mediates preconditioning-induced protection. Circulation 109(14):1714–1717. doi:10.1161/01.CIR.0000126294.81407.7D

62. Wang W, Fang H, Groom L, Cheng A, Zhang W, Liu J, Wang X, Li K, Han P, Zheng M, Yin J, Mattson MP, Kao JP, Lakatta EG, Sheu SS, Ouyang K, Chen J, Dirksen RT, Cheng H (2008) Superoxide flashes in single mitochondria. Cell 134(2):279–290. doi:10.1016/j.cell.2008.06.017

63. Korge P, Yang L, Yang JH, Wang Y, Qu Z, Weiss JN (2011) Protective role of transient pore openings in calcium handling by cardiac mitochondria. J Biol Chem 286(40):34851–34857. doi:10.1074/jbc.M111.239921

64. Bonora M, Pinton P (2014) Shedding light on molecular mechanisms and identity of mPTP. Mitochondrion. doi:10.1016/j.mito.2014.10.001

65. Baines CP (2011) The mitochondrial permeability transition pore and the cardiac necrotic program. Pediatr Cardiol 32(3):258–262. doi:10.1007/s00246-010-9880-9

66. Bonora M, Wieckowski MR, Chinopoulos C, Kepp O, Kroemer G, Galluzzi L, Pinton P (2014) Molecular mechanisms of cell death: central implication of ATP synthase in mitochondrial permeability transition. Oncogene. doi:10.1038/onc.2014.96

67. Szabo I, Bernardi P, Zoratti M (1992) Modulation of the mitochondrial megachannel by divalent cations and protons. J Biol Chem 267(5):2940–2946

68. Giorgio V, von Stockum S, Antoniel M, Fabbro A, Fogolari F, Forte M, Glick GD, Petronilli V, Zoratti M, Szabo I, Lippe G, Bernardi P (2013) Dimers of mitochondrial ATP synthase form the permeability transition pore. Proc Natl Acad Sci U S A 110(15):5887–5892. doi:10.1073/pnas.1217823110

69. Kowaltowski AJ, Naia-da-Silva ES, Castilho RF, Vercesi AE (1998) Ca2+-stimulated mitochondrial reactive oxygen species generation and permeability transition are inhibited by dibucaine or Mg2+. Arch Biochem Biophys 359(1):77–81. doi:10.1006/abbi.1998.0870

70. Giorgio V, Bisetto E, Soriano ME, Dabbeni-Sala F, Basso E, Petronilli V, Forte MA, Bernardi P, Lippe G (2009) Cyclophilin D modulates mitochondrial F0F1-ATP synthase by interacting with the lateral stalk of the complex. J Biol Chem 284(49):33982–33988

71. Abramov AY, Fraley C, Diao CT, Winkfein R, Colicos MA, Duchen MR, French RJ, Pavlov E (2007) Targeted polyphosphatase expression alters mitochondrial metabolism and inhibits calcium-dependent cell death. Proc Natl Acad Sci U S A 104(46):18091–18096. doi:10.1073/pnas.0708959104

72. Seidlmayer LK, Blatter LA, Pavlov E, Dedkova EN (2012) Inorganic polyphosphate--an unusual suspect of the mitochondrial permeability transition mystery. Channels 6(6):463–467. doi:10.4161/chan.21939

73. Holmstrom KM, Marina N, Baev AY, Wood NW, Gourine AV, Abramov AY (2013) Signalling properties of inorganic polyphosphate in the mammalian brain. Nat Commun 4:1362. doi:10.1038/ncomms2364

74. Stotz SC, Scott LO, Drummond-Main C, Avchalumov Y, Girotto F, Davidsen J, Gomez-Garcia MR, Rho JM, Pavlov EV, Colicos MA (2014) Inorganic polyphosphate regulates neuronal excitability through modulation of voltage-gated channels. Mol Brain 7:42. doi:10.1186/1756-6606-7-42

75. Elustondo PA, Angelova PR, Kawalec M, Michalak M, Kurcok P, Abramov AY, Pavlov EV (2013) Polyhydroxybutyrate targets mammalian mitochondria and increases permeability of plasmalemmal and mitochondrial membranes. PLoS One 8(9):e75812. doi:10.1371/journal.pone.0075812

76. Bernardi P (1999) Mitochondrial transport of cations: channels, exchangers, and permeability transition. Physiol Rev 79(4):1127–1155

77. Sorgato MC, Keller BU, Stuhmer W (1987) Patch-clamping of the inner mitochondrial membrane reveals a voltage-dependent ion channel. Nature 330(6147):498–500

78. Beal MF (2007) Mitochondria and neurodegeneration. Novartis Found Symp 287:183–192 discussion 192–186

79. Dodson MW, Guo M (2007) Pink1, Parkin, DJ-1 and mitochondrial dysfunction in Parkinson's disease. Curr Opin Neurobiol 17(3):331–337
80. Brand MD (2005) The efficiency and plasticity of mitochondrial energy transduction. Biochem Soc Trans 33(Pt 5):897–904
81. Hockenbery D, Nunez G, Milliman C, Schreiber RD, Korsmeyer SJ (1990) Bcl-2 is an inner mitochondrial membrane protein that blocks programmed cell death. Nature 348(6299):334–336
82. Alavian KN, Li H, Collis L, Bonanni L, Zeng L, Sacchetti S, Lazrove E, Nabili P, Flaherty B, Graham M, Chen Y, Messerli SM, Mariggio MA, Rahner C, McNay E, Shore GC, Smith PJ, Hardwick JM, Jonas EA (2011) Bcl-xL regulates metabolic efficiency of neurons through interaction with the mitochondrial F1FO ATP synthase. Nat Cell Biol 13(10):1224–1233. doi:10.1038/ncb2330
83. Chen YB, Aon MA, Hsu YT, Soane L, Teng X, McCaffery JM, Cheng WC, Qi B, Li H, Alavian KN, Dayhoff-Brannigan M, Zou S, Pineda FJ, O'Rourke B, Ko YH, Pedersen PL, Kaczmarek LK, Jonas EA, Hardwick JM (2011) Bcl-xL regulates mitochondrial energetics by stabilizing the inner membrane potential. J Cell Biochem 195(2):263–276
84. Jonas EA, Porter GA, Alavian KN (2014) Bcl-xL in neuroprotection and plasticity. Front Physiol 5:355. doi:10.3389/fphys.2014.00355
85. Park HA, Licznerski P, Alavian KN, Shanabrough M, Jonas EA (2014) Bcl-xL Is necessary for neurite outgrowth in hippocampal neurons. Antioxid Redox Signal. doi:10.1089/ars.2013.5570
86. Li H, Chen Y, Jones AF, Sanger RH, Collis LP, Flannery R, McNay EC, Yu T, Schwarzenbacher R, Bossy B, Bossy-Wetzel E, Bennett MV, Pypaert M, Hickman JA, Smith PJ, Hardwick JM, Jonas EA (2008) Bcl-xL induces Drp1-dependent synapse formation in cultured hippocampal neurons. Proc Natl Acad Sci U S A 105(6): 2169–2174
87. Li H, Alavian KN, Lazrove E, Mehta N, Jones A, Zhang P, Licznerski P, Graham M, Uo T, Guo J, Rahner C, Duman RS, Morrison RS, Jonas EA (2013) A Bcl-xL-Drp1 complex regulates synaptic vesicle membrane dynamics during endocytosis. Nat Cell Biol 15(7):773–785. doi:10.1038/ncb2791
88. Hom JR, Quintanilla RA, Hoffman DL, de Mesy Bentley KL, Molkentin JD, Sheu SS, Porter GA Jr (2011) The permeability transition pore controls cardiac mitochondrial maturation and myocyte differentiation. Dev Cell 21(3):469–478. doi:10.1016/j.devcel.2011.08.008
89. Beutner G, Eliseev RA, Porter GA Jr (2014) Initiation of electron transport chain activity in the embryonic heart coincides with the activation of mitochondrial complex 1 and the formation of supercomplexes. PLoS One 9(11):e113330. doi:10.1371/journal.pone.0113330
90. Cho SW, Park JS, Heo HJ, Park SW, Song S, Kim I, Han YM, Yamashita JK, Youm JB, Han J, Koh GY (2014) Dual modulation of the mitochondrial permeability transition pore and redox signaling synergistically promotes cardiomyocyte differentiation from pluripotent stem cells. J Am Heart Assoc 3(2): e000693. doi:10.1161/JAHA.113.000693
91. Fujiwara M, Yan P, Otsuji TG, Narazaki G, Uosaki H, Fukushima H, Kuwahara K, Harada M, Matsuda H, Matsuoka S, Okita K, Takahashi K, Nakagawa M, Ikeda T, Sakata R, Mummery CL, Nakatsuji N, Yamanaka S, Nakao K, Yamashita JK (2011) Induction and enhancement of cardiac cell differentiation from mouse and human induced pluripotent stem cells with cyclosporin-A. PLoS One 6(2):e16734. doi:10.1371/journal.pone.0016734
92. Drenckhahn JD (2011) Heart development: mitochondria in command of cardiomyocyte differentiation. Dev Cell 21(3):392–393. doi:10.1016/j.devcel.2011.08.021
93. Folmes CD, Dzeja PP, Nelson TJ, Terzic A (2012) Mitochondria in control of cell fate. Circ Res 110(4):526–529. doi:10.1161/RES.0b013e31824ae5c1
94. Halestrap AP, Davidson AM (1990) Inhibition of Ca2(+)-induced large-amplitude swelling of liver and heart mitochondria by cyclosporin is probably caused by the inhibitor binding to mitochondrial-matrix peptidyl-prolyl cis-trans isomerase and preventing it interacting with the adenine nucleotide translocase. Biochem J 268(1): 153–160
95. Crompton M, Virji S, Ward JM (1998) Cyclophilin-D binds strongly to complexes of the voltage-dependent anion channel and the adenine nucleotide translocase to form the permeability transition pore. Eur J Biochem 258(2):729–735
96. Beutner G, Ruck A, Riede B, Brdiczka D (1998) Complexes between porin, hexokinase, mitochondrial creatine kinase and adenylate translocator display properties of the

permeability transition pore. Implication for regulation of permeability transition by the kinases. Biochim Biophys Acta 1368(1):7–18

97. Beutner G, Ruck A, Riede B, Welte W, Brdiczka D (1996) Complexes between kinases, mitochondrial porin and adenylate translocator in rat brain resemble the permeability transition pore. FEBS Lett 396(2–3): 189–195

98. Leung AW, Halestrap AP (2008) Recent progress in elucidating the molecular mechanism of the mitochondrial permeability transition pore. Biochim Biophys Acta 1777(7–8): 946–952. doi:10.1016/j.bbabio.2008.03.009

99. Gutierrez-Aguilar M, Douglas DL, Gibson AK, Domeier TL, Molkentin JD, Baines CP (2014) Genetic manipulation of the cardiac mitochondrial phosphate carrier does not affect permeability transition. J Mol Cell Cardiol 72:316–325. doi:10.1016/j.yjmcc.2014.04.008

100. Kokoszka JE, Waymire KG, Levy SE, Sligh JE, Cai J, Jones DP, MacGregor GR, Wallace DC (2004) The ADP/ATP translocator is not essential for the mitochondrial permeability transition pore. [See comment]. Nature 427(6973):461–465

101. Kwong JQ, Davis J, Baines CP, Sargent MA, Karch J, Wang X, Huang T, Molkentin JD (2014) Genetic deletion of the mitochondrial phosphate carrier desensitizes the mitochondrial permeability transition pore and causes cardiomyopathy. Cell Death Differ 21(8): 1209–1217. doi:10.1038/cdd.2014.36

102. Gunter TE, Sheu SS (2009) Characteristics and possible functions of mitochondrial Ca(2+) transport mechanisms. Biochim Biophys Acta 1787(11):1291–1308. doi:10.1016/j.bbabio.2008.12.011

103. Kinnally KW, Peixoto PM, Ryu SY, Dejean LM (2011) Is mPTP the gatekeeper for necrosis, apoptosis, or both? Biochim Biophys Acta 1813(4):616–622. doi:10.1016/j.bbamcr.2010.09.013

104. Chen C, Ko Y, Delannoy M, Ludtke SJ, Chiu W, Pedersen PL (2004) Mitochondrial ATP synthasome: three-dimensional structure by electron microscopy of the ATP synthase in complex formation with carriers for Pi and ADP/ATP. J Biol Chem 279(30): 31761–31768

105. Bonora M, Bononi A, De Marchi E, Giorgi C, Lebiedzinska M, Marchi S, Patergnani S, Rimessi A, Suski JM, Wojtala A, Wieckowski MR, Kroemer G, Galluzzi L, Pinton P (2013) Role of the c subunit of the FO ATP synthase in mitochondrial permeability transition. Cell Cycle 12(4):674–683. doi:10.4161/cc.23599

106. Alavian KN, Beutner G, Lazrove E, Sacchetti S, Park HA, Licznerski P, Li H, Nabili P, Hockensmith K, Graham M, Porter GA Jr, Jonas EA (2014) An uncoupling channel within the c-subunit ring of the F1FO ATP synthase is the mitochondrial permeability transition pore. Proc Natl Acad Sci U S A 111(29):10580–10585. doi:10.1073/pnas.1401591111

107. Azarashvili T, Odinokova I, Bakunts A, Ternovsky V, Krestinina O, Tyynela J, Saris NE (2014) Potential role of subunit c of F0F1-ATPase and subunit c of storage body in the mitochondrial permeability transition. Effect of the phosphorylation status of subunit c on pore opening. Cell Calcium 55(2): 69–77. doi:10.1016/j.ceca.2013.12.002

108. Karch J, Molkentin JD (2014) Identifying the components of the elusive mitochondrial permeability transition pore. Proc Natl Acad Sci U S A 111(29):10396–10397. doi:10.1073/pnas.1410104111

109. Chinopoulos C, Szabadkai G (2014) What makes you can also break you, part III: mitochondrial permeability transition pore formation by an uncoupling channel within the C-subunit ring of the F1FO ATP synthase? Front Oncol 4:235. doi:10.3389/fonc.2014.00235

110. Carbajo RJ, Kellas FA, Runswick MJ, Montgomery MG, Walker JE, Neuhaus D (2005) Structure of the F1-binding domain of the stator of bovine F1Fo-ATPase and how it binds an alpha-subunit. J Mol Biol 351(4): 824–838. doi:10.1016/j.jmb.2005.06.012

111. Pedersen PL (1994) ATP synthase. The machine that makes ATP. Curr Biol 4(12): 1138–1141

112. Jonckheere AI, Smeitink JA, Rodenburg RJ (2012) Mitochondrial ATP synthase: architecture, function and pathology. J Inherit Metab Dis 35(2):211–225. doi:10.1007/s10545-011-9382-9

113. Walker JE (2013) The ATP synthase: the understood, the uncertain and the unknown. Biochem Soc Trans 41(1):1–16. doi:10.1042/BST20110773

114. Wittig I, Schagger H (2009) Supramolecular organization of ATP synthase and respiratory chain in mitochondrial membranes. Biochim Biophys Acta 1787(6):672–680. doi:10.1016/j.bbabio.2008.12.016

115. Pogoryelov D, Reichen C, Klyszejko AL, Brunisholz R, Muller DJ, Dimroth P, Meier T (2007) The oligomeric state of c rings from cyanobacterial F-ATP synthases varies from

13 to 15. J Bacteriol 189(16):5895–5902. doi:10.1128/JB.00581-07

116. Alavian KN, Beutner G, Lazrove E, Sacchetti S, Park HA, Licznerski P, Li H, Nabili P, Hockensmith K, Graham M, Porter GA Jr, Jonas EA (2014) An uncoupling channel within the c-subunit ring of the F1FO ATP synthase is the mitochondrial permeability transition pore. Proc Natl Acad Sci U S A. doi:10.1073/pnas.1401591111
117. Giorgio V, Bisetto E, Soriano ME, Dabbeni-Sala F, Basso E, Petronilli V, Forte MA, Bernardi P, Lippe G (2009) Cyclophilin D modulates mitochondrial F0F1-ATP synthase by interacting with the lateral stalk of the complex. J Biol Chem 284(49): 33982–33988
118. Giorgio V, von Stockum S, Antoniel M, Fabbro A, Fogolari F, Forte M, Glick GD, Petronilli V, Zoratti M, Szabo I, Lippe G, Bernardi P (2013) Dimers of mitochondrial ATP synthase form the permeability transition pore. Proc Natl Acad Sci U S A 110(15):5887–5892. doi:10.1073/pnas.1217823110
119. Roestenberg P, Manjeri GR, Valsecchi F, Smeitink JA, Willems PH, Koopman WJ (2012) Pharmacological targeting of mitochondrial complex I deficiency: the cellular level and beyond. Mitochondrion 12(1):57–65. doi:10.1016/j.mito.2011.06.011
120. Gomez L, Thibault H, Gharib A, Dumont JM, Vuagniaux G, Scalfaro P, Derumeaux G, Ovize M (2007) Inhibition of mitochondrial permeability transition improves functional recovery and reduces mortality following acute myocardial infarction in mice. Am J Physiol Heart Circ Physiol 293(3):H1654–H1661. doi:10.1152/ajpheart.01378.2006
121. Pedersen PL, Hullihen J (1978) Adenosine triphosphatase of rat liver mitochondria. Capacity of the homogeneous F1 component of the enzyme to restore ATP synthesis in urea-treated membranes. J Biol Chem 253(7):2176–2183
122. Kruse SE, Watt WC, Marcinek DJ, Kapur RP, Schenkman KA, Palmiter RD (2008) Mice with mitochondrial complex I deficiency develop a fatal encephalomyopathy. Cell Metab 7(4):312–320. doi:10.1016/j.cmet.2008.02.004
123. Oberfeld B, Brunner J, Dimroth P (2006) Phospholipids occupy the internal lumen of the c ring of the ATP synthase of *Escherichia coli*. Biochemistry 45(6):1841–1851. doi:10.1021/bi052304+
124. Meier T, Matthey U, Henzen F, Dimroth P, Muller DJ (2001) The central plug in the reconstituted undecameric c cylinder of a bacterial ATP synthase consists of phospholipids. FEBS Lett 505(3):353–356
125. Matthies D, Preiss L, Klyszejko AL, Muller DJ, Cook GM, Vonck J, Meier T (2009) The c13 ring from a thermoalkaliphilic ATP synthase reveals an extended diameter due to a special structural region. J Mol Biol 388(3):611–618. doi:10.1016/j.jmb.2009.03.052
126. Pavlov E, Zakharian E, Bladen C, Diao CT, Grimbly C, Reusch RN, French RJ (2005) A large, voltage-dependent channel, isolated from mitochondria by water-free chloroform extraction. Biophys J 88(4):2614–2625. doi:10.1529/biophysj.104.057281
127. McGeoch JE, McGeoch MW (2008) Entrapment of water by subunit c of ATP synthase. J R Soc Interface 5(20):311–318. doi:10.1098/rsif.2007.1146
128. Hamill OP, Marty A, Neher E, Sakmann B, Sigworth FJ (1981) Improved patch-clamp techniques for high-resolution current recording from cells and cell-free membrane patches. Pflugers Archiv 391(2):85–100
129. Brown MR, Sullivan PG, Dorenbos KA, Modafferi EA, Geddes JW, Steward O (2004) Nitrogen disruption of synaptoneurosomes: an alternative method to isolate brain mitochondria. J Neurosci Methods 137(2): 299–303
130. Brown MR, Sullivan PG, Geddes JW (2006) Synaptic mitochondria are more susceptible to Ca2+ overload than nonsynaptic mitochondria. J Biol Chem 281(17): 11658–11668
131. Chan TL, Greenawalt JW, Pedersen PL (1970) Biochemical and ultrastructural properties of a mitochondrial inner membrane fraction deficient in outer membrane and matrix activities. J Cell Biochem 45(2): 291–305
132. Martinez-Caballero S, Dejean LM, Jonas EA, Kinnally KW (2005) The role of the mitochondrial apoptosis induced channel MAC in cytochrome c release. J Bioenerg Biomembr 37(3):155–164

Chapter 12

Monitoring of Permeability Transition Pore Openings in Mitochondria of Cultured Neurons

Tatiana Brustovetsky and Nickolay Brustovetsky

Abstract

Mitochondrial permeability transition is a process marked by a significant increase in the permeability of the inner mitochondrial membrane that can be initiated by massive Ca^{2+} influx into mitochondria with or without accompanying oxidative stress. The increase in membrane permeability is due to induction of a proteinaceous pore in the inner mitochondrial membrane called the permeability transition pore (PTP). The detrimental role of PTP induction in neuronal mitochondria in various neuropathologies is well documented. There are different methods to study PTP in isolated mitochondria and in live cells. In the present chapter, we describe a method to monitor PTP induction in live cultured neurons based on monitoring changes in mitochondrial matrix pH. Although, this method is not often used, it could be an excellent addition to the arsenal of methodologies utilized in studies of PTP. In addition, we present here detailed descriptions of a method used by us to prepare neuronal cell culture from rat brain tissue and electroporation technique for transfection of cultured neurons.

Key words Permeability transition pore, Mitochondria, Neuron, Calcium, Glutamate, Mitochondrial matrix pH, Fluorescent microscopy

1 Introduction

Mitochondrial permeability transition is a process leading to induction of a gigantic proteinaceous pore in the inner mitochondrial membrane permeable for solutes with a molecular weight of up to 1.5 kDa called the mitochondrial permeability transition pore (PTP) [1]. Rapid and excessive Ca^{2+} accumulation in mitochondria and oxidative stress are the main inducers of the PTP [2]. The PTP can be induced and studied in isolated mitochondria, including brain synaptic and nonsynaptic mitochondria. Studies of PTP induction in isolated mitochondria generated significant amount of information about this process. An induction of the PTP in live cells was much less studied until in 1999, when Petronilli et al. introduced a method for monitoring PTP induction in live cells [3]. This method is based on loading cells with calcein-AM and subsequent entry of Co^{2+} into the cell and quenching of calcein

Stefan Strack and Yuriy M. Usachev (eds.), *Techniques to Investigate Mitochondrial Function in Neurons*, Neuromethods, vol. 123, DOI 10.1007/978-1-4939-6890-9_12, © Springer Science+Business Media LLC 2017

fluorescence. Co^{2+} quenches calcein in the cytosol but not in mitochondria unless mitochondria undergo PTP, thus allowing the mitochondrial matrix to become accessible to Co^{2+}. With this method PTP induction was demonstrated in different cell types, including neuronal cells [4, 5]. The main advantage of this method is that it allows detection of PTP induction in live cells. On the other hand, Co^{2+} inhibits Ca^{2+} uptake by mitochondria and decreases mitochondrial respiration [3] and consequently may interfere with Ca^{2+} signaling and mitochondrial bioenergetics. Recently, Bolshakov and colleagues developed a method based on acidification of mitochondrial matrix following PTP induction [6]. Under normal conditions, mitochondrial matrix is more alkaline than extra-mitochondrial space. Expression of mitochondrially targeted pH-sensitive enhanced yellow fluorescent protein (mito-eYFP) allows monitoring of changes in mitochondrial matrix pH. An induction of the PTP pore permeabilizes the inner membrane and results in H^+ influx into mitochondria and matrix acidification [6]. An increase in cytosolic Ca^{2+} induced by exposure to excitotoxic glutamate produces biphasic acidification of the mitochondrial matrix. The first phase is due to fast cytosolic acidification most likely produced by protons generated by activated plasma membrane Ca^{2+}-ATPase which translates into a decrease of mitochondrial matrix pH. The second phase is sensitive to cyclosporin A (CsA), an inhibitor of PTP, and therefore reflects opening of the PTP [6]. In our previous study, we demonstrated delayed onset of mitochondrial matrix acidification in neurons derived from cyclophilin D-knockout mice [7]. In this study, we simultaneously monitored changes in cytosolic Ca^{2+} and mitochondrial matrix pH by following the ratio of fluorescence signals of Fura-2FF excited at 340 and 380 nm as well as fluorescence of mito-eYFP, respectively. The delayed onset of matrix acidification correlated with deferred onset of Ca^{2+} dysregulation in neurons and due to PTP's dependence on cyclophilin D, a major regulatory component of the PTP [8, 9], was attributed to inhibition of PTP induction. Thus, this method appeared to be suitable for monitoring PTP induction in live cultured neurons. In this chapter, we describe methods for neuronal plating, neuronal transfection with electroporation technique, and fluorescent microscopy applied to cultured hippocampal neurons expressing mito-eYFP and loaded with calcium-sensitive dye Fura-2FF.

2 Materials

2.1 Surgical Tools and Supplies for Neuronal Isolation

The following items have to be prepared before brain dissection and neuronal isolation:

1. Autoclaved set of surgical tools: medium size 150 mm scissors, small size 110 mm scissors, two pairs of #5 forceps, round-end forceps, and dual-end spatula.

2. Six sterile 15 ml plastic test tubes.
3. 35 mm glass-bottomed petri dishes coated with rat tail collagen, Type 1 (BD Biosciences) and poly-L-lysine (Sigma).
4. Small plastic beaker with 50 ml of 70% ethanol.
5. 20 ml syringe.
6. 0.22 μm nitrocellulose membrane filter (Millipore) for sterilization of solutions.
7. Plastic bag for animal remains.

2.2 Solutions

The following solutions have to be prepared before the procedure:

1. L-15 solution (Life Sciences, Cat # 41300–039) at room temperature.
2. MEM solution (Life Sciences, Cat # 11430–030) supplemented with 10% NuSerum (BD Biosciences, Bedford, MA, Cat #355500) and 27 mM glucose and 26 mM $NaHCO_3$ (MEM+NuSerum solution) at room temperature.
3. 0.02% BSA solution, by adding 3 mg BSA (Sigma, Cat. #A1470) into 15 ml of L-15 solution.
4. 0.1% Papain solution, by adding 3 mg papain (Worthington, Lakewood NJ) into 3 ml of BSA solution and place in the incubator at 37 °C until dissection is complete.
5. 4% BSA solution, by adding 400 mg BSA (Sigma, Cat. # A1470) into 10 ml L-15 solution.

3 Methods

3.1 Brain Dissection and Neuronal Culture Preparation

This method is based on the procedure published previously [10] with some modifications.

1. Put 1.5 ml of 0.02% BSA solution into 35 mm petri dishes. Prepare four petri dishes: one for rinsing the brain, one for dissecting hippocampus, and two for rinsing the dissected hippocampus and rough mincing of hippocampus.
2. Decapitate a pup (1 day postnatal), remove brain from scull using spatula and round end forceps and then rinse in the first petri dish with 0.02% BSA solution and transfer into the second petri dish. For hippocampal culture, dissect hemispheres, remove meninges, scoop out hippocampus with #5 forceps and put into the third petri dish. Perform dissection under binocular microscope (we use Olympus C011). Continue until all pups are sacrificed.
3. Now, the third petri dish contains all dissected hippocampi. Remove remaining meninges, transfer cleaned tissue into the fourth petri dish and mince with small 110 mm scissors.

4. Sterilize 3 ml of 0.1% papain solution by passing through 0.22 μm filter into a fresh 15 ml plastic test tube and transfer minced tissue into this tube. Incubate for 30 min at 37 °C.
5. Discard from the tube as much of the papain solution as possible. Gently triturate with 2 ml of MEM+NuSerum solution at room temperature using fire-polished 145 mm Pasteur pipette. Use three pipettes with decreasing tip sizes. Triturate for 10 min with each pipette. At the end of trituration the solution should be cloudy. Let the tube stay for a few minutes at room temperature and to allow tissue chunks to settle. Transfer cloudy supernatant from the tube to another fresh one (collecting tube). Add 2 ml of fresh MEM + NuSerum solution at room temperature to remaining chunks of tissue and triturate again until all tissue is dissociated.
6. To remove tissue debris, layer cell suspension on the top of 4% BSA solution in the 15 ml plastic centrifuge tube. Spin at 1800 rpm for 5 min in Eppendorf centrifuge 5702, bucket rotor A-4-38, (500 × *g* for 5 min) at room temperature. Tissue debris will be in the BSA layer, neurons and glial cells will be in the pellet on the bottom of the tube. Discard supernatant and BSA layer.
7. Resuspend cells in 1 ml of MEM+NuSerum solution. Count cells in the aliquot of cell suspension using hemocytometer. Calculate volume of cell suspension to add to each petri dish for the final cell density of 200,000 cells per dish.
8. Plate cells by adding an aliquot of cell suspension to the center of a fresh collagen plus poly-L-lysine-coated glass-bottomed 35 mm petri dish filled with 1.5 ml MEM + NuSerum solution.
9. Next day, add 15 μl of FdU stock solution (uridine, 35 μg/ml (143 μM), plus 5-fluoro-2′-deoxyuridine, 15 μg/ml (61 μM)) into each 35 mm petri dish with plated cells filled with 1.5 ml of MEM + NuSerum solution. FdU is used to inhibit proliferation of non-neuronal cells, chiefly, microglia cells.
10. Neuronal cultures are maintained in a humidified air plus 5% CO_2 atmosphere at 37 °C in MEM+NuSerum solution.

3.2 Transfection of Primary Hippocampal Neurons Using Electroporation

To measure matrix pH (pH_m) in mitochondria within live cells, hippocampal neurons are transfected in suspension during plating with a plasmid encoding mito-eYFP (generously provided by Dr. Roger Tsien, University of California, San Diego) using electroporator BTX 630 ECM (Harvard Apparatus, Holliston, MA).

1. After removal of debris (centrifugation at 1800 rpm for 5 min), resuspend cell pellet in 0.1 ml PBS for the final cell density 2.5×10^6 cells per 0.1 ml. Add 4 μg DNA and transfer mixture into an electroporation cuvette with a 2 mm gap (Spectrocell, Oreland, PA, Cat # EL-102).

2. Electroporation is performed at room temperature in LV mode of BTX630 ECM electroporator at 125 V, capacitance 480, resistance 25.
3. After electroporation, transfer suspension into a sterile 1.5 ml Eppendorf tube filled with 1 ml MEM+NuSerum. Plate cells by an adding aliquot of cell suspension from the Eppendorf tube to the center of fresh collagen plus poly-L-lysine-coated glass-bottomed 35 mm petri dish filled with 1.5 ml MEM + NuSerum solution with the final cell density of 200,000 cells per dish.

This procedure usually provides 10–15% transfection efficacy, depending on plasmid quality, compared to <1% efficacy with commercial cationic lipid liposomes applied to primary neuronal cultures. Neurons can be imaged at 5–12 days after transfection. Figure 1 shows typical bright-field (A) and fluorescence images (B) of cultured hippocampal neurons expressing mito-eYFP at 12 DIV. Figure 1c shows 3D reconstruction of mitochondrial network in a live hippocampal neuron expressing mito-eYFP.

3.3 Fluorescence Microscopy, Calcium Imaging, and Mitochondrial Matrix pH Measurements

To monitor PTP openings in live neurons, we follow glutamate-triggered Ca^{2+}-dependent mitochondrial matrix acidification by measuring fluorescence of mito-eYFP.

1. In these experiments we use a Nikon Eclipse TE2000-U inverted microscope equipped with an oil-immersion Nikon CFI SuperFluor 40× 1.3 NA objective and a Photometrics cooled CCD camera CoolSNAP_{HQ} (Roper Scientific, Tucson, AZ) controlled by MetaMorph 6.3 software (Molecular Devices, Downingtown, PA).
2. Neurons are perfused using ValveBank 8 perfusion system (AutoMate Scientific, San Francisco, CA). The standard bath solution contains 139 mM NaCl, 3 mM KCl, 0.8 mM $MgCl_2$, 1.8 mM $CaCl_2$, 10 mM HEPES, pH 7.2, 5 mM glucose.
3. Neurons are illuminated sequentially at 340 and 380 nm to excite Fura-2FF, and 480 nm to excite mito-eYFP. The excitation filters (340 ± 5, 380 ± 7, and 480 ± 20 nm) are controlled by a Lambda 10-2 optical filter changer (Sutter Instruments, Novato, CA). The excitation light at 480 nm is usually attenuated to 10% with a neutral density filter. Fluorescence is recorded through a 505 nm dichroic mirror at 535 ± 25 nm by a CCD camera CoolSNAP_{HQ} (Photometrics, Tuscon, AZ).
4. To minimize photobleaching, take the images every 15 s during the time-course of the experiment using the minimal exposure time (50 ms) that provides acceptable image quality.
5. Matrix pH (pH_m) and cytosolic Ca^{2+} measurements are performed as described previously [6]. Cultured neurons, including

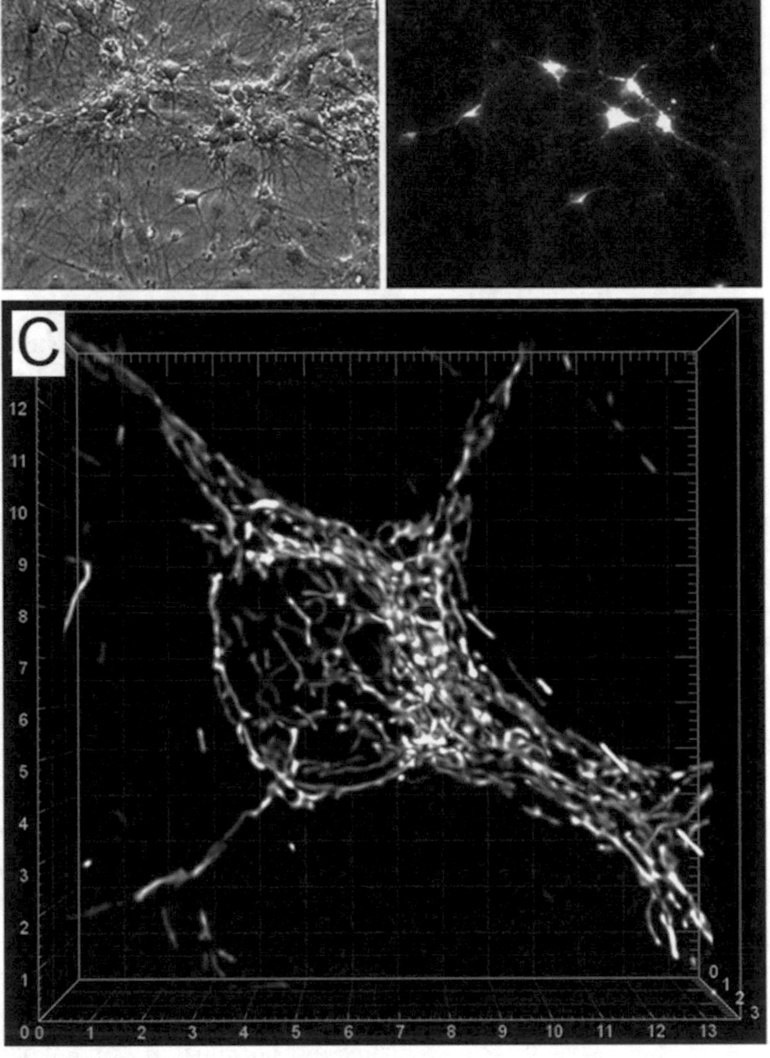

Fig. 1 Typical bright-field (**a**) and fluorescence images (**b**, **c**) of cultured hippocampal neurons (10 days in vitro) expressing mito-eYFP. In C, 3D reconstruction of mitochondrial network in live hippocampal neuron expressing mito-eYFP. (Republished with permission of Academic Press from Li et al. (2009) Role of cyclophilin D-dependent mitochondrial permeability transition in glutamate-induced calcium deregulation and excitotoxic neuronal death. Exp Neurol 218:171–182; permission conveyed through Copyright Clearance Center, Inc.)

those expressing mito-eYFP, are loaded withFura-2FF to follow pH_m and changes in cytosolic Ca^{2+} simultaneously.

6. Prepare 1 mM stock solution of Fura-2FF (Molecular Probes or Teflabs, 50 μg Fura-2FF/50 μl DMSO); keep it at −20 °C.

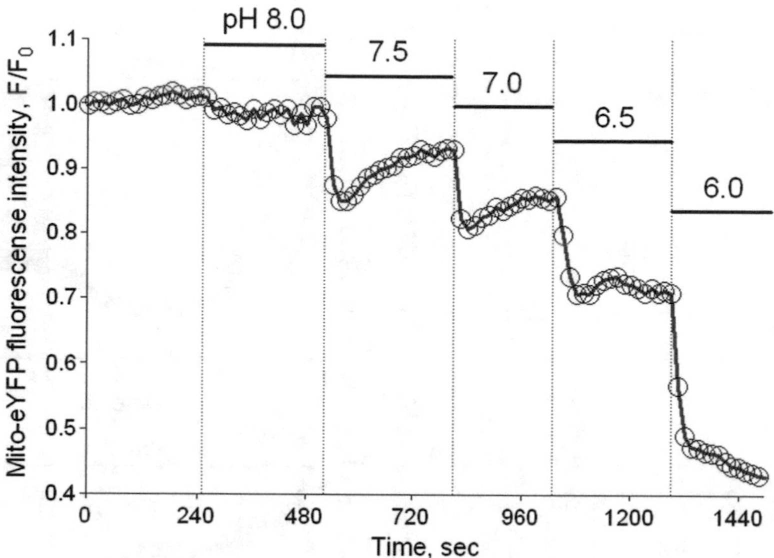

Fig. 2 The pH sensitivity of mito-eYFP expressed in mitochondria of hippocampal neurons. pH of perfusion solutions used for calibration is shown near the trace. pH was adjusted by adding NaOH or HCl

Before the experiment, mix 4 μl of Fura-2FF stock solution with 1 μl of 20% Pluronic-F127 (Sigma, Cat # 2443). Add 2.6 μM of Fura-2FF mixed with Pluronic-F127 to neurons and incubate for an hour at 37 °C.

7. Rinse neurons twice with a standard bath solution. Monitor the changes in $[Ca^{2+}]_c$ by following Fura-2FF fluorescence ratio F_{340}/F_{380} calculated after subtracting the background from both channels. Monitor the changes in mito-eYFP fluorescence as F/F_0 and then after background subtraction converted into pH_m using pH_m calibration procedure [6].

8. pH_m calibration is performed with the use of the following calibration solution [6]: 5 μM nigericin, 1 μM FCCP, 134 mM K-gluconate, 1 mM $MgCl_2$, and 20 mM HEPES. pH in calibration solutions is adjusted by the addition of HCl or KOH. Perfuse neurons sequentially with perfusion solutions with different pH and record mito-eYFP fluorescence. The pH sensitivity of mito-eYFP is shown in Fig. 2. According to these measurements, pH in the mitochondrial matrix is 8.0 ± 0.2.

9. Figure 3 shows simultaneous recordings of cytosolic Ca^{2+} monitored with Fura-2FF (red traces) and matrix pH (pH_m) followed with mito-eYFP (black traces with symbols) in cortical neurons from C57BL/6 mice (A) and cyclophilin D-knockout *Ppif–/–* mice (B, C).

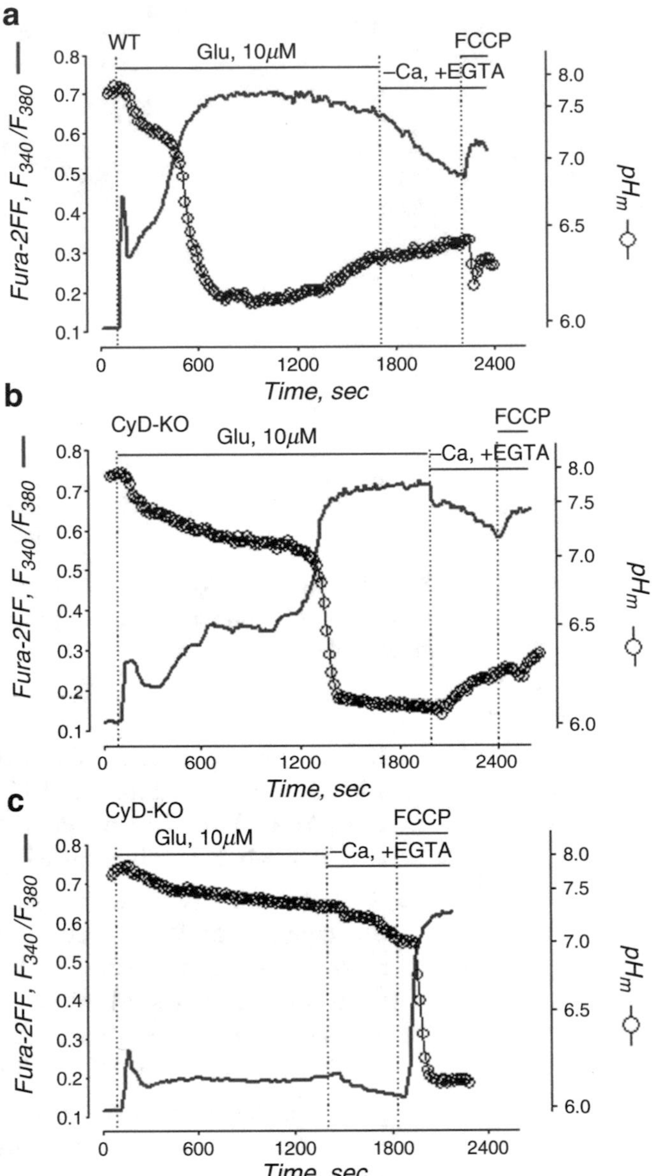

Fig. 3 Simultaneous recordings of cytosolic Ca^{2+} monitored with Fura-2FF (red traces without symbols) and matrix pH (pH_m) followed with mito-eYFP fluorescence (black traces with symbols). In (**a**), recordings from cortical neuron derived from wild-type C57BL/6 mouse. In (**b**) and (**c**), cortical neurons from cyclophilin D-knockout *Ppif–/–* mice. Where indicated, 10 μM glutamate plus 10 μM glycine (Glu), was applied. At the end of experiments, glutamate, glycine, and Ca^{2+} were removed from the bath solution and 1 μM FCCP was added to depolarize mitochondria (Republished with permission of Academic Press from Li et al. (2009) Role of cyclophilin D-dependent mitochondrial permeability transition in glutamate-induced calcium deregulation and excitotoxic neuronal death. Exp Neurol 218:171–182; permission conveyed through Copyright Clearance Center, Inc.)

4 Notes

4.1 Preparation of Neuronal Culture

1. We use the described above method (Sect. 3.1) to plate neurons together with astrocytes. The FdU treatment kills microglia, but does not eliminate astrocytes. Since a mixture of neurons with astrocytes represents more physiological neuronal model, we prefer to use coculture of neurons and astrocytes. However, if more pure neuronal culture is desired, embryonic day 18 animals should be used and 5 μM cytosine arabinofuranoside should be applied to cells on day 4 after plating [11].
2. We do not change MEM+NuSerum solution during cell culturing for 14–16 days.
3. We successfully use this method for preparing neuronal cell cultures from rat and mouse cortices and striata.

4.2 Transfection of Primary Neurons Using Electroporation

1. We successfully use the described above method (Sect. 3.2) to transfect cultured neurons from rat and mouse cortices and striata.
2. Neurons are the predominant cell type transfected by this method; transfection of astrocytes by this method is a rare event.
3. With time after transfection mito-eYFP fluorescence in neurons can dim. In this case, one can chose to image neurons at earlier ages (e.g., 8–10 DIV).

4.3 Ca^{2+} Imaging and Mitochondrial Matrix pH Measurements

1. PTP opening can be better detected when a majority of mitochondria in the cell respond to elevated cytosolic Ca^{2+} synchronously. This can be achieved with higher glutamate concentrations applied to neurons. The optimal glutamate concentration depends on species from which neurons have been derived. Usually, lower glutamate concentrations produce faster sustained elevation in cytosolic Ca^{2+} (Ca^{2+} dysregulation) in mouse cultured neurons compared with neurons from rats.
2. With increasing glutamate concentrations larger Ca^{2+} influx can trigger CsA-insensitive PTP and, consequently, sensitivity of mitochondrial acidification to CsA can decline.
3. All solutions used in experiments with cultured neurons have to be tested for osmolarity. We measure osmolarity of the solutions with the osmometer Osmette II™ (Precision Systems Inc., Natick, MA). Use sucrose (20–65 mM) in the standard bath solution to maintain osmolarity similar to that in the growth medium (300–340 mosm depending on duration of cell culturing).

Acknowledgment

This work was supported by Indiana Spinal Cord and Brain Injury Research Fund grant and NIH/NINDS grant R01 NS078008 to N.B.

References

1. Bernardi P, Krauskopf A, Basso E et al (2006) The mitochondrial permeability transition from in vitro artifact to disease target. FEBS J 273:2077–2099
2. Bernardi P, Rasola A (2007) Calcium and cell death: the mitochondrial connection. Subcell Biochem 45:481–506
3. Petronilli V, Miotto G, Canton M et al (1999) Transient and long-lasting openings of the mitochondrial permeability transition pore can be monitored directly in intact cells by changes in mitochondrial calcein fluorescence. Biophys J 76:725–734
4. Gillessen T, Grasshoff C, Szinicz L (2002) Mitochondrial permeability transition can be directly monitored in living neurons. Biomed Pharmacother 56:186–193
5. Quintanilla RA, Jin YN, von BR et al. (2013) Mitochondrial permeability transition pore induces mitochondria injury in Huntington disease. Mol Neurodegener 8:45
6. Bolshakov AP, Mikhailova MM, Szabadkai G et al (2008) Measurements of mitochondrial pH in cultured cortical neurons clarify contribution of mitochondrial pore to the mechanism of glutamate-induced delayed Ca(2+) deregulation. Cell Calcium 43:602–614
7. Li V, Brustovetsky T, Brustovetsky N (2009) Role of cyclophilin D-dependent mitochondrial permeability transition in glutamate-induced calcium deregulation and excitotoxic neuronal death. Exp Neurol 218:171–182
8. Schinzel AC, Takeuchi O, Huang Z et al (2005) Cyclophilin D is a component of mitochondrial permeability transition and mediates neuronal cell death after focal cerebral ischemia. Proc Natl Acad Sci U S A 102:12005–12010
9. Baines CP, Kaiser RA, Purcell NH et al (2005) Loss of cyclophilin D reveals a critical role for mitochondrial permeability transition in cell death. Nature 434:658–662
10. Dubinsky JM (1993) Intracellular calcium levels during the period of delayed excitotoxicity. J Neurosci 13:623–631
11. Polster BM, Nicholls DG, Ge SX et al (2014) Use of potentiometric fluorophores in the measurement of mitochondrial reactive oxygen species. Methods Enzymol 547: 225–250

Chapter 13

Protocols for Assessing Mitophagy in Neuronal Cell Lines and Primary Neurons

Ruben K. Dagda and Monica Rice

Abstract

Mitochondria are organelles that regulate essential eukaryotic functions including generating energy, sequestering excess calcium, and modulating cell survival. In order for neurons to thrive, mitochondria have to be continuously replenished by maintaining autophagic-lysosomal mediated degradation of mitochondria (mitophagy) and mitochondrial biogenesis. While a plethora of image- and biochemical-based techniques have been developed for measuring autophagy (macroautophagy) in eukaryotic cells, the molecular toolbox for quantifying and assessing mitophagy in neurons continues to evolve. Compared to proliferating cells, quantifying mitophagy in neurons poses a technical challenge given that mitochondria are predominantly present in neurites (axons and dendrites) and are highly dynamic.

In this chapter, we provide a brief overview on mitophagy and provide a list of validated fluorescence- and biochemistry-based techniques used for assessing mitophagy in neuronal cells and primary neurons. Secondly, we provide comprehensive guidelines for interpreting steady-state levels of mitophagy and mitophagic flux in neurons using modern fluorescence- and biochemistry-based techniques. Finally, we provide a comprehensive list of common pitfalls to avoid when assessing mitophagy and offer practical solutions to overcome technical issues.

Key words Autophagy, Mitophagy, GFP-LC3, RFP-LC3, Lysosome, Electron microscopy, Mitochondrial trafficking, Oxidative stress, Lysosome, Neurites

1 Introduction

Mitochondria are the main energy hubs of eukaryotic cells. To produce energy in the form of adenosine triphosphate (ATP), mitochondria employ oxygen, water, tricarboxylic cycle intermediates, and the electron transport chain. Other physiological functions of mitochondria include regulating cell death pathways that converge at the outer mitochondrial membrane (OMM), lipid synthesis (e.g., cardiolipin, steroids and phosphatidylethanolamine), sequestering calcium, heat production, and generating and sequestering reactive oxygen species (ROS).

The mitochondrion is a highly specialized organelle divided into four subcompartments: the OMM, the intermembrane space

Stefan Strack and Yuriy M. Usachev (eds.), *Techniques to Investigate Mitochondrial Function in Neurons*, Neuromethods, vol. 123, DOI 10.1007/978-1-4939-6890-9_13, © Springer Science+Business Media LLC 2017

(IMS), inner mitochondrial membrane (IMM), and the matrix. The OMM has emerged as a crucial platform by which a myriad of protein signaling pathways converge to regulate mitochondrial structure, metabolism, and oxidative phosphorylation [1–5].

There are several mitochondrial quality control and repair pathways that serve to maintain mitochondrial structure and function. At the OMM, oxidatively damaged proteins are targeted for localized degradation by the ubiquitin–proteasome system. In addition, the OMM serves as the "dumping ground" for IMS/IMM-localized proteins that were shuttled to the OMM so they can be tagged for degradation by E3 ubiquitin ligases [6]. The mitochondrial matrix has intrinsic quality control mechanisms for sensing and degrading damaged or unfolded proteins. Several proteases, including LON1, sense and proteolytically degrade damaged matrix-localized proteins. Therefore, the turnover of damaged matrix-localized proteins permits the efficient de novo import of new mitochondrial proteins [7].

On the other hand, when the whole organelle is extensively damaged, through exposure to ROS or through normal aging, mitochondria are sequestered by autophagic vacuoles (AVs) and targeted for lysosomal degradation through a catabolic process termed mitophagy.

Autophagy, or also known as macroautophagy, is a catabolic process by which oxidatively damaged and aged organelles are directed for lysosome-mediated degradation by acid hydrolases (reviewed in [8, 9]). Autophagy is a highly coordinated physiological process requiring several steps orchestrated by autophagy-related proteins that were initially identified in yeast as ATG proteins. Autophagy is initiated by the formation of isolation membranes (phagophores) that are derived from the endoplasmic reticulum (ER), mitochondria, and other organelles [10, 11] and are subsequently molded into double membrane organelles termed AVs. Microtubule-associated protein 1A/1B-light chain 3 (LC3) is a bona fide marker of AVs and the mammalian homolog of ATG8. ATG8 is posttranslationally processed by the protease ATG4 and it is subsequently C-terminally lipidated by ATG7 through the addition of phosphatidylethanolamine in a C-terminal glycine (G120 in the rat cDNA). This posttranslational modification permits ATG8/LC3 to physically associate with AVs. In cooperation with ATG8, the association of ATG5-ATG12 complexes—as formed through the sequential conjugation of ATG12 to ATG5—regulate the expansion, remodeling, and the successful closure of nascent AVs [12]. While ATG5-ATG12 complexes dissociate from the outer membrane of AV during the maturation phase of AVs, LC3 remains bound. The mature AVs eventually fuse with lysosomes to deliver the cargo for degradation by acidic hydrolases, a physiological process which allow for the eventual recycling of necessary nutrients in eukaryotic cells [8].

In mammalian tissues, basal autophagy is critical for development, differentiation of tissues (e.g., neurons, lymphocytes, and adipocytes), and for maintaining immunity [13, 14]. On the other hand, stimulated autophagy, such as induced by starvation, is critical for providing the necessary energy to allow a stressed eukaryotic cell to thrive. Autophagy was once believed to be dispensable in the central nervous system (CNS) given that neurons have privileged access to nutrients in vivo. However, this concept was challenged 7 years ago when two landmark studies demonstrated that specific gene deletions of ATG5 and ATG7 produce widespread degeneration in the brain. More importantly, these seminal studies suggested that autophagy predominantly plays a neuroprotective developmental role in the CNS [15, 16]. In other instances, autophagy is a double-edged sword that can be either beneficial or detrimental. During physiological conditions, autophagy contributes to the removal of large protein aggregates when the ubiquitin–proteasome system is overwhelmed. During neuronal development, autophagy participates in the pruning and remodeling of neurites [13, 17, 18]. On the other hand, high chronic levels of oxidative stress as induced by many neurodegenerative diseases—including amyotrophic lateral sclerosis (ALS), Alzheimer's disease (AD), Huntington's disease (HD), prion-related diseases (PRD), and Parkinson's disease (PD)—lead to early dysregulation of autophagy [13, 19–23]. While physiological mitophagy, as induced by starvation or by treating cells with the mTOR inhibitor rapamycin, permits the efficient recycling of nutrients for the formation of new mitochondria, pathological stimuli that tend to overactivate mitophagy in the absence of mitochondrial biogenesis—a concept coined "autophagic stress"—promotes neurodegeneration [24].

Once thought to be an unregulated physiological process, the signaling pathways and molecular players that govern mitophagy are beginning to be unveiled. In yeast, 36 autophagy-related genes (ATG) and 41 other non-ATG proteins initially identified to regulate autophagy. So far, only a few yeast proteins (Atg32 and AUP1P) are known to regulate mitophagy [25–27]. Three types of mitophagy have been well characterized in eukaryotic cells. In yeast, the OMM-localized Atg32p interacts with the Atg11p/Atg8p complex, leading to the recognition and engulfment of dysfunctional mitochondria by AVs. In mammalian cells, PINK1 and Parkin, two enzymes whose mutations are associated with familial forms of PD, link the ubiquitin–proteasome and autophagy machineries to promote mitophagy of severely damaged mitochondria [28]. In red blood cells, the OMM-localized protein NIX associates with LC3 to initiate mitophagy and allows for the maturation of red blood cells by eliminating mitochondria. In summary, mitophagy is now regarded as a highly selective physiological process that requires specific signals to trigger the highly coordinated temporal and spatial recruitment of a set of intermediate players and mitophagy effectors to the OMM [29–34].

In neurons, some mitochondrial toxins can concomitantly activate autophagy and mitophagy. For instance, treating neuronal cells or primary midbrain dopaminergic neurons with the Parkinsonian toxins 1-methyl-4-phenylpyridinium (MPP+), 6-hydroxydopamine (6-OHDA), or rotenone not only activates macroautophagy, but also promotes mitophagy in a concentration and time-dependent manner [30, 31, 35]. Other toxins such as staurosporine can also enhance the flux of mitochondria through the lysosomes (mitophagic flux) [30]. Recently, one landmark study showed that the externalization of cardiolipin (CL) from the IMM to the OMM leads to the recruitment of ATG8/LC3 to mitochondria by associating with CL. The externalization of CL in neurons appears to be a generalized mechanism for initiating mitophagy as induced by multiple mitochondria-directed toxins and apoptotic stimuli, a physiological process that does not require the recruitment of PINK1 and Parkin [30].

While neuronal mitophagy has been predominantly characterized in vitro (e.g., primary neurons), it is worth noting that a few protocols have been recently developed for analyzing mitophagy in vivo and ex vivo (e.g., brain slices) [36]. However, mitophagic flux ex vivo is challenging to assess given the poor permeation of reagents to manipulate autophagic flux (e.g., bafilomycin, chloroquine, and leupeptin) into tissue slices. Moreover, studying mitophagy in vivo requires sophisticated equipment to fluorescently monitor mitophagy. Moreover, there is a lack of reagents to induce flux in vivo. Given these aforementioned issues, this chapter provides a comprehensive list of fluorescent- and biochemical-based protocols for interpreting mitophagy in cultured neuronal cells and in primary neurons. Finally, we provide a thorough set of guidelines for interpreting mitophagy in neurons and troubleshooting various technical issues.

Key concepts: There are several key words and concepts to bear in mind in order to fully comprehend the mitophagy techniques presented in this chapter. **Steady-state autophagy** refers to rate of formation and turnover of AVs under physiological conditions whereas **induced autophagy** refers to the increased steady-state autophagy beyond baseline levels that occurs in response to an exogenous stimulus, chronic or acute stress. **Flux** is the fraction of the total pool of AVs that fuse with lysosomes to promote autophagic-lysosomal mediated degradation of cargo. **Mitophagic flux** is defined as the fraction of the total pool of AVs containing mitochondria that fuse with lysosomes to promote their degradation.

2 Materials

2.1 Key Fluorescently Tagged Reagents

Due to its intrinsic ability to bind to early AVs, ATG8/LC3 is a considered a bona fide marker for autophagy. Unlike Atg5-Atg12 complexes, the fact that LC3 remains associated to AVs until it is degraded by acid hydrolases in an autolysosome (autophagosomes fused to lysosomes) makes it an ideal AV marker [8]. Moreover, ever since the GFP fusion chimera of LC3 (GFP-LC3) was engineered by Dr. Tomatsu Yoshimori [37], this popular molecular biology reagent has been used by neurobiologists to monitor early AVs in live and fixed cells and for assessing autophagic flux as a tandem reporter version of LC3 [36]. Whereas the GFP moiety is labile in low pH, the RFP fusion chimeric version of LC3 (RFP-LC3) specifically labels autolysosomes due to the higher fluorescence stability of RFP inside the acidic environment of lysosomes [38]. Indeed, the tandem reporter construct which expresses a version of LC3 containing both fluorophores facilitates the analyses of autophagic flux in eukaryotic cells [38]. Overall, the ability to use the three fluorescent versions of LC3 gives the neurobiologist powerful reagents to quantify the initiation (GFP-LC3), progression (RFP-LC3), and flux (tandem reporter construct) using relatively simple, but powerful fluorescent image-based techniques.

2.2 Cell Culture

1. SH-SY5Y, a human dopaminergic neuroblastoma cell line, can be purchased from American Type Culture Collection (Rockville, MD), and primary cortical neurons are derived from embryos of 14–16 day-timed pregnant C57/BL6 mice (Charles River, Wilmington, MA) as previously described [39]. These cells (low passage number) are grown in Dulbecco's modified Eagle's medium (DMEM) with 4.5 g/l D-glucose (Life Technologies, Cat#10313021) and supplemented with the following additional components: 10% heat inactivated fetal bovine serum (SIGMA, St. Louis, MO, Cat#F2442), 10 mM HEPES, and 2 mM glutamine. Primary mouse cortical neurons are seeded in Neurobasal® medium (Life Technologies, Cat# 21103-049) containing 2% B27™ supplement (50× stock, Life Technologies, Cat#175044, stored at −20 °C until ready to use), 0.5% FBS and 2.0 mM L-glutamine/L-alanyl-L-glutamine (GlutaMAX, Thermo Fisher Scientific, Cat #35050061), preferably without serum or antibiotics. Three days after plating, 2/3 of the seeding media is replenished with maintenance media (same NB/B27 media as described above but without FBS to reduce growth of glia).
2. To study the effects of a stimulus on mitophagy in differentiated neuronal cells, SH-SY5Y cells can be differentiated with retinoic acid (RA) (10 μM; Sigma, Cat#R2625) or with dibutyryl cyclic

AMP (db-cAMP) (250 μM, Sigma, Cat# D0627) for 3–5 days. Stock solutions of 10 mM retinoic acid are prepared in DMSO (Sigma, St. Louis, MO, USA, Cat#D8418), and stored at −20 °C as 20–50 μl aliquots for up to 4 months or at 4 °C for 1 month and away from light. Db-cAMP (Sigma, St. Louis, MO, USA) is prepared as a 10 mM stock concentration in water and stored as 100 μl aliquots in −20 °C.

3. Tissue culture ready poly-L-lysine (0.01%, sterile-filtered) (Sigma, Cat# P-4707). Alternatively, poly-D-lysine hydrobromide (Sigma, Cat# P-6407) stocks can be prepared at 2 mg/ml (filter-sterilized) to coat chambered cover glasses at a final concentration of 150 μg/ml for 1 h prior to washing with water three times. Although poly-D-lysine is less expensive to purchase, it tends to degrade at a faster rate than poly-L-lysine which is stable at room temperature (RT).
4. 6-hydroxydopamine (1 mM stock, 6-OHDA; Sigma, Cat# H4381), freshly prepared in distilled water or media. Store powder stock in −20 °C in a desiccator jar and keep away from light. Use 6-OHDA at 75 μM (LD50) for 4 h in SH-SY5Y cells as a control to induce autophagy/mitophagy.
5. Rotenone (1 mM stock, Sigma, Cat# R8875), freshly prepared in water prior to each use in cells. Rotenone can be used at 1 μM for SH-SY5Y cells and at 100–250 nM for primary cortical neurons for 4–6 h to induce autophagy/mitophagy.
6. Staurosporine (1 mM ready-made solution stock, Sigma, Cat# S5921), can be used at 1 μM for 4 h for SH-SY5Y cells or at 100 nM for 4 h for primary cortical neurons to induce mitophagy.
7. E64-D (10 mM, Calbiochem, San Diego, CA, Cat# CAS 66701-25-5), stocks made in DMSO and stored at −20 °C. Use at 10 μM for 2 h for SH-SY5Y cells and primary cortical neurons for flux studies.
8. Pepstatin-A (25 mM, Calbiochem, Cat# 516481), dissolved in methanol or DMSO. Use at 25 μM for SH-SY5Y cells for 2–4 h for flux studies.
9. Bafilomycin A1 (10 μM, Sigma, Cat# B1793), dissolved in DMSO as a 32.1 mM stock solution. Use at 10 nM for 4 h for both SH-SY5Y cells and primary cortical neurons for flux studies.
10. Chambered Lab-Tek II cover glasses (#1.5 German borosilicate; Nalge Nunc International, Naperville, IL, USA, Cat# 155382). Alternatively, 35 mm pie-sectioned cell cultured dishes with glass bottoms (Cellview™, Greiner, Germany, Cat #627870), can also be used for analyzing autophagy/mitophagy by confocal microscopy. We have found that using 4-well

chambered Lab-Tek II cover glasses are better suited for running time-course experiments and for performing immunocytochemistry whereas Cellview™ culture dishes are optimal for confocal microscopy.

2.3 Electron Microscopy

1. 0.1 M phosphate buffered saline (PBS), pH 7.4.
2. 2.5% glutaraldehyde (Sigma, Cat# G5882) in 0.1 M PBS, pH 7.4, frozen as 30 ml aliquots and maintained in −20 °C or at 4 °C for 2 months.
3. 1% osmium oxide (Sigma, Cat# 419494) in 0.1 M PBS.
4. 2% uranyl acetate solution (Polysciences, Warrington, PA, Cat# 21447-25), prepared in distilled water. Stored at 4 °C.
5. 1% lead citrate solution (Sigma, Cat# 15326), prepared in distilled water and can be stored for 3–6 months at 4 °C.
6. Polybed 812 epoxy resin (Polysciences, Warrington, PA, Cat# 08792-1).
7. Phillips CM10 transmission electron microscope or equivalent electron microscope.
8. 15 mm cover glasses (German glass, #1 thickness, Electron Microscopy Sciences, Hatfield, PA, Cat# 72228-01).

2.4 Fluorescence Imaging-Based Analysis of Mitophagy

1. MitoTracker Red dye 580 (1 mM, MTR, 50 μg lyophilized pellet; Molecular Probes, Life Technologies, Cat# M22425). Pellet is dissolved in DMSO and stored at −20 °C.
2. Tetramethylrhodamine methyl-esther (10 mM, TMRM, powder, Sigma, Cat# T5428) is dissolved in DMSO as 10 mM stocks and frozen down as 20 μl aliquots at −20 °C. Use at 40 nM to stain mitochondria in SH-SY5Y cells and primary neurons.
3. MitoTracker Green FM dye (1 mM, Molecular Probes, Life Technologies, Cat# M7514) the pellet is dissolved as a 1 mM stock solution, and stored in small aliquots at −20 °C. Use at 250 nM to stain mitochondria in SH-SY5Y cells and primary cortical neurons.
4. LysoTracker Red DND-99 (1 mM, Molecular Probes, Life Technologies, Cat# L7528) is pre-prepared as a 1 mM cell culture ready stock solution, and stored in small aliquots at −20 °C. Use at 100 nM to stain lysosomes for both SH-SY5Y cells and primary neurons.
5. FluoView 1000 (Olympus America) or a Zeiss LSM 510 Meta laser-scanning confocal microscope (Carl Zeiss MicroImaging, Thornwood, NY) is used for imaging.
6. Mitochondria-targeted GFP (Mountain View, CA) or mtDsRed2 (CloneTech, MA, USA, Cat# 632421) is used to visualize mitochondria. GFP-LC3 (Addgene #11546, author: Dr. Tomatsu Yoshimori, Research Institute of Microbial Diseases,

Osaka, Japan) and RFP-LC3 (AddGene #21075, author: Dr. Tomatsu Yoshimori) plasmids for assessing initiation and maturation of mitophagy respectively.

7. Lipofectamine® 2000 (Life Technologies, Carlsbad, CA, Cat# 11688). Antibodies specific for mitochondrial markers: Anti-human mitochondrial antigen of 60 kDa antibody (Biogenex, San Ramon, CA, 1:1000), anti-human translocase of the outer mitochondrial membrane 20 kDa antibody (TOM20, Santa Cruz Biotechnologies, 1:1000, FL-145), anti-human cytochrome c oxidase antibody (Sigma, 1:100), and anti-human pyruvate dehydrogenase antibody (PDH, Molecular Probes, 1:1000).
8. Small interfering RNA (siRNA) reagents: To determine whether decreased mitochondrial levels in response to a stimulus is caused by mitophagy, Atg7 and Atg8 can be knocked down in SH-SY5Y cells using the following validated siRNAs: human Atg7 (human, 5′-GCCAGUGGGUUUGGAUCAA-3′ or Atg8 (LC3B, human, 5′ -GAAGGCGCUUACAGCUC AA-3′) and non-targeting/scrambled siRNA can be purchased from Life Technologies (40 nM stocks, Carlsbad, CA), transfect cells with 20 pmol of siRNA per well for cells grown in 4-well chambered slides for 72 h.

2.5 Western Blot Analysis of Mitophagy

1. Lysis buffer: 25 mM HEPES, pH 7.5; 150 mM NaCl; 1% Triton X-100; 10% glycerol containing freshly added proteinase and phosphatase inhibitors, including 100 μM E64, 1 mM sodium orthovanadate, 2 mM sodium pyrophosphate, and 2 mM PMSF.
2. 5–15% polyacrylamide gradient gels are used for the optimal separation of LC3-I and LC3-II.
3. Immobilon-PDVF membranes (Millipore, Bedford, MA, USA, Cat# IPVH00010).
4. Blocking solution: 5% nonfat dry milk in 20 mM potassium phosphate, 150 mM potassium chloride, pH 7.4, containing 0.3% (w/v) Tween 20 (PBST). Alternatively, 2% BSA (Fraction V) can be used to block PDVF membranes when immunodetecting for phosphorylated proteins in the same blot.
5. Antibodies: mouse-anti-LC3 (1:200, LC3-5F10: Nanotools, Cat# 0231-100), rabbit-anti-human LC3 antibody (1:1000, Pierce Biotechnologies), rabbit anti-outer mitochondrial membrane protein TOM20 Antibody (1:10,000, Santa Cruz Biotechnologies, Santa Cruz, CA), ATP Synthase subunit β antibody (1:1000, Complex V, MitoSciences, distributed by Fisher Scientific), anti-human mitochondrial antigen of 60 kDa antibody (1:1000, Biogenex, San Ramon, CA).

6. Small interfering RNA (siRNA) reagents: To measure mitophagic flux in SH-SY5Y cells using biochemical methods, siRNA targeting human ATG7 (5′-GCCAGUGGGUUUGGAUCAA-3′ or ATG8 (LC3B, human, 5′-GAAGGCGCUUACAGCUCAA-3′) and non-targeting/scrambled siRNA can be purchased from Life Technologies (prepared as 40 nM stocks). Transfect cells with 120 pmol of siRNA per well of a 6-well plate for 72 h.

3 Methods

3.1 Electron Microscopy

To date, electron microscopy (EM) remains the gold standard for assessing mitophagy in eukaryotic cells. Although EM is not considered a robust technique, this technique can be used to qualitatively assess mitophagy. At the ultrastructural level, mitophagy is characterized by the presence of early AVs (AVi), which contain mitochondria with other electron dense material. Given that neuronal cells and primary neurons have a high level of autophagic flux under basal conditions, it is worth noting that AVi are rarely detected by EM in neurons under basal conditions. Therefore, it is important to assess for mitophagy by exposing cells with pharmacological inhibitors of lysosomal fusion such as bafilomycin, an inhibitor of vacuolar-type H(+)-ATPase which inhibits the fusion of AVs with lysosomes [40], to increase the number of AVs containing mitochondria.

The following protocol has been optimized for assessing mitophagy in neuronal cells, particularly in SH-SY5Y cells:

1. SH-SY5Y cells are maintained in antibiotic-free, high-glucose (4.4 g/L) DMEM media containing 10% FBS and L-pyruvate. In their undifferentiated state, it is preferable to use early passages of SH-SY5Y cells (passages 15–25) which retain their characteristic fusiform bodies and neuronal-like processes. On the other hand, late passages of SH-SY5Y cells tend to lose their neuronal-like morphology and other non-neuronal like subpopulations—which contain a fibroblast-like morphology—tend to predominate.
2. For ultrastructural analyses of mitophagy, SH-SY5Y cells are plated at a density of 300,000 cells per well on a 6-well plate on either plastic or on glass coverslips kept in six well plates for 3 days. It is preferable to seed SH-SY5Y cells at this high cell density in order to include a sufficient number of cells when embedding the cell monolayer with Epon resin.
3. Given that neuronal cells have a high basal flux, it is challenging to visualize AVs with mitochondria contained inside them. Hence, to analyze for mitophagy, it is highly recommended to

pulse cells with 10 nM bafilomycin A1 for 4 h, or 20 nM bafilomycin A1 for 2 h prior to fixation. It is also advisable to check for cell morphology and appropriate density prior to fixing cells with glutaraldehyde as it is possible that co-treating cells with bafilomycin and a presumed mitophagy-inducing stimulus may enhance toxicity.

4. Following pharmacological treatments, cells are rinsed with 0.1 M PBS for 10 min per wash and fixed for at least 30 min in 2.5% glutaraldehyde at RT, or overnight at 4 °C.
5. After fixation, cell monolayers are washed three times in PBS (10 min per wash), followed by exposure to aqueous 1% OsO4 and 1% K3Fe(CN)6 for 1 h at 4 °C.
6. After three washes with PBS for 10 min per wash, the cultures are dehydrated through a graded series of 30–100% ethanol solutions (30%, 50%, 70%, and 90% ethanol for 10 min each wash and 100% ethanol for 10 min), and embedded in Polybed 812 epoxy resin contained in small capsules. The epoxy capsules containing cell monolayers are then transferred to a 37 °C incubator overnight to allow for the efficient preservation of samples followed by incubation at 60 °C for 48 h.
7. The preserved cell monolayers are cut as ultrathin sections and verified for cell morphology using a light microscope prior to mounting the samples on copper grids. The copper grids are sequentially stained with 2% uranyl acetate in 50% methanol for 10 min followed by a brief incubation in 1% lead citrate (7 min).
8. Cell monolayers are photographed, preferably at a magnification of 18,000×, to visualize autophagosomes using a Phillips CM10 transmission electron microscope or an equivalent electron microscope. As a general rule, AVs should have a size range of 1–2 μm under physiological conditions.

3.2 Fluorescence Imaging-Based Methods

1. SH-SY5Y cells are seeded on uncoated chambered Lab-Tek II #1.5 (Nunc) cover glasses (137,000 cells per well) in complete medium (high glucose DMEM with 10% FBS, L-glutamine, and sodium pyruvate). Primary cortical neurons are seeded on poly-L-lysine pre-coated Lab-Tek II (Nunc) cover glasses in complete medium (Neurobasal medium with 1× B27, 0.5% FBS and L-glutamine). To study the effects of a stimulus on mitophagy in differentiated neuronal cells, SH-SY5Y cells can be differentiated by exposing cells with a 3 day treatment of an acute dose of retinoic acid (10 μM) or with db-cAMP treatment (250 μM). RA-differentiated SH-SY5Y cells should extend neurites that are greater than one cell body length by 24 h of treatment whereas cAMP-differentiated SH-SY5Y tend to develop highly extended and complex neurite arbors. Primary cortical neurons should extend neurites immediately

1 day in vitro (DIV). Neurites of developing primary cortical neurons acquire polarity (axons or dendrites) by 3 DIV in culture and fully develop synapses by 7 DIV [41].

2. To monitor and quantify the initiation of mitophagy by confocal microscopy, 5 DIV mouse primary neurons or SH-SY5Y cells are transfected with 1 μg of GFP-LC3 (AddGene #11546, prepared as 100 ng/μl) in 0.07% and 0.10% Lipofectamine 2000 respectively. To monitor and quantify the progression of mitophagy induced by a presumed mitophagy-inducing stimulus, cells are transiently transfected with 1 μg of RFP-LC3 (AddGene #21075) in 0.10% for SH-SY5Y or in 0.07% Lipofectamine 2000 for primary neurons respectively. Four to six hours following transfection, it is critical to add one complete volume of media per well onto the transfection mix in order reduce any toxicity associated with any leftover Lipofectamine reagent present in the medium. It is important to carefully add the media on top of the transfection mix to avoid cell loss. Approximately two-thirds of the media is replaced 24 h after transfection. Transfected cells are then incubated for an additional 48 h in a 5% CO2-supported tissue culture incubator at 37 °C.

3. Three days post-transfection, the colocalization of AVs with mitochondria can be assessed in GFP-LC3 or RFP-LC3 expressing neurons by treating cells with 100 nM MitoTracker (MTR) Red CMXRos or with 250 nM MTR Green FM respectively for 45 min followed by one wash with warm media. The percentage of GFP-LC3 or RFP-LC3 puncta that colocalize with MTR Red or MTR Green-stained mitochondria respectively gives an overall assessment of the steady-state levels of mitophagy in neurons (See section 4.2).

4. Conditions that increase mitophagy, based on the colocalization of GFP-LC3 puncta with mitochondria, can be attributed to either increased or impaired mitophagic flux as both conditions can yield a high number and colocalization of GFP-LC3 puncta with mitochondria. Therefore, to analyze for flux, GFP-LC3 expressing cells can be treated with a single dose of bafilomycin A1 (10 nM for 4 h), a concentration that elicits a maximal accumulation of GFP-LC3 puncta without inducing significant cell death (<5%). To further corroborate the involvement of autophagy for delivering mitochondria to lysosomes, cells can be co-transfected with ATG7/ATG8 siRNA for 3 days to suppress autophagy. It is important to note that this procedure requires transfecting the cell cultures twice within a span of 2 days. For instance, GFP-LC3 or RFP-LC3 plasmids are transfected in SH-SY5Y cells 48 h after plating. The following day after the initial transfection, cells are re-transfected with ATG7/ATG8 siRNA at a 0.10% final concentration and

incubated for an additional 48 h (72 h total transfection time). Likewise, 5 DIV primary neurons can be sequentially transfected with LC3 plasmids and ATG7/ATG8 siRNA using the same transfection protocol as described above for SH-SY5Y cells. We have found that the sequential transfection of plasmids and siRNA do not significantly injure cells based on intact cell morphology.

5. Alternatively, cells can be co-transfected with mitochondrially targeted RFP (mito-RFP or mito-mCherry) or with mitochondrially targeted GFP (mito-GFP) and analyzed for the colocalization with GFP-LC3 and Lysotracker (LTR) Red respectively, an event defined as mitophagy maturation. It is highly recommended to allow neuronal cells/primary neurons to express mito-RFP or mito-mCherry for at least 2 days prior to mitophagy analyses in order to allow for sufficient import of mitochondrially targeted fluorophores through the TOM complex.
6. Alternatively, the colocalization of lysosomes with mitochondria in SH-SY5Y cells or mouse primary cortical neurons can be analyzed by co-staining neurons with 250 nM of MitoTracker Green and with 100 nM of LTR Red DND-99 in the same medium. Fluorescently labeled cells are incubated for at least 45 min in a 37 °C, 5% CO_2-mantained cell culture incubator. The colocalization of lysosomes, GFP-LC3 or RFP-LC3 with mitochondria can be analyzed by capturing high-resolution confocal slices (1024 × 1024 pixels) using a confocal microscope (Zeiss LSM510 or FluoView 1000) equipped with the appropriate set of filters (FITC and Texas Red-like filters) and a 60× objective. Based on experience, we typically analyze the colocalization of mitochondria with lysosomes, GFP-LC3 or with RFP-LC3 in at least 50 cells per condition in order to achieve statistically significant effects.

3.3 Biochemical Methods

Unlike the plethora of complementary imaging-based techniques available for assessing mitophagy, there is currently only one biochemical method used to corroborate mitophagy in response to a presumed mitophagy-inducing stimulus. This method, immunoblotting for LC3-II is described below:

1. SH-SY5Y cells (5×10^5 cells per well) or mouse primary neurons (1.2×10^6 cells per well) are seeded in 6-well tissue culture dishes. Cells are treated with a presumed inducer of mitophagy inducer (e.g., 2.5 mM MPP+, 75 μM 6-OHDA, or 1.0 μM rotenone for 4–6 h in SH-SY5Y cells) in the presence or absence of lysosomal fusion inhibitors (e.g., bafilomycin or chloroquine), or the cell permeable lysosomal protease inhibitors E64-D (40 μM for SH-SY5Y cells) and pepstatin (25 μM for SH-SY5Y cells) or transfected with siRNA targeting

autophagy-related genes (ATG7/8, 120 pmol per well, 6-well plate × 3 days) to assess for basal levels of mitophagy and mitophagic flux. Following treatments, cell lysates are collected in cell lysis buffers containing 1% Triton X-100. Approximately 30 μg of proteins are electrophoresed on 5–15% gradient gels and immuno-probed for the total levels of several OMM-, IMM-, and matrix-localized mitochondrial proteins (e.g., OMM-localized proteins: TOM20, mitofusin 2, and porin; IMM-localized proteins: complexes I–V, and matrix-localized proteins: human pyruvate dehydrogenase).

2. The lipidated form of LC3, termed LC3-II, exhibits a faster electrophoretic migration on a gradient gel (5–15%) compared to the cytosolic-localized LC3-I (17 kDa for LC3-I vs. 15 kDa for LC3-II). The quantitation of LC3-II/β-actin ratio is a valid index for assessing steady state levels of autophagy [36]. After blots have been probed for mitochondrial proteins, it is strongly recommended to verify the ability of ATG7/8 siRNA to suppress autophagy or of pharmacological and lysosomal fusion inhibitors (bafilomycin) for blocking the fusion of AVs with lysosomes by stripping and re-blotting the western blot membrane for LC3 and β-actin. Hence, a strong effect of bafilomycin on autophagy should increase the LC3-II/β-actin ratios compared to untreated cells when assessing for flux. Conversely, RNAi-mediated knockdown of ATG7/8 should decrease the LC3-II/β-actin ratio and reverse the loss of mitochondrial proteins in the case that a specific stimulus elicits mitophagy.

4 Data Analysis

4.1 Qualitative Assessment of Mitophagy Using GFP-LC3 and RFP-LC3

Generally, cells transiently transfected with GFP-LC3 show either diffuse cytosolic fluorescence or contain fluorescent puncta. It is worth noting that different morphological patterns of mitophagy can be identified when assessing for mitophagy using the GFP-LC3 construct as further described below.

AVs, as monitored by GFP-LC3, tend to show high mobility and have a propensity to coalesce near mitochondria during the initiation of mitophagy. AVs that target mitochondria for sequestration exhibit decreased mobility upon associating with mitochondria, followed by the eventual entrapment and delivery of mitochondria to lysosomes [42]. While it is not considered colocalization—in the strict sense of two perfectly overlapping fluorescently labeled organelles—the clustering of GFP-LC3 near mitochondria is interpreted as mitochondrial association, the first step of mitophagy. Secondly, the colocalization or overlap of GFP-LC3 or RFP-LC3 puncta with red and green fluorescently labeled mitochondria respectively is interpreted as direct association of an AV with mitochondrial membranes. Alternatively, fragmented mitochondria can be completely engulfed by an AV. This morphological pattern is typically characterized by the presence of large GFP-LC3 "rings" that can sequester several pieces of mitochondria or the whole organelle (Fig. 1b, inset). These large AVs that contain mitochondria have been previously termed "mitophagosomes" [43]. As with EM analyses, it is critical to categorize and classify these three morphological patterns of mitophagy in order to gain a comprehensive understanding on the mechanisms by which a specific stimulus or protein promotes the initiation of mitophagy.

4.2 Quantitative Analysis of Autophagy and Mitophagy Using GFP-LC3, RFP-LC3, and Fluorescent Lysosome-Specific Dyes

To quantify mitophagy using fluorescence-based imaging protocols, the percentage of LC3 puncta (endogenous, GFP-LC3 or RFP-LC3) that colocalize with mitochondria in the soma is a valid method for assessing steady-state levels of mitophagy in primary neurons. The percentage of LC3 puncta that colocalize with mitochondria for a certain segment of proximal or distal neurites (axons or dendrites, 50–100 μm) is a valid index for assessing mitophagy in neurites. Alternatively, the number of mitochondrial colocalizing LC3 puncta per soma or neurite segment can also be quantified in conjunction with the percentage of LC3 puncta that colocalize with mitochondria to assess for mitophagy. Given that mitochondria are highly dynamic in neurites, it is necessary to rule out transient colocalization of LC3 with mitochondria. To overcome this issue, it is highly recommended to monitor mitophagy at least two different time points in live or fixed cells. Data on autophagy/mitophagy can be represented as bar graphs that display the number of GFP-LC3 or RFP-LC3 puncta per neuron as well as the percentage of GFP-LC3 or RFP- LC3 puncta that colocalize with mitochondria in response to a presumed mitophagic inducer (Figs. 1, 2, and 3). Likewise, to study lysosomal sequestration of mitochondria in response to a specific stimulus, the percentage of lysosomes that colocalize with mitochondria can be quantified to analyze late-stage mitophagy (Fig. 2).

Thus far, we have presented several methods and guidelines to measure mitophagy using LC3 as a marker of AVs. However, measuring the content of autophagosomes such as mitochondria in different neuronal compartments is a direct assessment of mitophagy that can further corroborate analyses involving mitochondrial colocalization of LC3. As numbers of mitochondria in neurites are governed by different factors including mitochondrial biogenesis, trafficking, and mitophagy, it is imperative to assess the contribution of mitophagy to mitochondrial content in the soma or neurites. Therefore, the percentage of cytosolic area occupied by mitochondria in the presence or absence of lysosomal fusion inhibitors or in cells transfected with ATG7/8 siRNA is a valid method

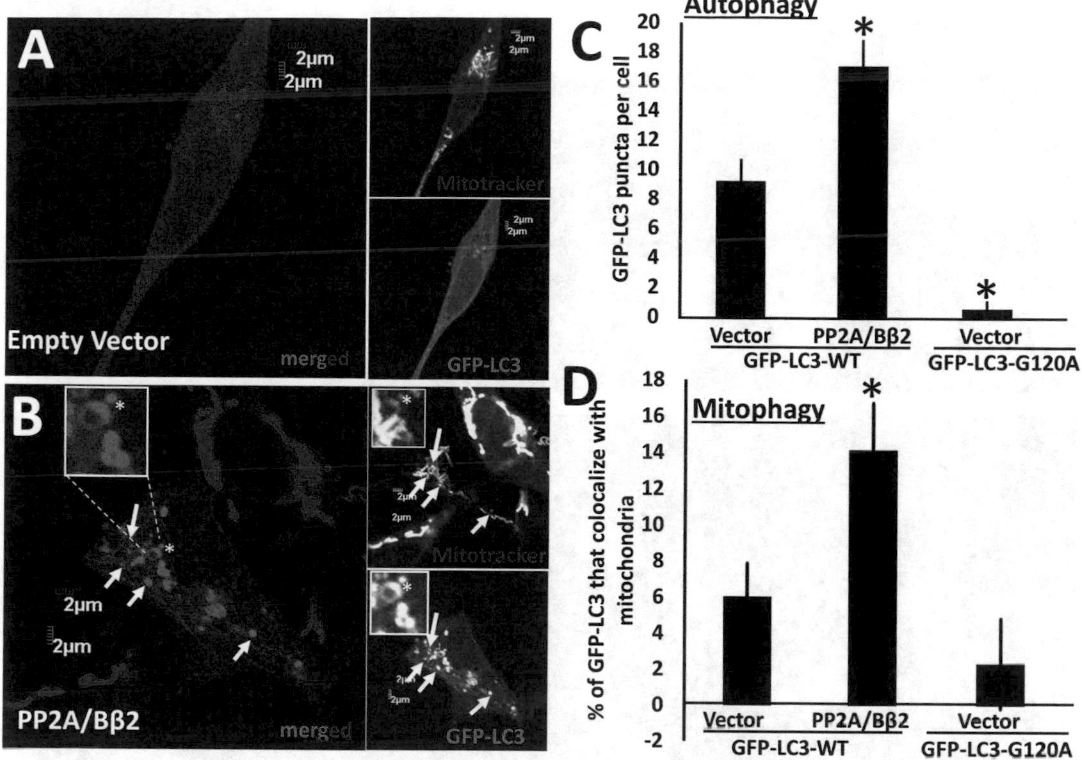

Fig. 1 Example of autophagy/mitophagy analyses using the GFP-LC3 construct in live SH-SY5Y cells. The effects of overexpressing Bβ2 on autophagy/mitophagy were analyzed in SH-SY5Y cells. Representative epifluorescence images of SH-SY5Y cells transiently co-expressing GFP-LC3 and either an empty vector (**a**) or Flag-tagged Bβ2 (**b**) and stained with MitoTracker Red to visualize mitochondria. Note that forced overexpression of PP2A/Bβ2 increases the number and mitochondrial colocalization of GFP-LC3 puncta (*arrows*). Some Bβ2-expressing neuronal cells showed the presence of large GFP-LC3 ring structures containing mitochondria (*asterisk*, *inset* in **b**). Scale bars: 2 μm. (**c**, **d**) Representative quantifications of GFP-LC3 puncta per cell (**c**) and the percentage of GFP-LC3 puncta that colocalize with mitochondria per cell (**d**). Note that forced overexpression of PP2A/Bβ2 increases the number and mitochondrial colocalization of GFP-LC3 puncta per cell. As a negative control, the number and mitochondrial colocalization of GFP-LC3 puncta were quantified in cells transiently expressing the lipidation deficient mutant construct of GFP-LC3 (GFP-LC3 G120A). (*$p < 0.05$ vs. empty vector, ±SEM, $n = 22$–30 cells per condition, Student's *t*-test)

for assessing mitophagy. As with western blots, it is imperative to immunostain for at least two different mitochondrial markers when assessing mitophagy using image-based methods. Secondly, to facilitate the concomitant analyses of mitochondrial content and mitophagy, cells can be co-transfected with mito-RFP and with GFP-LC3 to demarcate the cell boundaries given the mixed diffused and punctate staining pattern observed when cells express GFP-LC3.

Quantifying mitophagic flux: To facilitate the analyses of flux, treating neuronal cells with bafilomycin can be used to trap mitochondria in AVs. **Mitophagic flux** = (AVmbaf)/AVm)/Tbaf) [43].

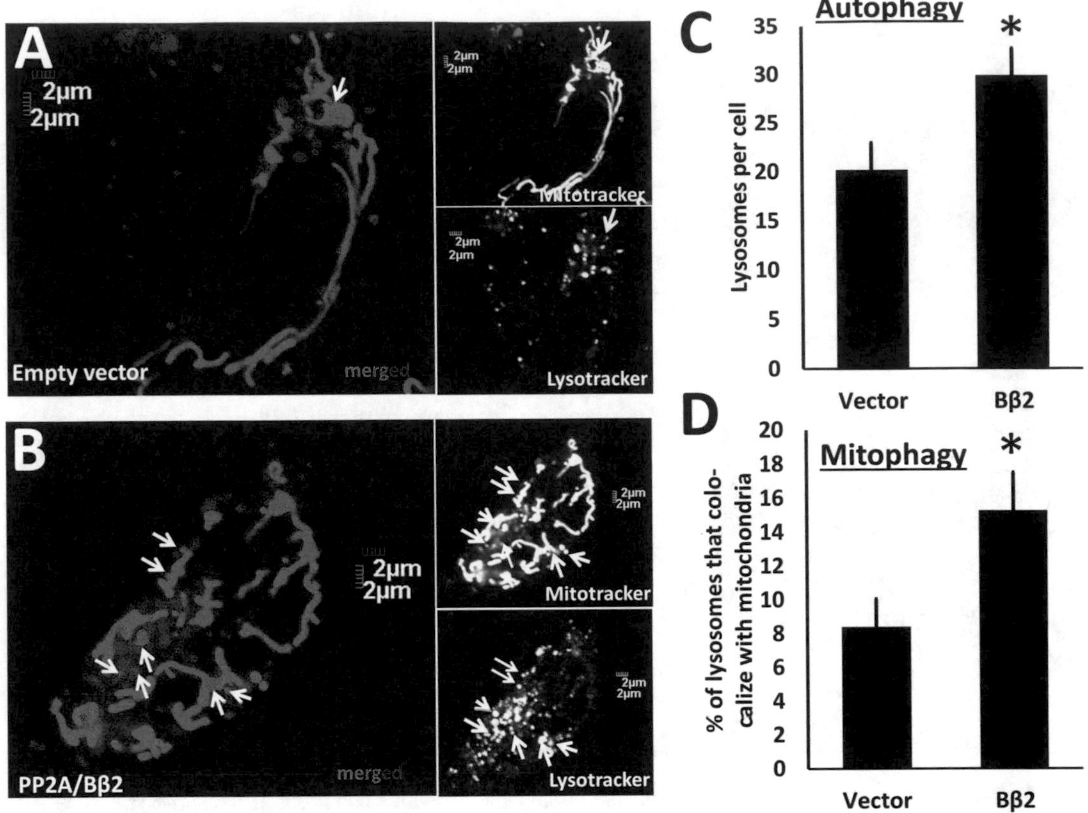

Fig. 2 Example of autophagy/mitophagy analyses using lysosomal and mitochondrial specific fluorescent dyes in live SH-SY5Y cells. The effects of forced overexpression of Bβ2 on lysosomal-mediated sequestration of mitochondria were analyzed in SH-SY5Y cells. Representative epifluorescence images of SH-SY5Y cells transiently expressing empty vector (**a**) or Flag-tagged Bβ2 (**b**) were co-stained with Lysotracker Red and MitoTracker Green FM to visualize lysosomes and mitochondria respectively. Note that forced overexpression of Bβ2 increases the number and the colocalization of lysosomes with mitochondria (*white arrows*). Scale bars: 2 μm. (**c**, **d**) Quantifications of the number Lysotracker (LTR) stained structures per cell and the average percentage of LTR puncta that colocalize with mitochondria (mitophagy). (*$p < 0.05$ vs. empty vector, ±SEM, $n =$ 22–30 cells per condition, Student's t-test)

AVmbaf: # of AVs per cell containing mitochondria in the presence of bafilomycin; AVm: # of AVs per cell containing mitochondria in the absence of bafilomycin; Tbaf: time in bafilomycin treatment. Therefore, a high mitophagic flux ratio suggests that a specific stimulus or protein of interest enhances mitophagic flux, whereas a low ratio (~1.0) indicates impaired flux.

4.3 Guidelines for Measuring Mitophagy in Neurites

Mitochondria are highly dynamic organelles that show bidirectional movement in neurites [44]. When assessing mitophagy in presynaptic and postsynaptic compartments, live imaging can be performed in cells co-expressing mito-GFP and RFP-LC3. Mitophagy in neurites is assessed as the percentage of AVs that contain mitochondria at least two different time points. Hence,

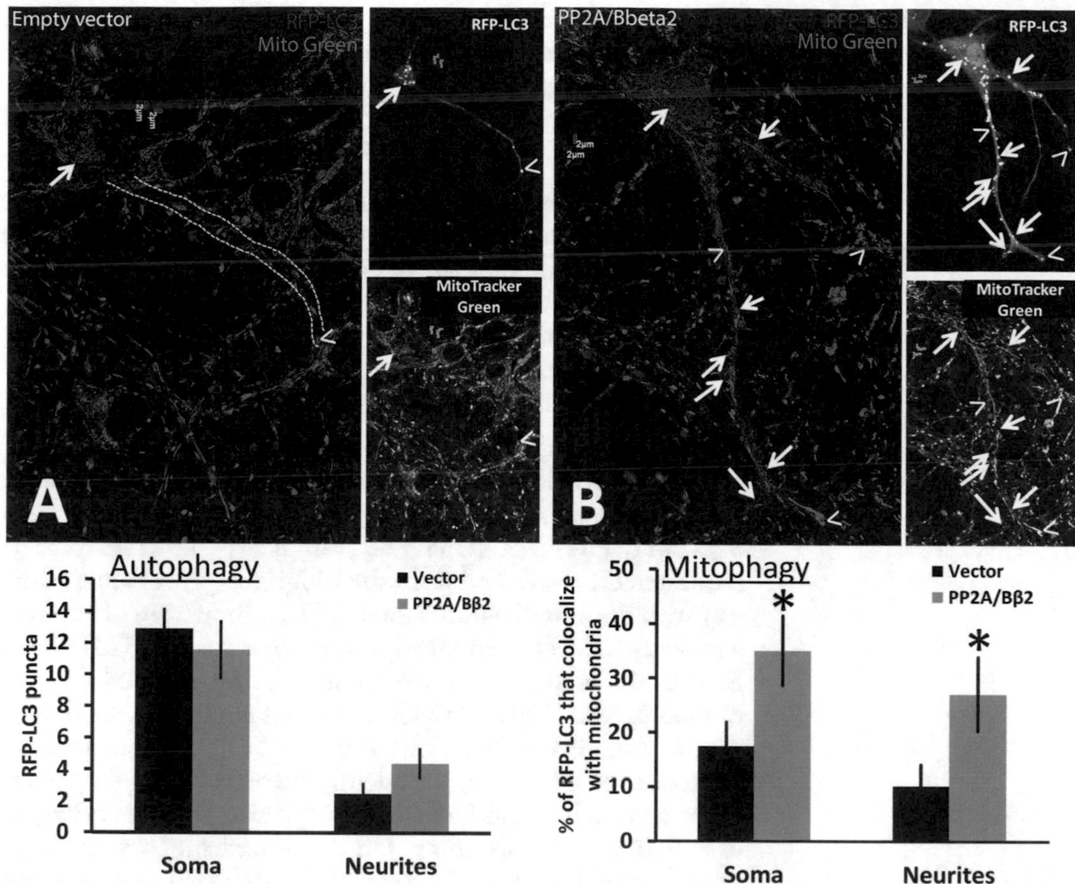

Fig. 3 Example of autophagy/mitophagy analyses in the soma and neurites of primary cortical neurons expressing the RFP-LC3 construct. The effects of forced overexpression of PP2A/Bβ2 on autophagy/mitophagy were further confirmed in 8 DIV primary cortical neurons. Representative epifluorescence images of primary cortical neurons transiently co-expressing RFP-LC3 and either an empty vector (**a**) or Flag-tagged Bβ2 (**b**) and stained with MitoTracker Green (MTRG) to visualize mitochondria. Scale bars: 2 μm. Note that forced overexpression of Bβ2 increases the mitochondrial colocalization of RFP-LC3 puncta (*white arrows*) or its association with mitochondria (*arrow heads*). For clarity, a dendrite was traced with *white hatched lines* in the representative neuron shown in panel **a**. (**c**, **d**) Quantification of RFP-LC3 puncta per cell and the average percentage of RFP-LC3 that colocalize with mitochondria (mitophagy) in the soma or neurites. While overexpression of Bβ2 induces a nonsignificant increase in RFP-LC3 puncta per neurite (50 μm), forced overexpression of Bβ2 significantly increases the mitochondrial colocalization of RFP-LC3 puncta in the soma and neurites of neurons (*$p < 0.05$ vs. empty vector, ±SEM, $n = 22$–30 cells analyzed per condition as shown in panel **c** or $n = 20$–45 neurites analyzed per condition as shown in panel **d**, Student's *t*-test)

the colocalization of two dynamic organelles labeled with two different fluorophores can be analyzed by generating two 5-min kymographs for mitochondria (e.g., mito-RFP) and AVs (e.g., GFP-LC3). Individual kymographs for each organelle can then be assembled and overlaid to analyze for colocalized objects using Image J and the Multiple Kymograph" plug-in (J. Rietdorf, A. Seitz, EMBL, Heidelberg, http://www.embl.de/eamnet/html/body_kymograph.html).

4.4 Western Blot Quantification of Mitophagy

Unlike imaged-based analysis, one major caveat is that western blots do not measure mitophagy in distinct neuronal compartments but represent an average of heterogeneous autophagic/mitophagic responses for a population of cells. We provide the following guidelines to measure mitophagy using western blot assays.

1. To corroborate mitophagy, it is important to perform densitometric quantification of at least two different mitochondrial proteins as measured by the integrated density of immunoreactive bands. Moreover, it is important to analyze the levels of mitochondrial proteins in cells treated in the presence or absence of bafilomycin or in cells transfected with ATG7/8 siRNA to assess for mitophagy.
2. An increase in the LC3-II/β-actin ratios (not the LC3- I/LC3-II levels) along with a concomitant decrease in the level of mitochondrial proteins suggests enhanced mitophagy. SQSTM1/P62 is a bona fide autophagy substrate that is sequestered by AVs and links the ubiquitin–proteasome pathway with the autophagic machinery [18]. Induction of autophagy leads to decreased steady-state levels of SQSTM1/P62. Hence, the western blot membrane can be stripped and reblotted for SQSTM1/P62 levels to further confirm for induction of autophagy. It is important to bear in mind that an induction of mitophagy transiently increases LC3-II/β-actin ratios while a block in flux will show persistent increased levels of SQSTM1/P62 over time. On the other hand, a decrease in mitochondrial markers, increased LC3-II/β-actin ratios along with decreased SQSTIM1/P62 levels for at least two time points is a strong indication that mitophagy is induced by a specific stimulus or protein of interest.
3. If possible, it is helpful to include a positive control for all biochemical-based assessments of mitophagy such as treating cells with rapamycin (an inducer of mild, physiological mitophagy) or with staurosporine (a pathological inducer of mitophagy).

5 Results

In this section, we present several experimental examples that show how autophagy and mitophagy are properly analyzed in neuronal cell lines and primary neurons using image-based methods.

Also, as mitophagy can occur in the context of both physiological and pathological conditions, several caveats and potential pitfalls that can lead to a misinterpretation of the results will be indicated appropriately for each example.

5.1 Image-Based Analysis of Mitophagy

Figure 4 shows an example of how mitophagy is properly analyzed at the ultrastructural level in SH-SY5Y cells treated with rotenone, a pathological inducer of mitophagy [30]. While untreated cells show elongated mitochondria and occasional AVs, a 4 h treatment of rotenone (1 μM) robustly swells/fragments mitochondria and increases the number of AVi in SH-SY5Y cells (Fig. 4a, b). Co-treating cells with a short pulse of bafilomycin (10 nM, 4 h), leads to a further increase in the number of AVs per cell as well as the accumulation of swollen/fragmented mitochondria engulfed in AVi compared to cells treated with rotenone alone (Fig. 4c–e). These results suggest that rotenone alters mitochondrial structure and promotes mitophagy in SH-SY5Y cells. We calculated the mitophagic flux induced by rotenone in the same experiment using the formula described in Sect. 4.2. Although rotenone increases the mean number of mitophagosomes per cell (0.22 vs. 0.62 mitophagosomes per cell in untreated or rotenone treated cells respectively), mitophagic flux of SH-SY5Y cells exposed to rotenone (0.15 mitophagosomes/h) is similar to mitophagic flux in the absence of rotenone (0.15 mitophagosomes/h). These results suggest that rotenone elevates autophagic sequestration of mitochondria without inducing flux.

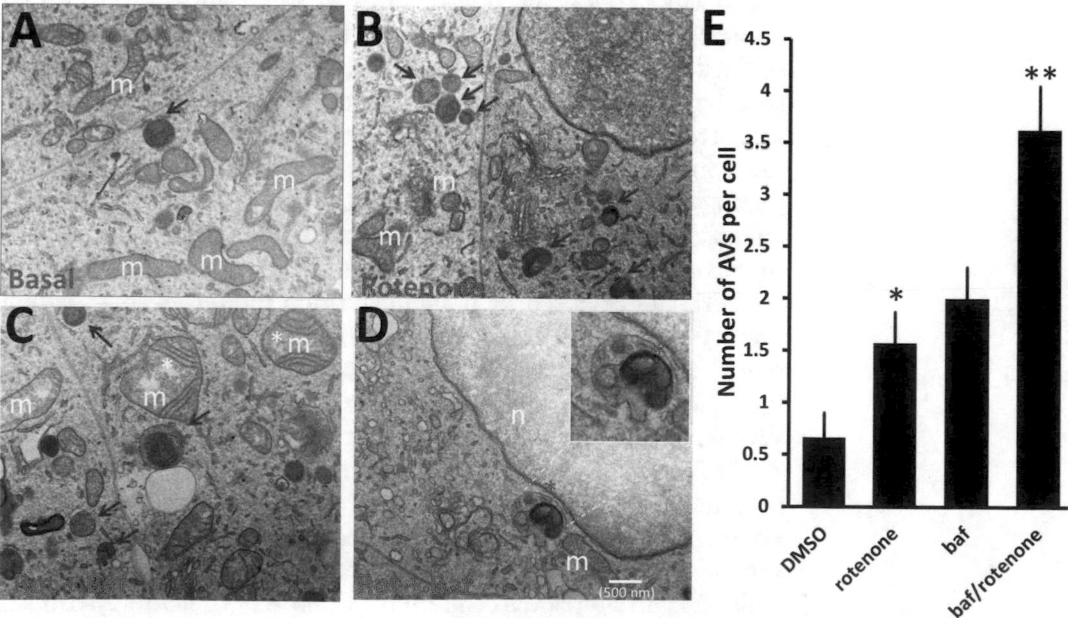

Fig. 4 Example of ultrastructural analyses of autophagy/mitophagy in SH-SY5Y cells. a. SH-SY5Y cells were treated with rotenone (**b**) or DMSO (**a**) as a vehicle control in the absence (**b**) or presence of bafilomycin (**c**, **d**) for 4 h prior to processing for ultrastructural analyses of mitophagy. (**e**) The number of AVs (*arrows*) per cell was quantified for the indicated conditions. An *asterisk* in panel (**d**) shows an example of an AV containing mitochondria (electron dense structure with noticeable cristae). *m* mitochondria, *n* nuclei, *arrows* AVs, *asterisk* mitophagosome. Scale bar: 500 nm (*$p < 0.05$ vs. DMSO, **$p < 0.05$ vs. rotenone, ±SEM, $n = 15–22$ cells per condition, Student's *t*-test)

The next three experimental examples demonstrate how studying the effects of overexpressing one protein on mitophagy can be assessed using multiple but complementary fluorescence-based techniques. The neuronal regulatory subunit of protein phosphatase 2A (PP2A), termed Bβ2, targets the PP2A holoenzyme to the OMM to promote mitochondrial fission (fragmentation) and accelerate apoptosis in oxidatively stressed neurons [45, 46]. As mitochondrial fragmentation precedes mitophagy, we sought to determine whether PP2A/Bβ2 modulates autophagy and mitophagy in SH-SY5Y neuronal cells. Overexpression of PP2A/Bβ2 in SH-SY5Y neuronal cells significantly increased the number of GFP-LC3 puncta per cell as well as the percentage of GFP-LC3 that colocalize with mitochondria suggesting that overexpression of PP2A/Bβ2 concomitantly upregulates autophagy and mitophagy (Rice M, Strack S, and Dagda RK, unpublished observations). As a specificity control for autophagy, cells transiently expressing a C-terminal lipid deficient mutant of GFP-LC3 (G120A) yields a significantly low number of GFP-LC3 puncta per cell. The presence of large GFP-LC3 rings engulfing mitochondria were also observed in some Bβ2-expressing cells (Fig. 1b).

Alternatively, mitochondrial content in response to mitophagic stimuli can be measured at the single cell level as a valid index of mitophagy. Following treatment with a presumed mitophagy-inducing stimulus, in the presence or absence of flux inhibitors, live or fixed cells can be analyzed for the percentage of cytosol occupied by mitochondrial-specific fluorescent pixels by using a pre-validated Image J macro ("Mitophagy" macro, http://imagejdocu.tudor.lu/doku.php?id=macro:mitpophagy_mitochondrial_ morphology_content_lc3_colocalization_macro) [39, 46]. Therefore, enhanced mitophagy in response to a specific stimulus is expected to decrease the percentage of the cytosolic area occupied by mitochondria but reversed by lysosomal fusion inhibitors (e.g., bafilomycin) or by ATG7/8 siRNA. Conversely, an increase in mitochondrial content in response to a stimulus, which is not further increased by lysosomal fusion inhibitors or by ATG7/8 siRNA, suggest an impairment of mitophagy [31].

It is important to determine whether a specific stimulus not only increases the initiation of mitophagy but also leads to the eventual sequestration and degradation of mitochondria by lysosomes. In conjunction with mitophagy, lysosomal expansion—defined as an increase in the number and size of acidic lysosomes per cell—is an indication of enhanced autophagy [31, 47]. In addition to stimulating mitophagy initiation, to ascertain whether forced overexpression of PP2A/Bβ2 promotes the lysosomal-mediated sequestration of mitochondria, SH-SY5Y cells were transiently transfected with an empty vector or Flag-tagged Bβ2 for 3 days and analyzed for the number and percentage of lysosomes that colocalize with mitochondria per cell. Compared to empty vector expressing cells, transient expression of PP2A/Bβ2 increases both the number and percentage of lysosomes that colocalize with mitochondria in SH-SY5Y cells (Fig. 2). Overall, these results suggest that PP2A/Bβ2 not only promotes the initiation of mitophagy, but elicits lysosomal expansion and the active lysosomal-mediated degradation of mitochondria (Rice M, Strack S, and Dagda RK, unpublished observations).

5.2 Imaged-Based Analysis of Mitophagy in Neurites

The quantification of the number of GFP-LC3 or RFP-LC3 puncta (granules) per segment of a neurite (50–100 μm) in either paraformaldehyde fixed or live neurons is a valid method to measure autophagy in neurites as previously published [17]. Mitochondria in neurites play critical physiological roles such as maintaining the structural stability of neurites and providing the necessary energy required for remodeling dendritic spines, synaptogenesis, and for enabling the release of synaptic vesicles [39, 48–51]. Therefore, quantitating mitophagy in neurites in response to a specific stimulus can yield valuable insight into the physiological role that mitophagy plays in neuronal development and neurodegeneration. To quantitate mitophagy in neurons, the percentage of the area per neurite occupied by mitochondria (also known as mitochondrial content) in the presence or absence of flux inhibitors or ATG siRNA, can be used to assess mitophagy in neurites [52].

An experimental example of how mitophagy is measured in neurites is presented in Fig. 3. To study the effects of overexpressing PP2A/Bβ2 on autophagy and mitophagy in neurites, 5 DIV primary cortical neurons were co-transfected with RFP-LC3 and with Flag-tagged Bβ2 for 3 days prior to analyzing for both macroautophagy and mitophagy by confocal microscopy. While PP2A/Bβ2 nonsignificantly increased the number of RFP-LC3 puncta per neurite (RFP-LC3 puncta/50 μm of neurite segment), transient expression of PP2A/Bβ2 significantly elevated the percentage of RFP-LC3 puncta that colocalize with mitochondria per neurite (% of RFP-LC3 that colocalize with mitochondria/50 μm of neurite segment) and at the soma as well. Overall, these results suggest that PP2A/Bβ2 upregulates the steady-state levels of mitophagy in neurites and soma of primary neurons (Rice M, Strack S, and Dagda RK, unpublished observations).

5.3 Biochemical Quantification of Mitophagy

The lipidated form of LC3, termed LC3-II, exhibits a much faster electrophoretic migration on a gradient gel (5–15%) compared to LC3-I [36]. As with image-based analysis of LC3 puncta, macroautophagy and mitophagy can be analyzed in the same western blot. A specific effect of a stimulus on macroautophagy should transiently increase the LC3-II/β-actin ratios and return to baseline levels, an indication of autophagic flux. On the other hand, a block on autophagic flux should lead to a persistent increase in the LC3-II/β-actin ratio which is not further increased with bafilomycin treatment.

In conjunction to enhanced autophagy, a specific effect of a stimulus on mitophagy should promote a decrease in the levels of at least two mitochondrial proteins (e.g., ATP synthase V, Mitochondrial antigen of 60 kDa, and cytochrome c oxidase) and this effect should be blocked, or partially suppressed by co-treating cells with bafilomycin or by ATG7/8 siRNA.

6 Troubleshooting

In this section, we present the most common pitfalls that neurobiologists encounter when assessing mitophagy in neurons and provide solutions to overcome the following issues:

1. Pathological stimuli that elicit mitophagy, especially mitochondria-directed toxins, tend to damage and depolarize mitochondria. Hence, mitophagy assessments may be underestimated under these circumstances as oxidatively damaged and fragmented mitochondria do not stain well with mitochondrial membrane potential-dependent dyes (e.g., MitoTracker Red dye series) and may appear not to colocalize with AVs. Therefore, it is highly advisable to analyze mitophagy using live fluorescent reagents that do not depend on the membrane potential (e.g., MitoTracker Green) or transfect with mitochondria- targeted fluorescent proteins.
2. There are several caveats to consider when analyzing mitophagy by transient vs. stable expression of fluorescent LC3 fusion proteins (e.g., GFP-LC3). For instance, a major caveat of transiently co-expressing a protein of interest and fluorescent LC3 fusion proteins is that only short-term mitophagy responses can be monitored. On the other hand, using stable cell lines affords the ability to analyze mitophagy for prolonged periods of time. Secondly, the use of stable cell lines eliminates potential toxicity of liposome-mediated transient transfection. Third, it facilitates biochemical analysis of mitophagy in response to stably overexpressing or down-regulating a protein of interest. However, a major caveat of stable cell lines is that GFP-LC3 may be incorporated into aggregates (not AVs), or the GFP moiety may aggregate itself, which can confound analyses of autophagy and mitophagy [53].
3. It is important to check the quality of cells prior to starting an experiment as a high baseline level of autophagy and mitophagy in primary neurons likely indicates an unhealthy, stressed or contaminated culture.
4. Primary neurons transfected with GFP-LC3 or cultured from transgenic GFP-LC3 mouse embryos [42] predominantly show a diffuse staining pattern and a low number of AVs

which precludes analyzing steady-state basal levels of mitophagy using this construct. On the other hand, primary neurons tend to display a higher number of RFP-LC3 puncta compared to GFP-LC3 puncta. Given that the RFP moiety is fluorescently stable in the acidic environment of the lysosome compared to GFP [38], a high number of RFP-LC3 puncta relative to the number of GFP-LC3 puncta per cell is indicative of high autophagic flux. Neurons display a high autophagic flux under basal conditions as indicated by the high number of RFP-LC3 puncta per cell relative to the GFP-LC3 puncta. To this end, mitophagy in the soma and neurites of primary neurons should be assessed by analyzing for the colocalization of RFP-LC3 puncta with MitoTracker Green-labeled mitochondria (Fig. 4).

5. Although analyses of mitophagy in presynaptic or postsynaptic compartments is technically challenging in primary neurons, it is highly advisable to use a spinning disc confocal microscope equipped with high performance camera (e.g., EMCCD Hamamatsu) to perform time-lapse imaging in order to prevent phototoxicity-induced mitophagy [54] and to have the ability to analyze the dynamic movement and colocalization of dynamic organelles (AVs and mitochondria) at a high frame rate (40–62 frames/s). Alternatively, the mitochondrial content in the neurites in response to a specific stimulus can be assessed to further confirm mitophagy as described in section 4.2.

6. Unlike IMM-localized proteins, OMM-localized proteins can undergo fast turnover through the ubiquitin–proteasome pathway. Therefore, a major pitfall when assessing mitophagy by western blot is that the protein levels of OMM-localized proteins may not change in parallel with IMM- or matrix-localized proteins [36]. To overcome this issue, it is important to strip and re-blot a western blot membrane for multiple mitochondrial markers, preferably IMM- and matrix-localized proteins. Secondly, a decrease in IMM- and matrix-localized proteins in response to a specific stimulus can be attributed to proteolytic processing, impaired mitochondrial import or impaired biogenesis. Therefore, it is critical to determine whether specific decreases in mitochondrial proteins can be reversed by RNAi-mediated knockdown of ATG7/8 or by short-term treatment of cells with bafilomycin. In addition, biochemical assessments of mitophagy should be qualitatively confirmed by EM as well.

7. Another possibility is that treatments that promote high mitophagic flux may not change the total levels of mitochondrial proteins due to increased compensatory mitochondrial biogenesis. To overcome this issue, it is important to

analyze for potential compensatory mitochondrial biogenesis by re-blotting the western blot membrane for the total protein levels of peroxisome proliferator-activated receptor gamma coactivator 1-alpha (PGC1-α) and mitochondrial transcription factor A (TFAM), two regulators of mitochondrial biogenesis [55, 56].

8. When performing LC3 shift assays, LC3-II may not be very visible in treatments that modestly induce autophagy. In addition, unlike other tissue types, brain tissue contains a high amount of LC3-I. Hence, high levels of LC3-I may block the detection of LC3-II in a western blot [51]. Hence, to ascertain whether a specific treatment induces macroautophagy, it may be necessary to probe for LC3-II in membrane fractions obtained by centrifugation steps as previously described [51]. Finally, immunodetection of endogenous LC3 by immunofluorescence or by western blot can be difficult at times. Some good commercial antibodies available include mouse anti-human LC3 monoclonal antibodies (Nanotools, clone 5F1 or clone 2G6 for western blot). In addition, good antibodies that can be used to detect the GFP moiety to resolve GFP-LC3-II from GFP-LC3-I by western blot have been used as well (Rabbit anti-GFP serum, Life Technologies, Cat# A-6455) [17].

7 Conclusions

Although a vast amount of knowledge has been garnered over the last decade regarding the role of mitophagy in the soma of neurons, studies that address the physiological roles of mitophagy in other neuronal compartments have only recently taken center stage. Indeed, the seminal discovery that LC3 is a bona fide marker of AVs and the development of sophisticated, yet affordable live cell imaging equipment have allowed neuroscientists to assess the role of mitophagy in the context of neuronal development and neurodegeneration. It is important to keep in mind that the image-based techniques presented in this chapter may allow for the simultaneous analyses of autophagy, mitophagy, mitochondrial morphology, bioenergetics and mitochondrial dynamics in neurons when multiplexed with other techniques described in this book. Secondly, it is important to keep in mind that a single technique is not sufficient to establish mitophagy induced by a specific physiological/pathological stimulus. Therefore, we encourage the use of complementary techniques, including electron microscopy, fluorescence- and biochemistry-based techniques to analyze mitophagy in neurons as described in this chapter. Finally, the techniques outlined in this chapter are meant to encourage the development of new methodology and reagents to assess mitophagy in ex vivo and in in vivo models in the near future. However, it should be noted that regardless of the experimental paradigm(s) used to analyze mitophagy in neurons, the concepts and guidelines for assessing mitophagy should remain the same.

8 Notes

1. It is important to freeze down a high number of vials containing early passages of SH-SY5Y cells that respond well to neurotrophic/differentiation factors (cyclic AMP and retinoic acid). Approximately 1.0×10^6 cells are frozen in 1 ml of cell-freezing media containing 10% DMSO (e.g., CryoStor, CS10, STEMCELL Technologies).
2. It is important to isolate embryonic primary neurons and dissociate them preferably through mechanical means (e.g., passing cells through Pasteur glass pipettes with decreasing bore sizes) without the need to enzymatically digest the tissue with papain or trypsin as these treatments tend to significantly reduce neuronal viability.
3. It is critical to routinely test different stocks and each lot of RA for its differentiation potential in SH-SY5Y cells as RA tends to get inactivated by light. SH-SY5Y cells should respond well to RA by spreading out on plastic dishes, adopting fusiform neuronal cell bodies, and typically send out neurites that are as long as two cell body lengths within 24 h of exposure to RA.
4. Materials transfer agreement needs to be processed prior to purchasing reagents from AddGene (www.addgene.com).
5. One Lipofectamine 2000 reagent should be purchased for transfecting DNA plasmids in neuronal cells/primary neurons whereas a second vial of Lipofectamine 2000 should be purchased for transfecting cells with siRNA to avoid cross-contaminating siRNA reagents with RNAases.
6. The anti-TOM20 and anti-cytochrome C oxidase antibodies can label mitochondria in both SH-SY5Y cells and primary cortical neurons.
7. Unlike cells grown on plastic, we have noticed that growing SH-SY5Y cells on coverslips pre-coated with poly-L-lysine can facilitate the removal of the Epon-imbedded cell monolayer by exposing the Epon capsules through sequential freeze/thaw cycles (Agarwal A, Rice M, and Dagda RK 2014, unpublished observations).
8. It is important to test the epoxy resin, at least 3 days prior to processing the experimental samples, as the quality of the epoxy resin decreases with time. Therefore, it is advisable to

freeze a large working stock (~50 ml) of epoxy resin at −20 °C. Also, to remove the epoxy resin from plastic for cells seeded on 6-well plastic dishes, it is important to snap off the epoxy capsule from the bottom using a small pliers to prevent shattering the capsule if not handled correctly.

9. To analyze mitophagy in neurons, it is imperative to obtain high resolution and high quality images by confocal microscopy as it is conceivable that a cluster of several overlapping bright fluorescent GFP-LC3 or RFP-LC3 puncta can be misinterpreted as a single LC3 granule in low quality epifluorescence micrographs, a very common but avoidable pitfall.

10. It is important to keep in mind that a decrease in the protein levels of one mitochondrial protein is not sufficient evidence for mitophagy. Hence, it is highly advisable to immunoblot for several mitochondrial markers at different time points in order to determine whether a decrease in mitochondrial levels is caused by mitophagy as opposed to impaired mitochondrial biogenesis or ubiquitin-mediated degradation of OMM-localized mitochondrial proteins. A final word of caution is to avoid immunoblotting for apoptogenic factors (e.g., cytochrome C) when assessing mitophagy as cytochrome C is released from mitochondria in response to apoptotic stimuli.

11. Depending on the level of transient expression of GFP-LC3, an important caveat to consider when overexpressing GFP-LC3 or RFP-LC3 is that these LC3 chimeras can associate with large ubiquitinated protein aggregates or aggresomes, a phenomenon that does not reflect autophagosome formation or flux and likely represents an artifact of overexpression [53]. Hence, it is conceivable that a small percentage of GFP-LC3 puncta is not associated with AVs in any experimental context. To overcome this issue, cells can be transfected with the lipid binding deficient version of GFP-LC3 (G120A) to determine the percentage of GFP-LC3 puncta that is not associated with AVs.

Acknowledgments

The method development and the research data presented in this book chapter were supported by an NIH-NIGMS grant (GM103554) and by a University of Pittsburgh Pathology Postdoctoral Research Training Program Grant awarded to R.K.D. We give special thanks to Dr. Stefan Strack (Department of Pharmacology, University of Iowa College of Medicine) for graciously providing the mito-GFP and Flag-tagged Bβ2 constructs.

References

1. Pagliarini DJ, Dixon JE (2006) Mitochondrial modulation: reversible phosphorylation takes center stage? Trends Biochem Sci 31(1):26–34
2. Pekkurnaz G, Trinidad JC, Wang X, Kong D, Schwarz TL (2014) Glucose regulates mitochondrial motility via Milton modification by O-GlcNAc transferase. Cell 158(1):54–68. doi:10.1016/j.cell.2014.06.007
3. Gawlowski T, Suarez J, Scott B, Torres-Gonzalez M, Wang H, Schwappacher R, Han X, Yates JR III, Hoshijima M, Dillmann W (2012) Modulation of dynamin-related protein 1 (DRP1) function by increased O-linked-beta-N-acetylglucosamine modification (O-GlcNAc) in cardiac myocytes. J Biol Chem 287(35): 30024–30034. doi:10.1074/jbc.M112.390682
4. Geisler S, Holmstrom KM, Skujat D, Fiesel FC, Rothfuss OC, Kahle PJ, Springer W (2010) PINK1/Parkin-mediated mitophagy is dependent on VDAC1 and p62/SQSTM1. Nat Cell Biol 12(2):119–131
5. Benischke AS, Hemion C, Flammer J, Neutzner A (2014) Proteasome-mediated quality control of S-nitrosylated mitochondrial proteins. Mitochondrion 17:182–186. doi:10.1016/j.mito.2014.04.001
6. Taylor EB, Rutter J (2011) Mitochondrial quality control by the ubiquitin-proteasome system. Biochem Soc Trans 39(5):1509–1513. doi:10.1042/BST0391509
7. Hamon MP, Bulteau AL, Friguet B (2015) Mitochondrial proteases and protein quality control in ageing and longevity. Ageing Res Rev 23(Pt A):56–66. doi:10.1016/j.arr.2014.12.010
8. Klionsky DJ, Emr SD (2000) Autophagy as a regulated pathway of cellular degradation. Science (New York, NY) 290(5497): 1717–1721
9. Levine B, Klionsky DJ (2004) Development by self-digestion: molecular mechanisms and biological functions of autophagy. Dev Cell 6(4):463–477
10. Cook KL, Soto-Pantoja DR, Abu-Asab M, Clarke PA, Roberts DD, Clarke R (2014) Mitochondria directly donate their membrane to form autophagosomes during a novel mechanism of parkin-associated mitophagy. Cell Biosci 4(1):16. doi:10.1186/2045-3701-4-16
11. Hayashi-Nishino M, Fujita N, Noda T, Yamaguchi A, Yoshimori T, Yamamoto A (2010) Electron tomography reveals the endoplasmic reticulum as a membrane source for autophagosome formation. Autophagy 6(2): 301–303
12. Geng J, Klionsky DJ (2008) The Atg8 and Atg12 ubiquitin-like conjugation systems in macroautophagy. 'Protein modifications: beyond the usual suspects' review series. EMBO Rep 9(9):859–864. doi:10.1038/embor.2008.163
13. Cherra SJ, Chu CT (2008) Autophagy in neuroprotection and neurodegeneration: A question of balance. Fut Neurol 3(3):309–323
14. Cuervo AM (2004) Autophagy: in sickness and in health. Trends Cell Biol 14(2):70–77
15. Komatsu M, Wang QJ, Holstein GR, Friedrich VL Jr, Iwata J, Kominami E, Chait BT, Tanaka K, Yue Z (2007) Essential role for autophagy protein Atg7 in the maintenance of axonal homeostasis and the prevention of axonal degeneration. Proc Natl Acad Sci U S A 104(36):14489–14494
16. Nishiyama J, Miura E, Mizushima N, Watanabe M, Yuzaki M (2007) Aberrant membranes and double-membrane structures accumulate in the axons of Atg5-null Purkinje cells before neuronal death. Autophagy 3(6):591–596
17. Plowey ED, Cherra SJ III, Liu YJ, Chu CT (2008) Role of autophagy in G2019S-LRRK2-associated neurite shortening in differentiated SH-SY5Y cells. J Neurochem 105(3): 1048–1056
18. Bjorkoy G, Lamark T, Brech A, Outzen H, Perander M, Overvatn A, Stenmark H, Johansen T (2005) p62/SQSTM1 forms protein aggregates degraded by autophagy and has a protective effect on huntingtin-induced cell death. J Cell Biol 171(4):603–614
19. Nixon RA, Wegiel J, Kumar A, Yu WH, Peterhoff C, Cataldo A, Cuervo AM (2005) Extensive involvement of autophagy in Alzheimer disease: an immuno-electron microscopy study. J Neuropathol Exp Neurol 64(2):113–122
20. Zhang Y, Murshid A, Prince T, Calderwood SK (2011) Protein kinase A regulates molecular chaperone transcription and protein aggregation. PLoS One 6(12):e28950. doi:10.1371/journal.pone.0028950
21. Ravikumar B, Duden R, Rubinsztein DC (2002) Aggregate-prone proteins with polyglutamine and polyalanine expansions are degraded by autophagy. Hum Mol Genet 11(9):1107–1117
22. Yu WH, Kumar A, Peterhoff C, Shapiro Kulnane L, Uchiyama Y, Lamb BT, Cuervo AM, Nixon RA (2004) Autophagic vacuoles are enriched in amyloid precursor protein-secretase activities: implications for beta-amyloid peptide

over-production and localization in Alzheimer's disease. Int J Biochem Cell Biol 36(12): 2531–2540

23. Jeffrey M, Scott JR, Williams A, Fraser H (1992) Ultrastructural features of spongiform encephalopathy transmitted to mice from three species of bovidae. Acta Neuropathol (Berl) 84(5):559–569
24. Zhu J, Wang KZ, Chu CT (2013) After the banquet: Mitochondrial biogenesis, mitophagy and cell survival. Autophagy 9(11)
25. Kanki T, Klionsky DJ (2009) Atg32 is a tag for mitochondria degradation in yeast. Autophagy 5(8):1201–1202
26. Tal R, Winter G, Ecker N, Klionsky DJ, Abeliovich H (2007) Aup1p, a yeast mitochondrial protein phosphatase homolog, is required for efficient stationary phase mitophagy and cell survival. J Biol Chem 282(8):5617–5624
27. Dagda RK, Das Banerjee T, Janda E (2013) How Parkinsonian toxins dysregulate the autophagy machinery. Int J Mol Sci 14(11): 22163–22189. doi:10.3390/ijms141122163
28. Youle RJ, Narendra DP (2011) Mechanisms of mitophagy. Nat Rev 12(1):9–14
29. Lemasters JJ (2005) Selective mitochondrial autophagy, or mitophagy, as a targeted defense against oxidative stress, mitochondrial dysfunction, and aging. Rejuvenation Res 8(1):3–5
30. Chu CT, Ji J, Dagda RK, Jiang JF, Tyurina YY, Kapralov AA, Tyurin VA, Yanamala N, Shrivastava IH, Mohammadyani D, Qiang Wang KZ, Zhu J, Klein-Seetharaman J, Balasubramanian K, Amoscato AA, Borisenko G, Huang Z, Gusdon AM, Cheikhi A, Steer EK, Wang R, Baty C, Watkins S, Bahar I, Bayir H, Kagan VE (2013) Cardiolipin externalization to the outer mitochondrial membrane acts as an elimination signal for mitophagy in neuronal cells. Nat Cell Biol. doi:10.1038/ncb2837
31. Dagda RK, Zhu J, Kulich SM, Chu CT (2008) Mitochondrially localized ERK2 regulates mitophagy and autophagic cell stress: implications for Parkinson's disease. Autophagy 4(6): 770–782
32. Geisler S, Holmstrom KM, Skujat D, Fiesel FC, Rothfuss OC, Kahle PJ, Springer W (2010) PINK1/Parkin-mediated mitophagy is dependent on VDAC1 and p62/SQSTM1. Nat Cell Biol 12(2):119–131. doi:10.1038/ncb2012
33. Gusdon AM, Chu CT (2011) To eat or not to eat: neuronal metabolism, mitophagy, and Parkinson's disease. Antioxid Redox Signal 14(10):1979–1987. doi:10.1089/ars.2010.3763
34. Whitworth AJ, Pallanck LJ (2009) The PINK1/Parkin pathway: a mitochondrial quality control system? J Bioenerg Biomembr 41(6): 499–503
35. Zhu J, Horbinski C, Guo F, Watkins S, Uchiyama Y, Chu CT (2006) Regulation of autophagy by extracellular signal regulated protein kinases during 1-methyl-4-phenylpyridinium induced cell death. Am J Pathol 170(1):75–86
36. Klionsky DJ, Abdelmohsen K, Abe A, Abedin MJ, Abeliovich H, Acevedo Arozena A, Adachi H, Adams CM, Adams PD, Adeli K, Adhihetty PJ, Adler SG, Agam G, Agarwal R, Aghi MK et al (2016) Guidelines for the use and interpretation of assays for monitoring autophagy (3rd edn). Autophagy 12(1):1–222. doi:10.1080/15548627.2015.1100356
37. Kabeya Y, Mizushima N, Ueno T, Yamamoto A, Kirisako T, Noda T, Kominami E, Ohsumi Y, Yoshimori T (2000) LC3, a mammalian homologue of yeast Apg8p, is localized in autophagosome membranes after processing. EMBO J 19(21):5720–5728
38. Kimura S, Noda T, Yoshimori T (2007) Dissection of the autophagosome maturation process by a novel reporter protein, tandem fluorescent-tagged LC3. Autophagy 3(5): 452–460
39. Dagda RK, Pien I, Wang R, Zhu J, Wang KZ, Callio J, Banerjee TD, Dagda RY, Chu CT (2014) Beyond the mitochondrion: cytosolic PINK1 remodels dendrites through protein kinase A. J Neurochem 128(6):864–877. doi:10.1111/jnc.12494
40. Yamamoto A, Tagawa Y, Yoshimori T, Moriyama Y, Masaki R, Tashiro Y (1998) Bafilomycin A1 prevents maturation of autophagic vacuoles by inhibiting fusion between autophagosomes and lysosomes in rat hepatoma cell line, H-4-II-E cells. Cell Struct Funct 23(1):33–42
41. Kaech S, Banker G (2006) Culturing hippocampal neurons. Nat Protoc 1(5):2406–2415. doi:10.1038/nprot.2006.356
42. Kim I, Lemasters JJ (2011) Mitochondrial degradation by autophagy (mitophagy) in GFP-LC3 transgenic hepatocytes during nutrient deprivation. Am J Physiol Cell Physiol 300(2): C308–C317. doi:10.1152/ajpcell.00056.2010
43. Zhu J, Dagda RK, Chu CT (2011) Monitoring mitophagy in neuronal cell cultures. Methods Mol Biol 793:325–339. doi:10.1007/978-1-61779-328-8_21
44. Schwarz TL (2013) Mitochondrial trafficking in neurons. Cold Spring Harb Perspect Biol 5(6). doi:10.1101/cshperspect.a011304
45. Dagda RK, Zaucha JA, Wadzinski BE, Strack S (2003) A developmentally regulated, neuron-specific splice variant of the variable subunit

Bbeta targets protein phosphatase 2A to mitochondria and modulates apoptosis. J Biol Chem 278(27):24976–24985

46. Dagda RK, Merrill RA, Cribbs JT, Chen Y, Hell JW, Usachev YM, Strack S (2008) The spinocerebellar ataxia 12 gene product and protein phosphatase 2A regulatory subunit Bbeta 2 antagonizes neuronal survival by promoting mitochondrial fission. J Biol Chem 283: 36241–36248
47. Rodriguez-Enriquez S, Kim I, Currin RT, Lemasters JJ (2006) Tracker dyes to probe mitochondrial autophagy (mitophagy) in rat hepatocytes. Autophagy 2(1):39–46
48. Dickey AS, Strack S (2011) PKA/AKAP1 and PP2A/Bbeta2 regulate neuronal morphogenesis via Drp1 phosphorylation and mitochondrial bioenergetics. J Neurosci 31(44):15716–15726
49. Li Z, Okamoto K, Hayashi Y, Sheng M (2004) The importance of dendritic mitochondria in the morphogenesis and plasticity of spines and synapses. Cell 119(6):873–887
50. Verstreken P, Ly CV, Venken KJ, Koh TW, Zhou Y, Bellen HJ (2005) Synaptic mitochondria are critical for mobilization of reserve pool vesicles at Drosophila neuromuscular junctions. Neuron 47(3):365–378
51. Chu CT, Plowey ED, Dagda RK, Hickey RW, Cherra SJ III, Clark RS (2009) Autophagy in neurite injury and neurodegeneration: in vitro and in vivo models. Methods Enzymol 453: 217–249
52. Cherra SJ III, Steer E, Gusdon AM, Kiselyov K, Chu CT (2013) Mutant LRRK2 elicits calcium imbalance and depletion of dendritic mitochondria in neurons. Am J Pathol 182(2):474–484. doi:10.1016/j.ajpath.2012.10.027
53. Kuma A, Matsui M, Mizushima N (2007) LC3, an autophagosome marker, can be incorporated into protein aggregates independent of autophagy: caution in the interpretation of LC3 localization. Autophagy 3(4):323–328
54. Ashrafi G, Schlehe JS, LaVoie MJ, Schwarz TL (2014) Mitophagy of damaged mitochondria occurs locally in distal neuronal axons and requires PINK1 and Parkin. J Cell Biol 206(5):655–670. doi:10.1083/jcb.201401070
55. Rantanen A, Jansson M, Oldfors A, Larsson NG (2001) Downregulation of Tfam and mtDNA copy number during mammalian spermatogenesis. Mamm Genome 12(10): 787–792
56. Wang KZ, Zhu J, Dagda RK, Uechi G, Cherra SJ III, Gusdon AM, Balasubramani M, Chu CT (2014) ERK-mediated phosphorylation of TFAM downregulates mitochondrial transcription: implications for Parkinson's disease. Mitochondrion 17:132–140. doi:10.1016/j.mito.2014.04.008

Chapter 14

Examining Mitochondrial Function at Synapses In Situ

Gregory T. Macleod and Maxim V. Ivannikov

Abstract

Synaptic mitochondria are exposed to an environment quite unlike any other subcellular environment. The cytosolic proton concentration ($[H^+]_c$) in the presynaptic compartment can increase several fold within seconds during a burst of action potentials (APs) while the cytosolic free Ca^{2+} concentration ($[Ca^{2+}]_c$) can increase several fold within milliseconds of a single AP. To understand how mitochondria function under such dynamic conditions they must be examined in the context of these small synaptic compartments. Here, we describe the application of fluorescent reporters to examine mitochondrial function in situ at the *Drosophila melanogaster* (fruit fly) larval neuromuscular junction (NMJ). Emphasis is placed on genetically encoded fluorescent (GEF) probes, rather than chemical fluorescent probes, due to the large range of GEF-probes now available and their specificity of targeting. We describe how best to prepare NMJs for ex vivo interrogation, apply stimuli to motor axons, and collect and analyze fluorescence data. We summarize the probes that have been used successfully at the *Drosophila* NMJ to monitor changes in mitochondrial $[Ca^{2+}]$, $[H^+]$, $[O_2 \cdot^-]$, [ATP] and voltage across the inner mitochondrial membrane (IMM). Lastly, for those who wish to generate transgenic *Drosophila* to express other GEF-probes, we list useful genetic reagents and signal sequences known to be effective at targeting GEF-probes to mitochondria in *Drosophila*. Overall, the techniques described here should provide a starting point for a diverse array of researchers who may wish to use GEF-probes targeted to mitochondria in *Drosophila* either as screening tools or as reporters of mitochondrial function at synapses in situ.

Key words Mitochondria, Synapse, Motor neuron, Neuromuscular junction, Calcium, Mitochondrial inner membrane potential, Mitochondrial matrix pH

1 Introduction and Background

1.1 Introduction

As an endosymbiont, the mitochondrion has become fully integrated with the biology of the host cell. However, the biology of the cell is far from uniform in spatiotemporal domains and the mechanisms that integrate mitochondria in one subcellular domain may not be the same in another subcellular domain. This diversity in domains is readily evident in highly polar cells such as neurons which form synapses—specialized sites for communication with other cells [1, 2]. The archetypal synapse is a small swelling immediately adjacent to a similar structure on another cell. Due to the high density of channels and transporters, the alacrity with

Stefan Strack and Yuriy M. Usachev (eds.), *Techniques to Investigate Mitochondrial Function in Neurons*, Neuromethods, vol. 123, DOI 10.1007/978-1-4939-6890-9_14, which they can be activated and inactivated, substantial transmembrane ionic gradients, and readily apparent limitations to solute diffusion, mitochondria in these small synaptic compartments are exposed to a highly dynamic biochemical environment unlike any other subcellular cytosolic domain. For example, $[Ca^{2+}]_c$ may increase several fold within milliseconds of an AP [3, 4], and $[H^+]_c$ may increase several fold within seconds during sustained activity [5]. Even the best microfluidics device would not be able to replicate such a dynamic environment. Presynaptic ATP consumption is also believed to be highly volatile [6, 7] due to the rapid onset of AP-triggered activity and a high concentration of ATP- and GTP-ases [8] in a very small volume. Therefore, to understand the ways in which mitochondria integrate with synaptic function, and ultimately to understand the ways in which they empower and possibly restrain neural function, synaptic mitochondria must be studied in the context of their extreme environment—in situ. Such an understanding may reveal the ways in which mitochondrial biology could be targeted as part of a therapeutic approach to neurological disorders and neurodegenerative diseases.

1.2 A Brief History

Excluding studies that have examined ensemble mitochondrial responses from synapses in cell culture, and synaptosomal preparations, the number of laboratories that have examined mitochondrial function at single synapses is exceedingly small. Pioneering studies of mitochondrial physiology at individually identified synapses in situ were performed by Gavriel David and Ellen Barrett on lizard [9, 10] and mouse [11–13] motor neuron terminals, where they examined mitochondrial Ca^{2+} uptake using chemical fluorescent probes. Billups and Forsythe [14] also directly examined mitochondrial Ca^{2+} uptake in the calyx of Held presynaptic terminal. Mitochondrial Ca^{2+} uptake in *Drosophila* motor nerve terminals was first reported in 2005 [15]. Subsequently, in the laboratory of Gavriel David, changes in the electrical gradient across the IMM (ψ_m) and mitochondrial NADH levels were measured at lizard motor nerve terminals [16], and more recently changes in ψ_m were measured in mouse motor nerve terminals [17]. The first application of GEF-probes at synapses in situ was performed in *Drosophila* by Chouhan and others [18, 19], closely followed by Fumiko Kawasaki [20]. GEF-probes have also been targeted to mouse motor neurons [21, 22].

1.3 Method Overview

The methods described here exploit the genetic tractability of *Drosophila* and the physical accessibility of motor nerve terminals in the larva. Perhaps the biggest impediment to anyone wishing to adopt these techniques is the dissection itself, but with perseverance, and the right tools, the adopter will find that the dissection requires no more dexterity than nerve-muscle dissections in any other organism. Importantly, with practice, this can be a quick dissection.

Table 1
Important steps in the method for examining mitochondria in *Drosophila* larval motor nerve terminals

Time	Protocol step	Equipment and materials
−9 days	Set adult *Drosophila* to mate and deposit embryos on food	Stereomicroscope, CO_2 perfused fly pushing equipment, fly food and fly vials
0 min	Select a larva	Fluorescence stereomicroscope
2 min	Dissect and pin larva	Stereomicroscope, dissection bath, fine scissors and forceps, dissection solution
7 min	Draw up hemi-segment nerve *en passant*	Micropipette, suction tube and filling filament
10 min	Exchange solution for final physiological solution and equilibrate preparation for 15 min, or begin incubation with chemical probes	Physiological solution, chemical probes
25 min	Move preparation onto the imaging rig	Compound fluorescence microscope, camera
26 min	Attach stimulus wires	Pulse stimulator and stimulus isolator
28 min	Find/focus an NMJ	High NA water-dipping objective
30 min	Run imaging protocol	Data acquisition system
32 min	Analyze image data	Image processing software
35 min	Export, collate, and analyze numerical data	Statistical analysis software

The time stamp on the *left hand side* refers to timing for an experienced practitioner when using GEF-probes. Essential equipment and materials are listed on the *right hand side*. As the equilibration time, from time = 10 to time = 25 min, represents down time when using GEF-probes, the total time investment for obtaining an analyzed record can be as little as 20 min, allowing for records from numerous animals in a day

The protocol is broken into sections to facilitate different points of entry, depending on your level of familiarity with each of *Drosophila* genetics, molecular biology, the larval dissection, and live fluorescence imaging. A list of the important steps in the method is given in Table 1 below to allow an appreciation of the scope of the protocol. While the protocol in the following sections is comprehensive, where possible we refer to a greater depth of detail that might be provided in other readily accessed sources.

2 Equipment and Materials

2.1 Major Equipment

The major piece of equipment required is a fluorescence compound microscope, which can be either widefield or confocal. The lasers and/or filter sets need to match the spectra of the fluorescent probes. If using a non-confocal microscope a fluorescence excitation system, camera and image acquisition system will be required. The best data will be collected using a water-dipping objective with a high numerical aperture (NA), e.g., ≥0.9. Most of the probes described in this protocol are bright and dynamic, and, when using a high NA objective, they yield a good signal such that a standard CCD camera is adequate to capture mitochondrial fluorescence. However, if long periods of monitoring are anticipated, and/or small numbers of mitochondria are being interrogated, then CCD cameras with high quantum efficiency are desirable. Numerous commercial image processing packages are available, but ImageJ in the public domain will also suffice as it is particularly flexible and offers many useful plugins (imagej.nih.gov/ij/). A stereomicroscope is essential for dissections, but it only needs to have an epi-fluorescence capacity if GEF-probe expressing larvae need to be screened before dissection. For dissections, dual goose-neck fiber optic illumination is preferable to a fiber optic ring illumination. The same stereomicroscope, fitted with a CO_2 line and a CO_2 pad to anaesthetize the adult flies, would be used for "fly-pushing." An isolated pulse stimulator can be used to deliver electrical pulses to the nerve held in the micropipette. AWG (gauge) 18 insulated copper strand wires connect the 26 gauge chlorided silver wires that enter the bath and micropipette to the poles of the isolated pulse stimulator.

2.2 Minor Equipment and Materials

For the dissection we use Vannas spring scissors from Fine Science Tools (FST; Cat.No.15000-08) and two pairs of fine forceps (Dumont #3 and #5). The baths that contain the dissection and in which all procedures are performed are custom made, Sylgard lined, and relatively disposable (Fig. 1a, and see ref. 23). Entomological pins used in the dissection are also available from FST (Cat.No.26002-10). Dental wax keeps the micropipettes in place and both are modified to suit as described in Rossano and Macleod [23]. All chemicals used for the physiological solutions described below are readily available from either Sigma-Aldrich or Life Technologies.

3 Methods

3.1 Experimental Animals and Probe Selection

A choice is available between chemical and GEF probes. Examples of data collected using chemical and GEF probes at *Drosophila* motor nerve terminals are shown in Fig. 2. Chemical probes can be applied regardless of the genotype, but an optimized protocol is required for their loading (concentration, vehicle, incubation time, temperature, wash time), with typical incubation and/or wash-out times of several hours. When using a GEF probe a lead time that incorporates a breeding program of more than a week is usually required, but the lead time before individual experiments is short as the mitochondria are preloaded with the GEF-probe. Also, a large

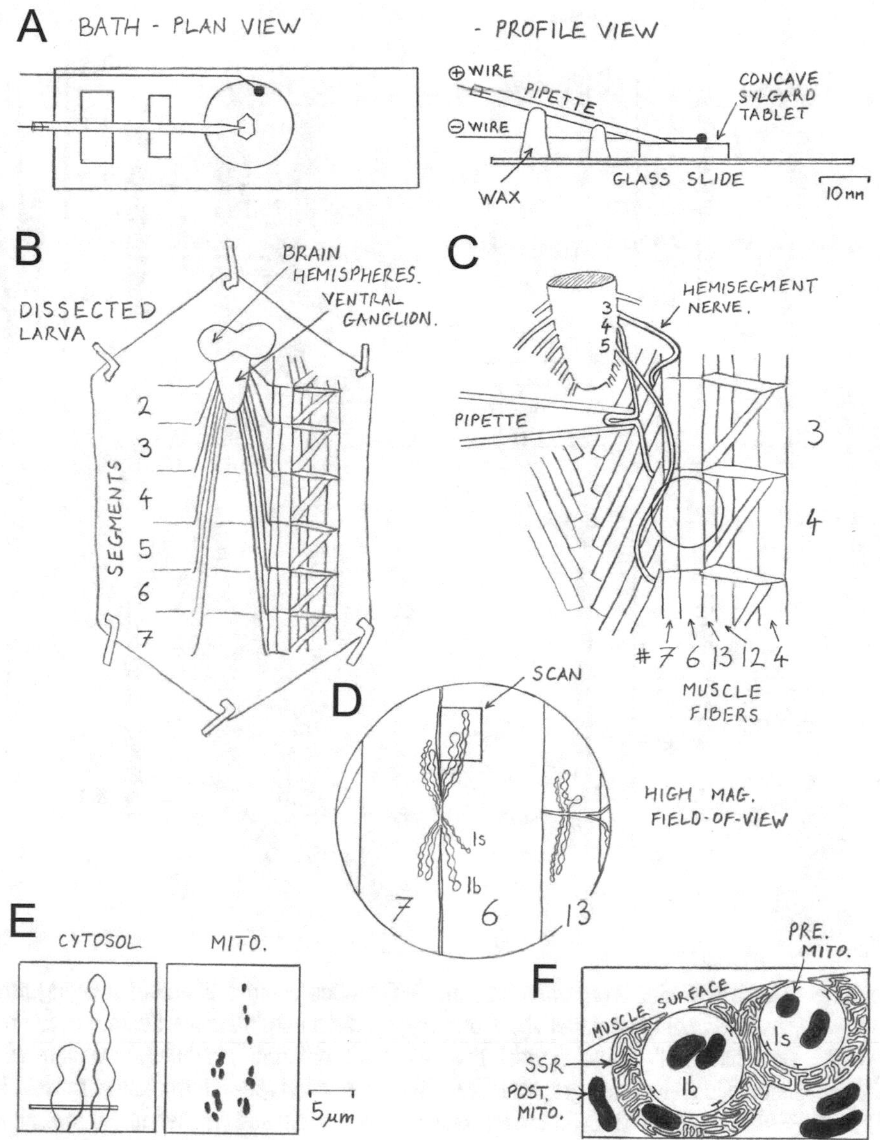

Fig. 1 Sketches depicting aspects of the bath and detail of the *Drosophila* larval fillet preparation and its neuromuscular junctions (NMJs). (**a**) Plan and profile views of a dissection dish (bath) consisting of a shallow concave Sylgard tablet on a 1″ × 3″ glass slide, adorned with a two-piece dental wax ramp supporting a glass micropipette that is used to draw up and stimulate a loop of a nerve. (**b**) A plan view of the *Drosophila* larval fillet preparation emphasizing the relative locations of the central nervous system, radiating hemi-segment nerves, abdominal segments and some body-wall muscle fibers on the *right hand side*. (**c**) A view of the ventral ganglion (severed from the brain hemispheres) with a nerve to abdominal segment #4 still intact and drawn into a glass micropipette tip in a short loop. In this sketch the oblique muscle fibers that terminate at the ventral midline feature prominently, but they were not shown in (**b**). (**d**) Three commonly examined muscle fibers and their innervating motor neuron terminals, shown in a field of view through a high NA objective (indicated with a *circle* in **c**). Type-II and type-III motor neuron terminals are not shown. (**e**) Further detail of nerve terminals on muscle fiber #6 within a nominal laser scan area, or cropped field of view of a CCD, revealed by fluorophores in the presynaptic cytosol versus fluorophores in the mitochondria (observed through different filter sets). (**f**) A cutaway view through the NMJ (location indicated in **e**) showing presynaptic and postsynaptic mitochondria, and the highly elaborated postsynaptic folds referred to as the sub-synaptic reticulum (SSR)

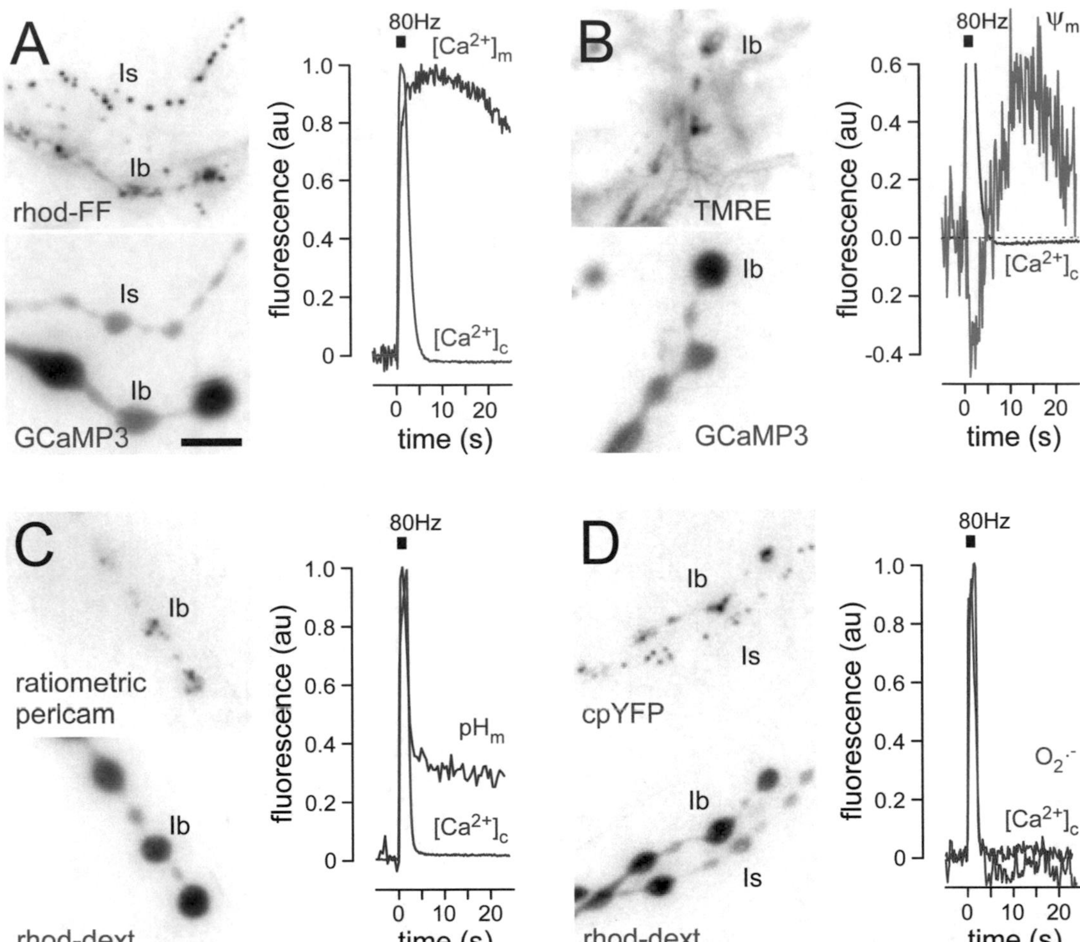

Fig. 2 Examples of data collected from chemical and GEF-probes using a standard imaging protocol. The images on the *left hand side* of each panel show presynaptic mitochondria loaded with a probe relative to a chemical or GE Ca^{2+} indicator in the cytosol revealed with different excitation/emission wavelengths. Fluorescence is rendered in an inverted grayscale. (**a**) Mitochondria in type-Ib terminals of muscle #13 loaded using 5 μM rhod-FF AM in 0.25% DMSO/0.05% Pluronic acid for 10 min and washed for 90 min, show a strong response to nerve stimulation (80 Hz/2 s). GCaMP3 in the cytosol of the same terminal shows the period of elevated $[Ca^{2+}]_c$. (**b**) Mitochondria in type-Ib terminals of muscle #13, loaded using 50 nM TMRE in 0.0005% DMSO/0.0001% Pluronic acid for 7 min at ~6 °C and washed for 120 min at room temperature, show a strong response to nerve stimulation (80 Hz/2 s). This trace indicates rapid depolarization followed by repolarization then hyperpolarization. GCaMP3 in the cytosol of the same terminal shows the period of elevated $[Ca^{2+}]_c$. (**c**) Mitochondria in type-Ib terminals of muscle #13 targeted with the ratiometric pericam, a GEF Ca^{2+}-indicator responsive to both Ca^{2+} change (420 nm ex.) and pH change (490 nm ex.—data shown in this panel), show a strong response to nerve stimulation (80 Hz/2 s). Rhod-dextran loaded into the cytosol of the same terminal by forward-filling [24] shows the period of elevated $[Ca^{2+}]_c$. (**d**) Mitochondria in type-Ib terminals of muscle #13 targeted with cpYFP, a putative superoxide anion reporter [25], show a strong response to nerve stimulation (80 Hz/2 s). Rhod-dextran loaded into the cytosol of the same terminal shows the period of elevated $[Ca^{2+}]_c$. All data were collected with an Andor Technology DU860 EMCCD camera on a widefield fluorescence microscope (Olympus BX51WI) fitted with a Sutter Instrument Company 150 W Xenon bulb DG4, and using a 100× 1.0NA water-dipping objective. Each trace represents a single trial without data averaging or smoothing

range of GEF probes are now available and they can be targeted with exquisite specificity at cellular and subcellular levels. For a recent review of the GEF probes that have been used in mitochondria see De Michele and others [26]. A prime consideration is that the signal of a GEF probe not be perturbed by changes in the concentration of entities that the experimenter does not intend to measure. The pH sensitivity of GFP-based probes is well known and this is an important consideration when choosing a GEF-probe for any mitochondrial compartment, especially in the matrix where we have reported pH changes of as much as 0.7 of a $\log_{10}$ unit within less than a second [27]. Secondly, ratiometric probes are preferable, as they allow for calibrations that are independent of GEF probe expression levels in different animals and the amount of protein present in an individual mitochondrion. Other considerations include, whether the GEF probe has the appropriate affinity for the range of concentrations that one expects to encounter, and whether there is spectral overlap with other fluorophores present. A summary of chemical and GEF probes that have been used successfully in *Drosophila* motor nerve terminals can be found in Table 2.

3.2 Larva Selection

The wandering 3rd instar is the preferred experimental larval stage as it is the largest and most easily dissected, and distinguishes itself for easy extraction (using a fine wetted paintbrush) from a culture by climbing out of the food. Where genetic crosses have yielded larvae with different genotypes a fluorescence stereomicroscope can be used to screen for those expressing the GEF-probe, providing the protein expresses in an anatomically large region and its resting fluorescence is high. For example, larvae expressing GEF-probes in motor neurons will be revealed by a detectable level of fluorescence in the ventral ganglion when the larva is examined through its ventral surface.

3.3 Dissection

The *Drosophila* larval NMJ preparation is best accessed in the fillet preparation. The dissection has been described or demonstrated in numerous manuscripts and video presentations, and will not be repeated here. We recommend a dissection as demonstrated by Rossano and Macleod [23]. For the uninitiated, a successful dissection will take several days to master. A successful dissection has no muscle fibers damage between the midline and body-wall muscle fiber #2, and all nerves are intact between the ventral ganglion and abdominal segment 8 (Fig. 1b). Damaged muscle fibers, which appear opaque, whitish or torn under a dissection stereomicroscope, will cause ongoing mechanical instability during fluorescence recordings.

The preparation should be allowed to equilibrate in the final physiological saline for 20 min between the time of the last shock (nerve "pinching" or solution exchange) and the time of data acquisition. Where chemical probes are applied and washed out over a long incubation period, the nerve should only be drawn up 20 min prior to the time of data acquisition.

Table 2
A summary of fluorescent probes that have been used to investigate mitochondrial function at the *Drosophila* neuromuscular junction (NMJ)

Chemical							
Measure	Name	K_d	$\lambda_{ex}/\lambda_{em}$	Load	Solvent	Wash	References
Ca^{2+}	Rhod-2 (AM)	0.57 μM	552/581	10 min@5 μM	0.25% DMSO	>30 min	[19]
	Rhod-FF (AM)	19 μM	552/580	"	"	"	[18, 19, 27]
	Mag-fluo-4 (AM)	22 μM	488/620	"	"	"	[18]
	Rhod-5N (AM)	320 μM	551/576	"	"	"	[19, 27]
ψm	TMRE	NA	546&573/590	7 min@50 nM	0.0005% DMSO	>30 min	[18, 27]
	$DiOC_2(5)$	NA	579/603	40s@40nM	0.0004% DMSO	>10 min	[15, 34]*
	2-Di-4-Asp	NA	488/607	5 min@5 μM	–	–	[33, 39]*
	Rhodamine-123	NA	511/534	10 min@26 μM	–	–	[32, 34]*
	JC-1	NA	490/527&590	5 min@6 μM	–	–	[15, 35]*
Genetically encoded							
Measure	Name	K_d/pKa	$\lambda_{ex.}/\lambda_{em.}$	Sig. seq.	Copy number	Gal4	References
Ca^{2+}	Ratiometric pericam	1.7 nM	415&494/517	2MT8	One (hetero)	OK6	[18, 19, 53]*
	Camgaroo 2	5.3 μM	490/528	"	"	"	[19, 54]*
	Yellow cameleon 2	1.2 μM	430/475&535	"	"	"	[19, 55]*
	TN-XXL	0.8 μM	430/475&528	"	"	"	[27, 56]
	D4cpv	64 μM	436/475&535	"	"	"	[27, 57]*
	D2cpv	0.3/3 μM	436/475&535	"	"	"	Unpubl., [57]*
	GCaMP3	0.66 μM	485/518	MT8	"	elav	[20, 21]
pH	mtAlpHi	8.5	498/522	2MT8	One (hetero)	OK6	[18, 58]*
	SypHer	8.7	431&482/535	"	"	"	Unpubl., [59]*
$O_2^{\bullet -}$	cpYFP	8.6	488/520	2MT8	One (hetero)	OK6	Fig. 2, [25]*
ATP	ATeams (AT1.03)	3.3 mM	435/475&525	2MT8	One (hetero)	OK6	Unpubl., [60]*

Chemical: K_d *values* indicate in vitro dissociation constants provided by the manufacturer. Wavelengths (λ) are reported in nm. The surfactant Pluronic acid is present with DMSO, at 20% of the concentration of DMSO (F127, Life Technologies). *Load/solvent/wash* refer to the loading conditions in situ only. *Genetically encoded*: K_d *values* indicate dissociation constants and pK_a *values* indicate acid dissociation constants provided by the original publisher of the GEF-probe. Filters are specified where absorbance and emission maxima are not given. *Gal4* refers to the driver line, e.g., OK6-Gal4 [61]. References that do not refer to *Drosophila* are indicated with an asterisk (*). *Unpubl.* refers to an unpublished transgenic fly containing the construct in the Macleod laboratory and data derived from the fly line

3.4 Solutions and Reagents

The signs of NMJ function such as postsynaptically recorded excitatory junction potentials (EJPs) are surprisingly tolerant of a range of solution compositions. Solutions used in *Drosophila* range from the standard solution [28], which closely resembles a frog Ringer, to those that attempt to mimic some of the distinctive properties of insect hemolymph where K^+ and sugar concentrations are high and Na^+ is low. The latter so-called hemolymph-like (HL) solutions—HL3 [29], HL3.1 [30] and HL6 [4] have been used successfully to preserve and monitor physiological properties of *Drosophila* NMJs in situ (see Table 3 for their detailed compositions). Providing your protocol does not require extended post dissection

Table 3

Chemical compositions of physiological solutions typically used for dye loading and imaging of *Drosophila* larval NMJs. Concentrations are shown in mM units

	Standard solution	HL3	HL3.1	HL6
1× stock solution components:				
NaCl	128.0	70.0	70.0	23.7
KCl	2.0	5.0	5.0	24.8
$MgCl_2$	4.0	20.0	4.0	15.0
$NaHCO_3$	–	10.0	10.0	10.0
Trehalose	–	5.0	5.0	80.0
Sucrose	35.5	115	115	–
HEPES[a]	5.0	–	5.0	–
BES[a]	–	5.0	–	5.0
Isethionic acid Na-salt	–	–	–	20.0
L-Alanine	–	–	–	5.7
L-Arginine HCl-salt	–	–	–	2.0
Glycine	–	–	–	14.5
L-Histidine	–	–	–	11.0
L-Methionine	–	–	–	1.7
L-Proline	–	–	–	13.0
L-Serine	–	–	–	2.3
L-Threonine	–	–	–	2.5
L-Tyrosine	–	–	–	1.4
L-Valine	–	–	–	1.0
Trolox[a]	–	–	–	1.0
TPEN[a]	–	–	–	0.0001
pH	7.3	7.3	7.1	7.2
Components added prior use:				
$CaCl_2$	1.8	1.5	1.5	0.5
L-Glutamate Na-salt	7.0	7.0	7.0	7.0

Note: 1× stock solutions are typically prepared in 0.5–1 L batches, aliquoted into 15 ml polyethylene conical tubes, and stored at −80 °C until use

[a]Full chemical names: BES: (*N,N*-bis[2-hydroxyethyl]-2-aminoethanesulfonic acid); HEPES: (*N*-[2-hydroxyethyl]piperazine-*N'*-[2-ethanesulfonic acid]); Trolox: 6-hydroxy-2,5,7,8-tetramethylchroman-2-carboxylic acid; TPEN: tetrakis-(2-pyridylmethyl) ethylenediamine

times, any of the solutions above would suffice assuming controls are performed in the same solution. We recommend using HL6 as a physiological solution, based on its abilities to support NMJ function over extended periods of time and to stave off signs of muscle deterioration. Schneider's Insect Medium (available from many chemical vendors) is probably more supportive of the fillet preparation, but it is excluded as a physiological solution as it is manufactured with $CaCl_2$ added and shows some autofluorescence. Techniques that require >45 min of preparation time would benefit from incubation in more supportive solutions such as HL6 or Schneider's before transfer into the final physiological solution.

3.5 Chemical Fluorescent Dye Loading of Mitochondria

A large range of chemical fluorescent dyes can be loaded with specificity into mitochondria in situ. The acetoxymethyl (AM) ester forms of small organic compounds with a net positive charge are particularly amenable to loading through application in the presence of DMSO (0.25%) and Pluronic acid (0.05%), e.g., the Ca^{2+} indicators; rhod-2, rhod-FF, rhod-5N, mag-fluo-4 [18, 19, 27]. Prior dye reduction with $NaBH_4$ may help [31] specificity of loading but is not always necessary. When applied at 5 μM, load times may be as short as 10 min, but to ensure that all of the excess dye has cleared from the cytosol wash times may extend to several hours. Dyes that report ψ_m through their Nernstian accumulation across the IMM can be loaded into the mitochondria even through exposure over a short period of time at concentrations in bathing media as low as 40 nM: TMRE, TMRM and rhodamine 123 [32], 4-Di-2-Asp [33], $DiOC_2(5)$ [34], and JC-1 [15, 35]. With these dyes it is always essential to know whether you are working in quench (high dye loading concentrations) or non-quench mode [36]. Also interpretation of any data collected from Nernstian dyes requires an understanding of how the dye exchanges between the IMM and plasma membrane [37].

Postsynaptic mitochondria can be loaded at the *Drosophila* NMJ with the same dyes that load presynaptic mitochondria (personal observation), but greater load times are required due to the much lower surface-to-volume ratio of muscle relative to axon terminals. The low surface-to-volume ratio also mitigates against clearing the muscle cytosol of dye even with the longest wash times. Fluorescence responses from postsynaptic mitochondria are generally not problematic, as the same solution additive that prevents muscle concentration during nerve stimulation and imaging (L-glutamic acid), also prevents Ca^{2+} entry through postsynaptic glutamate receptors. Although mitochondria might be found in the postsynaptic folds close to presynaptic terminal boutons (Fig. 1f), presynaptic versus postsynaptic loading of mitochondria is rarely problematic when intending to collect data from presynaptic mitochondria using GEF-probes as the genetics define the cell type in which the GEF-probe is expressed.

3.6 Preparation for Nerve Stimulation

Nerve transection is common, but we prefer to draw a short loop of the nerve into a glass micropipette with a 12 μm tip (inner diameter) (Fig. 1c). Application of short (0.4 ms) pulses of <5 V will still evoke an action potential that propagates into the nerve terminal. Capture of the nerve requires manual manipulation of a glass micropipette positioned on a dental wax ramp immediately adjacent to the dissected preparation (Fig. 1a) while applying suction via a syringe attached to the blunt end of the micropipette via a length of tubing (see ref. 23 for a demonstration of the technique). The entire lumen of the micropipette must be filled with solution from the bath to provide a conductor for nerve stimulation. This is achieved using a fine hollow filament (70 μm outer, 40 μm inner diameter) that can reach down the length of the micropipette. Filling the lumen without bubbles is important to ensure electrical connectivity.

3.7 Imaging Rig

Where the footprint of the dissection chamber (bath) is a microscope slide (Fig. 1a) it is trivial to attach to a microscope stage, but where the bath is larger, or the stage is unadorned, then two-sided adhesive tape represents a simple solution. If using a dental wax "ramp" on the side of the bath to support the micropipette that contains the nerve, then a separate micromanipulator is not required.

A critical component of the imaging "rig" is the quality of the objective, but with a high NA objective comes a limited field of view and this will make it difficult for most to find and identify nerve terminals. Apart from familiarity with the anatomy of muscle fibers and nerve branches in the larval fillet preparation (Fig. 1b), we would suggest finding the micropipette under the objective (possibly by first centering it in the field of view with a low magnification non-dipping objective) and then tracking along the micropipette to the nerve and the appropriate muscle fiber. Lastly, it is important to ensure that there are no bubbles under the objective, and this is usually done by lifting the objective out of the bath solution and dipping it back several times.

3.8 Nerve Stimulation & Stimulus Bus

Attachment of stimulation wires can be as simple as placing the tip of one chlorided silver wire in the bath and inserting the tip of the other chlorided silver wire into the blunt end of the micropipette, and taping the connecting insulated wires to the microscope stage. Alternatively a coarse manipulator or a wiring bus can be used to swing the wires into position and secure them in place. An essential step is to touch the wire destined for the micropipette lumen on the wire already in the bath to discharge any static electricity. Electrical pulse stimulators with built in isolation units (e.g., A-M Systems, Model 2100) represent a simple all-in-one solution to nerve stimulation, especially when they can be triggered by the data acquisition software. However, where multiple pulse trains of different frequencies and/or durations are required in the same imaging episode more flexible control may be required and for this we have adopted a Master-8 pulse stimulator (A.M.P.I.) to drive a separate stimulus isolator (Iso-Flex, A.M.P.I.).

3.9 Choosing a Synapse

While most motor neurons form glutamatergic NMJs in the periphery of *Drosophila* larvae, they are quite diverse in their morphology and function [18, 19, 38, 39]. The NMJ is best considered a relay synapse with a high neurotransmission safety factor which it achieves though a battery of release sites (active zones: AZs), each with a relatively high probability of release, and all in the one large synaptic connection or NMJ. The NMJs most accessible to interrogation are those formed by monosynaptic (innervating only one muscle fiber) motor neurons producing type-Ib (big bouton) terminals. Those on muscles fibers #13 and #4 generally form on the top of their respective muscle fibers - close to the objective. Polysynaptic motor neurons produce type-Is (small bouton) terminals immediately adjacent to type-Ib terminals and manifest slow endogenous firing rates. A third type of NMJ, referred to as type-II, is formed by a polysynaptic neuron, and is characterized by small presynaptic swellings without postsynaptic folds in the adjacent muscle [40]. These NMJs appear to release both glutamate and octopamine, and generally contain only a single AZ which may or may not be accompanied by a mitochondrion. Finally, a fourth type of NMJ is formed by a monosynaptic neuron on muscle #12, and appears to be peptidergic rather than glutamatergic. An excellent overview is provided by Hoang and Chiba [41].

3.10 Imaging Protocols

Your imaging protocol will largely depend on your question and the specifics of your experimental design, but a standard protocol used in our laboratory may be a useful starting point. Images (or pairs of images if using a ratiometric probe) are collected at a rate of 5 Hz [or, frames per second (fps)], and after a short period collecting baseline data (e.g., 10 s), a challenge is applied and data continues to be collected well into the post-challenge period (e.g., 30 s). A challenge of relevance to presynaptic mitochondria in *Drosophila* larva is a train of action potentials arriving in the presynaptic compartment at a rate close to the motor neuron's endogenous firing frequency [19] and for a period of time consistent with the peristalsis of larval locomotion [42]. The rate and duration for the type-Ib terminal on muscle 13 would be 42 Hz for 0.83 s. Each action potential in the nerve can be induced by the pulse stimulator, which should be triggered (with the appropriate delay) by the data acquisition software. Alternatively, the pulse stimulator may trigger the data acquisition software.

Several principles will assist in gathering data that most closely reflect physiological conditions:

1. Reduce illumination as much as possible to reduce photo-damage. For example maximize gain for fluorescence emission detection and minimize illumination (excitation) light intensity. Do not illuminate unnecessarily, e.g., when the preparation is not being viewed. Avoid oversampling. Reprogram software protocols to remove all illumination that is not accompanied by data collection.
2. Circulate the bath solution. Whether by the influence of a K^+ layer effect, anoxia, or nutrient depletion of the solution immediately adjacent to the NMJ, we have observed an increase in resting $[Ca^{2+}]_c$ when the solution between the preparation and the objective is allowed to remain unstirred for more than 10 min. Therefore, we recommend that the objective be lifted out of the solution and replaced within several minutes of acquiring data.
3. Beware of preconditioning. Even if the preparation is challenged without data collection it may constitute preconditioning. It may take up to 3 min for presynaptic mitochondria to lose Ca^{2+} subsequent to nerve stimulation [19, 43].

3.11 Analysis

The procedure for analysis of the fluorescence data will be dictated by capabilities of the software available, but there are some generic steps which are common to analyzing changes in the fluorescence of objects in a larger field of view. The protocol outlined for the analysis of a cytosolic Ca^{2+} reporter in *Drosophila* motor nerve terminals [44] can also be used for presynaptic mitochondria. This recommendation is based on the understanding that the only probe fluorescing with given spectra in the terminal is that targeted to the mitochondria. It is particularly difficult to deconvolve challenge-specific signals coming from the same probe in different compartments when they cannot be spatially resolved. Should a signal emanate from mis-targeted probe in the cytosol, more work should be put into loading mitochondria with specificity, rather than trying to deconvolve signals (as roughly demonstrated [45]). Movement is often the bane of imaging synapses on contractile tissue such as skeletal muscle, and movements of as little as 1–2 μm over a period of several seconds can degrade data quality. Small movements in the *X*–*Y* dimension can be corrected using image processing algorithms such as ImageJ plugins, e.g., Image Stabilizer, but large movements in the *Z* dimension are more difficult to correct especially in confocal imaging systems.

A standard that we have adhered to for discriminating test from control, or similarly, phenomenon from chance event, is a single trial/measurement from each of six different animals and using standard statistical methods. For this purpose, numerical data should be exported from your image processing software and collated within a statistics capable spreadsheet.

3.12 Calibration Protocols for GEF-Probes

The purpose of calibration is to establish a probe's apparent affinity for a metabolite and its dynamic range in the conditions that closely approximate the probe's intracellular environment. One approach is to perform calibration in-vitro in a cuvette or capillary when sufficient quantities of the probe are available. When attempting to calibrate GEF Ca^{2+}-indicators, calibration parameters obtained from an in vitro calibration may be useful as a reference, but they generally underestimate Ca^{2+} concentration when applied to in-situ measurements. The more practical approach, particularly for GEF Ca^{2+}-indicators, is to calibrate the expressed indicator directly in the tissue of interest by selectively permeabilizing the cell membrane with Ca^{2+} ionophores such as ionomycin or A23187. Unlike calibration of cytosolic GEF Ca^{2+}-indicators, calibration of mitochondrial GEF Ca^{2+}-indicators presents more challenges as it requires an efficient permeabilization of the plasma membrane and both mitochondrial membranes. Some success in calibrating mitochondrial GEF Ca^{2+}-indicators has been achieved by using high concentrations of ionomycin (>10 μM) together with mitochondrial uncouplers [27]. Mitochondrial uncouplers are added to prevent active potential-driven Ca^{2+} accumulation in mitochondria.

Below is a detailed description of a typical calibration procedure for mitochondrially expressed FRET based GEF Ca^{2+}-indicators such as D4cpV and TN-XXL in *Drosophila* motor nerve terminals:

(a) Preparation of standard Ca^{2+} solutions: Standard Ca^{2+} concentration solutions are prepared by mixing various volumes of two different HL3 (HL6 solution could be used in place of HL3) solutions, one Ca^{2+}-free supplemented with 5 mM Na_2EGTA and the other one containing 5 mM Na_2EGTA and 5 mM $CaCl_2$. Mixing of Ca^{2+} with EGTA results in acidification, which requires pH adjustment back to 7.3 in the final solutions unless strong pH buffers (MOPS, BTP), or BAPTA, a less pH sensitive Ca^{2+} chelator, is used. Free Ca^{2+} concentration in the final solution can be calculated using standard equations [46] or more conveniently, using a web-based calculator Maxchelator (http://maxchelator.stanford.edu). Calibration solutions with free Ca^{2+} concentrations >5 μM used for low-affinity Ca^{2+} indicators are prepared without adding EGTA, as it has a limited Ca^{2+} buffering capacity in that range.

(b) Calibration procedure: Before calibration, 5 mM ionomycin and 10 mM CCCP stock solutions in DMSO are added to all standard Ca^{2+} solutions to the final concentrations of 50 μM and 10 μM, respectively. It is recommended to start calibration by determining indicator's minimum F_{EYFP}/F_{ECFP} emission ratio in Ca^{2+}-free solution (R_{min}). To determine R_{min}, *Drosophila* larval fillet preparations are incubated with Ca^{2+}-free HL3 solution supplemented with 5 mM Na_2EGTA, 50 μM ionomycin and 10 μM CCCP for 30 min to release and bind internally stored Ca^{2+}, followed by an additional 10 min incubation in a fresh aliquot of the same solution and fluorescent measurements. Intermediate *R* values for each preparation are obtained after ~15 min incubation with successively increasing Ca^{2+} concentrations. R_{max} values are typically measured last by adding HL3 containing 10 mM $CaCl_2$ and supplemented with 50 μM ionomycin and 10 μM CCCP. It should be noted that calibration solutions with free Ca^{2+} concentration >1 μM result in considerable muscle contraction, which can make identification of presynaptic mitochondria more difficult.

(c) Calibration curve plotting and fitting: Obtained ratios or FRET fraction values $((R - R_{min})/(R_{max} - R_{min}))$ are typically plotted against standard Ca^{2+} concentrations converted to a logarithmic scale (base 10). Fluorescent responses of calmodulin-based FRET Ca^{2+} indicators follow a simple sigmoid relationship with Ca^{2+} and can be well fitted with one binding site four parameter Hill's equation:

$$(R - R_{min}) = (R_{max} - R_{min}) \times \left(\left[Ca^{2+}\right]^n / \left[Ca^{2+}\right]^n + K^n_{d,app}\right)$$

n—Hill's coefficient and $K_{d,\,app}$ indicator's apparent Ca^{2+} dissociation constant are easily derived by fitting data using SigmaPlot or other curve fitting software. Average R_{max} and R_{min} values derived from calibrations, and calculated *n* and $K_{d,\,app}$ are used in the equation above to calculate Ca^{2+} concentration from experimentally measured *R*.

3.13 Molecular Biology & Genetics

GEF-probes can be targeted to mitochondria in defined cell types using molecular biological and genetic techniques that are well established for *Drosophila*. Binary expression systems allow for a relatively straightforward means of expressing transgenes under cell specific and temporal control [47, 48]. The short period of time needed to make transgenic flies (2–3 months) along with their low cost has made this technology attractive for many laboratories. Tens of thousands of *Drosophila* "driver" lines are available off-the-shelf. These lines specify the identity of the cells that will express the first part of the binary system, the transcriptional activator. Therefore, the challenge is only to make the second part of the binary system, i.e., the "reporter" line. This line is made by cloning the cDNA of a GEF-probe, along with a mitochondrial targeting sequence, into a transformation vector and stably transform an animal. Once the reporter line is established, it is just a single cross with the appropriate driver line to obtain larvae with the GEF-probe targeted to mitochondria in the cell type of interest. Transformation vectors for making reporter lines in both the Gal4/UAS system (e.g., JFRC14) and the LexA/LexAop system (e.g., JFRC19) are available from either the Drosophila Genome Resource Center (DGRC) or the HHMI Janelia Farm Research Campus. Commercial injection services are available for a reasonable fee to inject plasmids into embryos for stable germ line transformation.

Matrix targeting: We have had success targeting GEF-probes to mitochondrial matrices using both a tandem repeat of the first 36 amino acids of subunit VIII of human cytochrome oxidase (COX) (2MT8; [49]), and the first 12 amino acids of subunit IV of yeast COX (MT4). 2MT8 was designed to increase the mitochondrial localization efficiency of GEF-probe in cultured cells (in vitro), but we have rarely been troubled by signals that appear to derive from non-mitochondrially located GEF-probe in nerve terminals. However, this may simply be the happy result of most mitochondrial proteins being translated and translocated into mitochondria close to the soma, and only arriving at the nerve terminals if transported within a mitochondrion.

Mitochondrial intermembrane space (IMS) targeting: Although we have attempted to target GEF-probes to the IMS using glycerol phosphate dehydrogenase leading sequence (GPD/MIMS), the same signal sequence as used by Porcelli and others [50] in cultures, we obtained no progeny expressing the GEF-probe, and concluded that targeting with GPD to this space was lethal for the organism. We have heard reports that targeting to the IMS using part of a mitochondrial AAA protease (iAAA) is effective [51], but success has so far eluded us with this approach.

Outer mitochondrial membrane (OMM) targeting: As the mitochondrion might be viewed as a signaling hub with the outer surface of the OMM harboring signaling complexes, it may be informative to place GEF-probes in this domain. This has been achieved in cultured cells by György Hajnóczky and colleagues [52] and the same construct (AKAP-FKBP-ratiometric pericam) expresses in *Drosophila* motor nerve terminals with an apparent OMM localization (unpublished data).

4 Notes/Troubleshooting Tips

1. Problem: Preparation shows no fluorescence response during stimulation.

   Potential solutions: Verify electrical circuit continuity: the wire is immersed in the solution inside the stimulating micropipette and no bubbles are present.

2. Problem: Probe resting fluorescence/expression levels are weak.

   Potential solutions: GEF probe expression can be moderately increased by maintaining larvae at a higher temperature. Alternatively, broadband rather than narrowband emission filters might be used to increase emission light intensity. Overexposure by increasing illumination light intensity is generally not recommended.

3. Problem: The nerve is unresponsive to nerve stimulation and muscles show continual, slow movements.

   Potential solutions: The nerve might have been damaged by static electricity during stimulating wire insertion. Ensure that the output wires from the isolated-pulse stimulator are earthed and that the wire to be inserted into the stimulating micropipette is discharged by touching it to the reference wire in the bath.

Acknowledgments

G.T.M. was supported by NIH NINDS awards NS061914 and NS083031.

References

1. Szabadkai G, Simoni A et al (2006) Mitochondrial dynamics and Ca^{2+} signaling. Biochim Biophys Acta 1763(5):442–449
2. Lee CW, Peng HB (2008) The function of mitochondria in presynaptic development at the neuromuscular junction. Mol Biol Cell 19(1):150–158
3. Helmchen F, Borst J et al (1997) Calcium dynamics associated with a single action potential in a CNS presynaptic terminal. Biophys J 72(3):1458
4. Macleod G, Hegström-Wojtowicz M et al (2002) Fast calcium signals in *Drosophila* motor neuron terminals. J Neurophysiol 88(5):2659–2663
5. Rossano AJ, Chouhan AK et al (2013) Genetically encoded pH-indicators reveal activity-dependent cytosolic acidification of *Drosophila* motor nerve termini in vivo. J Physiol 591(7):1691–1706
6. Harris JJ, Jolivet R et al (2012) Synaptic energy use and supply. Neuron 75(5):762–777
7. Rangaraju V, Calloway N et al (2014) Activity-driven local ATP synthesis is required for synaptic function. Cell 156(4):825–835
8. Weingarten J, Laßek M et al (2014) The proteome of the presynaptic active zone from mouse brain. Mol Cell Neurosci 59:106–118
9. David G, Barrett JN et al (1998) Evidence that mitochondria buffer physiological Ca^{2+} loads in lizard motor nerve terminals. J Physiol 509(1):59–65
10. David G (1999) Mitochondrial clearance of cytosolic Ca^{2+} in stimulated lizard motor nerve terminals proceeds without progressive elevation of mitochondrial matrix [Ca^{2+}]. J Neurosci 19(17):7495–7506
11. David G, Barrett EF (2000) Stimulation-evoked increases in cytosolic [Ca^{2+}] in mouse motor nerve terminals are limited by mitochondrial uptake and are temperature-dependent. J Neurosci 20(19):7290–7296
12. David G, Barrett EF (2003) Mitochondrial Ca^{2+} uptake prevents desynchronization of quantal release and minimizes depletion during repetitive stimulation of mouse motor nerve terminals. J Physiol 548(2):425–438
13. David G, Talbot J et al (2003) Quantitative estimate of mitochondrial [Ca2+] in stimulated motor nerve terminals. Cell Calcium 33(3): 197–206
14. Billups B, Forsythe ID (2002) Presynaptic mitochondrial calcium sequestration influences transmission at mammalian central synapses. J Neurosci 22(14):5840–5847
15. Guo X, Macleod GT et al (2005) The GTPase dMiro is required for axonal transport of mitochondria to *Drosophila* synapses. Neuron 47(3):379–393
16. Talbot J, Barrett JN et al (2007) Stimulation-induced changes in NADH fluorescence and mitochondrial membrane potential in lizard motor nerve terminals. J Physiol 579(3): 783–798

17. Nguyen KT, García-Chacón LE et al (2009) The Ψ_m depolarization that accompanies mitochondrial Ca^{2+} uptake is greater in mutant SOD1 than in wild-type mouse motor terminals. Proc Natl Acad Sci 106(6):2007–2011
18. Chouhan AK, Ivannikov MV et al (2012) Cytosolic calcium coordinates mitochondrial energy metabolism with presynaptic activity. J Neurosci 32(4):1233–1243
19. Chouhan AK, Zhang J et al (2010) Presynaptic mitochondria in functionally different motor neurons exhibit similar affinities for Ca^{2+} but exert little influence as Ca^{2+} buffers at nerve firing rates in situ. J Neurosci 30(5):1869–1881
20. Lutas A, Wahlmark CJ et al (2012) Genetic analysis in *Drosophila* reveals a role for the mitochondrial protein p32 in synaptic transmission. G3 2(1):59–69
21. Tian L, Hires SA et al (2009) Imaging neural activity in worms, flies and mice with improved GCaMP calcium indicators. Nat Methods 6(12):875–881
22. Chen Q, Cichon J et al (2012) Imaging neural activity using Thy1-GCaMP transgenic mice. Neuron 76(2):297–308
23. Rossano AJ, Macleod GT (2007) Loading *Drosophila* nerve terminals with calcium indicators. J Vis Exp 6:250
24. Macleod GT (2012) Forward-filling of dextran-conjugated indicators for calcium imaging at the *Drosophila* Larval NMJ. Cold Spring Harb Protoc 7:791–796
25. Wang W et al (2008) Superoxide flashes in single mitochondria. Cell 134:279–290
26. De Michele R, Carimi F et al (2014) Mitochondrial biosensors. Int J Biochem Cell Biol 48:39–44
27. Ivannikov MV, Macleod GT (2013) Mitochondrial free Ca2+ levels and their effects on energy metabolism in *Drosophila* motor nerve terminals. Biophys J 104(11): 2353–2361
28. Jan L, Jan Y (1976) Properties of the larval neuromuscular junction in *Drosophila melanogaster*. J Physiol 262(1):189–214
29. Stewart B, Atwood H et al (1994) Improved stability of *Drosophila* larval neuromuscular preparations in haemolymph-like physiological solutions. J Comp Physiol A 175(2):179–191
30. Feng Y, Ueda A et al (2004) A modified minimal hemolymph-like solution, HL3.1, for physiological recordings at the neuromuscular junctions of normal and mutant *Drosophila* larvae. J Neurogenet 18(2):377–402
31. Hajnóczky G, Robb-Gaspers LD et al (1995) Decoding of cytosolic calcium oscillations in the mitochondria. Cell 82(3):415–424
32. Scaduto RC, Grotyohann LW (1999) Measurement of mitochondrial membrane potential using fluorescent rhodamine derivatives. Biophys J 76(1):469–477
33. Magrassi L, Purves D et al (1987) Fluorescent probes that stain living nerve terminals. J Neurosci 7(4):1207–1214
34. Yoshikami D, Okun LM (1984) Staining of living presynaptic nerve terminals with selective fluorescent dyes. Nature 310:53–56
35. Reers M, Smiley ST et al (1995) Mitochondrial membrane potential monitored by JC-1 dye. Methods Enzymol 260:406–417
36. O'Reilly CM, Fogarty KE et al (2003) Quantitative analysis of spontaneous mitochondrial depolarizations. Biophys J 85(5): 3350–3357
37. Gerencser AA, Chinopoulos C et al (2012) Quantitative measurement of mitochondrial membrane potential in cultured cells: calcium-induced de- and hyperpolarization of neuronal mitochondria. J Physiol 590(12):2845–2871
38. Atwood H, Govind C et al (1993) Differential ultrastructure of synaptic terminals on ventral longitudinal abdominal muscles in *Drosophila* larvae. J Neurobiol 24(8):1008–1024
39. Kurdyak P, Atwood H et al (1994) Differential physiology and morphology of motor axons to ventral longitudinal muscles in larval *Drosophila*. J Comp Neurol 350(3):463–472
40. Koon AC, Ashley J et al (2011) Autoregulatory and paracrine control of synaptic and behavioral plasticity by octopaminergic signaling. Nat Neurosci 14(2):190–199
41. Hoang B, Chiba A (2001) Single-cell analysis of *Drosophila* larval neuromuscular synapses. Dev Biol 229:55–70
42. Klose MK, Chu D et al (2005) Heat shock-mediated thermoprotection of larval locomotion compromised by ubiquitous overexpression of Hsp70 in *Drosophila melanogaster*. J Neurophysiol 94(5):3563–3572
43. García-Chacón LE, Nguyen KT et al (2006) Extrusion of Ca^{2+} from mouse motor terminal mitochondria via a Na^+–Ca^{2+} exchanger increases post-tetanic evoked release. J Physiol 574(3):663–675
44. Macleod GT (2012) Imaging and analysis of nonratiometric calcium indicators at the *Drosophila* larval neuromuscular junction. Cold Spring Harb Protoc 7:802–809
45. Macleod GT (2012) Topical application of indicators for calcium imaging at the *Drosophila* larval NMJ. Cold Spring Harb Protoc 7:786–790
46. Bers DM, Patton CW et al (1994) A practical guide to the preparation of Ca^{2+} buffers. Methods Cell Biol 40:3–29

47. Brand AH, Perrimon N (1993) Targeted gene expression as a means of altering cell fates and generating dominant phenotypes. Development 118(2):401–415
48. Venken KJ, Simpson JH et al (2011) Genetic manipulation of genes and cells in the nervous system of the fruit fly. Neuron 72(2): 202–230
49. Filippin L, Abad MC et al (2005) Improved strategies for the delivery of GFP-based Ca^{2+} sensors into the mitochondrial matrix. Cell Calcium 37(2):129–136
50. Porcelli AM, Ghelli A et al (2005) pH difference across the outer mitochondrial membrane measured with a green fluorescent protein mutant. Biochem Biophys Res Commun 326(4):799–804
51. Rainey RN, Glavin JD et al (2006) A new function in translocation for the mitochondrial i-AAA protease Yme1: import of polynucleotide phosphorylase into the intermembrane space. Mol Cell Biol 26(22):8488–8497
52. Csordás G, Várnai P et al (2010) Imaging interorganelle contacts and local calcium dynamics at the ER-mitochondrial interface. Mol Cell 39(1):121–132
53. Nagai T (2001) Circularly permuted green fluorescent proteins engineered to sense Ca^{2+}. Proc Natl Acad Sci U S A 98:3197–3202
54. Griesbeck O, Baird GS et al (2001) Reducing the environmental sensitivity of yellow fluorescent protein. J Biol Chem 276: 29188–29194
55. Arnaudeau S, Kelley WL et al (2001) Mitochondria recycle Ca^{2+} to the endoplasmic reticulum and prevent the depletion of neighboring endoplasmic reticulum regions. J Biol Chem 276:29430–29439
56. Mank M, Santos AF et al (2008) A genetically encoded calcium indicator for chronic in vivo two-photon imaging. Nat Methods 5(9):805–811
57. Palmer A, Tsien RY (2006) Measuring calcium signaling using genetically-targetable fluorescent indicators. Nat Protoc 1:1057–1061
58. Abad MCF, Di Benedetto G et al (2004) Mitochondrial pH monitored by a new engineered green fluorescent protein mutant. J Biol Chem 279:11521–11529
59. Poburko D et al (2011) Dynamic regulation of the mitochondrial proton gradient during cytosolic calcium elevations. J Biol Chem 286:11672–11684
60. Imamura et al. (2009). Visualization of ATP levels inside single living cells with fluorescence resonance energy transfer-based genetically encoded indicators. Proc Natl Acad Sci U S A 106: 15651–15656.
61. Aberle H, Haghighi AP et al (2002) wishful thinking encodes a BMP type II receptor that regulates synaptic growth in *Drosophila*. Neuron 33(4):545–558

Chapter 15

Proteomic Analysis of Neuronal Mitochondria

Kelly L. Stauch and Howard S. Fox

Abstract

Neurons critically depend on mitochondrial function to establish membrane excitability and to execute the complex processes of neurotransmission and plasticity. Although mitochondria are central for various cellular processes important for neuronal function that include ATP production, intracellular calcium signaling, and generation of reactive oxygen species, our understanding of the roles of mitochondria in neurons and intact brain tissue are limited. Proteins carry out an array of mitochondrial functions, with structural, transporter, hormone, enzyme, and receptor protein classes working in complexes and pathways to serve these critical functions. Knowledge of the protein composition of mitochondria and how it changes under different conditions will yield important information to further our insight into mitochondrial processes. Here we describe the implementation of mass spectrometry (MS)-based proteomics to the study of neuronal mitochondria, with emphasis on quantitative approaches. This chapter describes the isolation of mitochondria from primary neuronal cultures and brain tissue nerve terminals, sample preparation for liquid chromatography dual MS (LC-MS/MS), the LC-MS/MS procedure, and the subsequent data processing steps to compare global mitochondrial proteomic alterations. Identifying mitochondrial proteins that are altered within neurons during development and disease is of central importance for understanding the multifaceted roles of mitochondria.

Key words Mass spectrometry, Mitochondria, Neurons, Proteomics, Synaptosomes

1 Introduction

Mitochondria are ubiquitous organelles, critical for cellular survival and function [1]. The functions of these specialized organelles vary according to the cell type or tissue in which they reside and take part in many critical functions in neurons, where their dysfunction is recognized to play an important role in neurodegenerative disorders. Further, mitochondria dynamically undergo morphology changes through regulated processes of fusion and fission as well as actively trafficking between cellular locations such as the soma, axon, and presynaptic terminals of neurons [2]. In fact, mitochondria respond acutely to the metabolic environment by undergoing morphological and functional changes [3]. Due to their metabolic demands neurons are particularly sensitive to mitochondrial

Stefan Strack and Yuriy M. Usachev (eds.), *Techniques to Investigate Mitochondrial Function in Neurons*, Neuromethods, vol. 123, DOI 10.1007/978-1-4939-6890-9_15, © Springer Science+Business Media LLC 2017

alterations, thus changes in mitochondrial proteins affecting their function may influence fundamental aspects of brain physiology.

Our understanding of the involvement of mitochondrial proteomic changes in neuronal processes is limited. While it is clear that mitochondria have roles important for neuronal activity and health, these organelles are highly dynamic and the molecular mechanisms underlying their function are not completely understood. Identifying key proteins and alterations in their expression is an important step towards functional characterization. Studying the quantitative changes in the neuronal mitochondrial proteome during different biological events is one way to elucidate this. Several groups have shown both temporal and spatial regulation of the mitochondrial proteome in different tissues, suggesting the expression of mitochondrial proteins may be an important determinant of their function within distinct tissues, cells, and cellular locations [4–6].

Here, we discuss how MS can be used to determine the identity of mitochondrial proteins that change significantly throughout a given cellular event. Specifically, we introduce MS-based quantitative proteomics approaches and describe the experimental and analytical steps to study the mitochondrial proteome in primary neuronal cultures and brain tissue nerve terminals—from sample preparation (Fig. 1) to data analysis using two distinct techniques, stable-isotope labeling by amino acids in cell culture (SILAC)-MS and sequential window acquisition of all theoretical spectra (SWATH)-MS (Fig. 2) [7, 8]. These techniques provide reproducible detection and accurate quantification of mitochondrial proteins from neuronal samples [6, 9, 10].

1.1 Super-SILAC Mix for Quantitative Proteomics of Neuronal Mitochondria

Technological advancement in the field of mass spectrometry has lead to improvements in the identification and quantification of tissue proteomes. One example of such developments is the use of a mix of SILAC-labeled cell lines as an internal standard for the quantification of a tissue of interest, a technique termed super-SILAC [7, 11]. The mixture of SILAC-labeled cell lines (the super-SILAC mix) must be chosen to provide labeled peptides that accurately represent the proteome of the tissue of interest. The application of the super-SILAC technique has allowed for the analysis of mitochondria from isolated nerve terminals (synaptosomes, which contain presynaptic mitochondria) to study the effects of aging on the presynaptic mitochondrial proteome in mice as well as the proteome differences between presynaptic and nonsynaptic mitochondria [6, 10]. The super-SILAC technique can be utilized for the application of mass spectrometry-based proteomics for the study of mitochondria from isolated nerve terminals and primary neuronal cultures derived from different animal models.

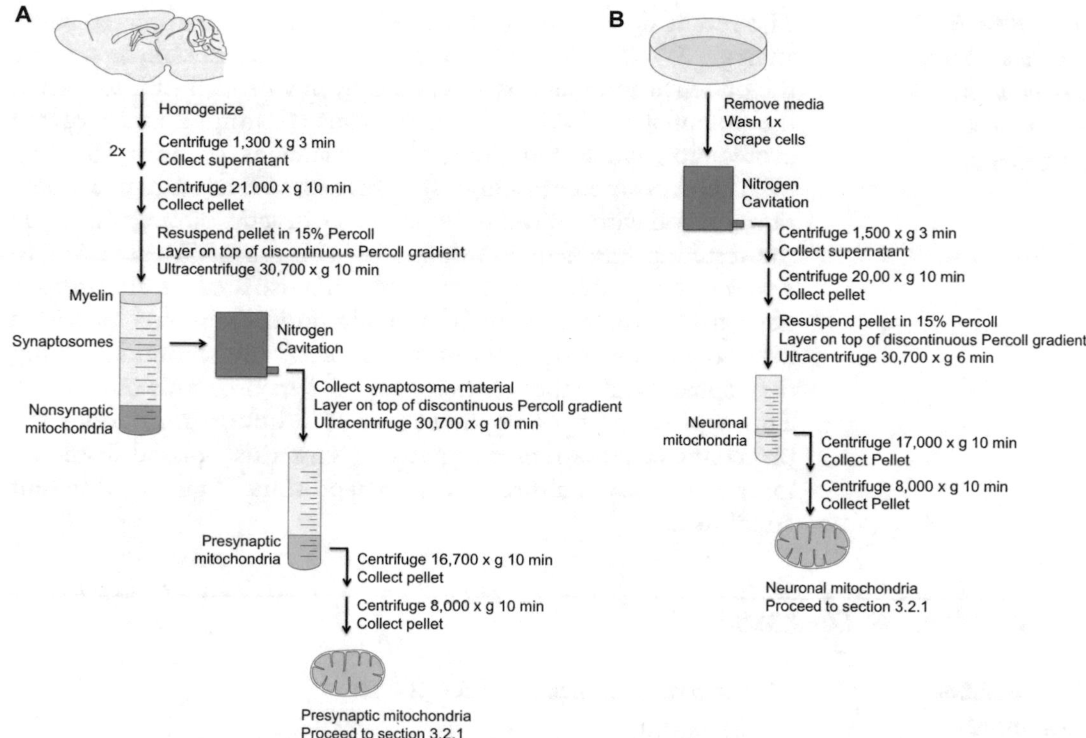

Fig. 1 Sample preparation. (**a**) Isolation of presynaptic mitochondria from nerve terminals obtained from brain tissue (Sect. 3.1.1). (**b**) Isolation of mitochondria from primary cultures of neurons (Sect. 3.1.2)

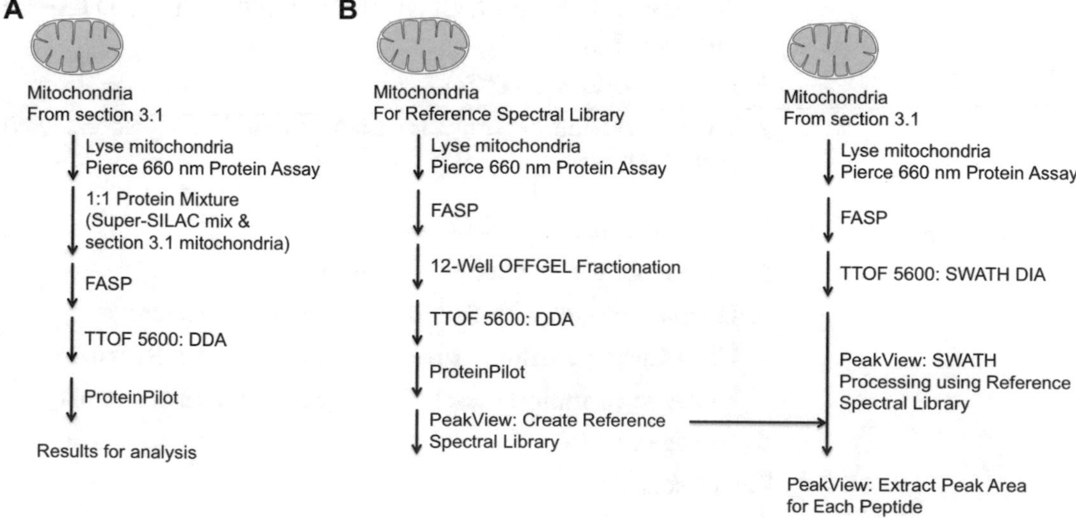

Fig. 2 Mass spectrometry sample and data processing. (**a**) Workflow for super-SILAC sample preparation (Sects. 3.2 and 3.3) and data processing (Sect. 3.4). (**b**) Workflow for SWATH-MS sample preparation (Sects. 3.2 and 3.3) and data processing (Sects. 3.5 and 3.6)

1.2 SWATH-MS for Quantitative Proteomics of Neuronal Mitochondria

The recent development of a label-free mass spectrometry-based strategy, SWATH-MS, that systematically queries sample sets for the presence and quantity of essentially any protein of interest has the potential to alleviate the limitations of shotgun and targeted proteomics, while exploiting the advantages of both [8, 12]. SWATH-MS combines a high specificity data-independent acquisition method with a novel targeted data extraction strategy to mine the resulting fragment ion data sets. The use of SWATH-MS to analyze the proteome of mitochondria isolated from primary neuronal cultures and brain mitochondria from rats provided invaluable information about the role of mitochondria during development and in the context of familial Parkinson's disease [9, 13]. This approach can be applied to acquire information about the proteomic changes that occur in mitochondria isolated from primary neuronal cultures and synaptosomes under different conditions.

2 Materials and Equipment

2.1 Isolation of Neuronal Mitochondria

2.1.1 Reagents

1. Sucrose (also need in Sect. 2.2).
2. Mannitol.
3. Ethylene glycol tetraacetic acid (EGTA).
4. 4-(2-hydroxyethyl)-1-piperazineethanesulfonic acid (HEPES).
5. Fatty-acid-free bovine serum albumin (BSA).
6. Percoll (GE Healthcare).
7. Protease inhibitor cocktail tablets (cOmplete Mini, EDTA-free from Roche).
8. Hydrochloric acid (HCl).
9. Tris(hydroxymethyl)aminomethane (Tris)-base (also need in Sect. 2.2).

2.1.2 Equipment

1. Dounce homogenizer (2 mL).
2. Centrifuge with a fixed rotor bucket.
3. Ultracentrifuge (with SW40 Ti rotor from Beckman).
4. Ultra-Clear centrifuge tubes (14 × 95 mm from Beckman).
5. Nitrogen cavitation vessel (45 mL from Parr Instruments).
6. Nitrogen tank.
7. Cell scraper.
8. Polypropylene tubes (15 mL).
9. Magnetic stir bar and stir plate.

2.1.3 Solutions to Prepare

1. *Isolation Media (IM)*: 70 mM sucrose, 210 mM mannitol, 1 mM EGTA, 5 mM HEPES, 0.5 % (w/v) fatty-acid-free BSA, made in deionized, distilled water, pH to 7.2 with Tris-base.
   (a) For brain tissue homogenation and subfractionation, add protease inhibitor cocktail tablets (1 tablet per 10 mL).
2. *100 % Percoll solution*: 70 mM sucrose, 210 mM mannitol, 1 mM EGTA, 5 mM HEPES, 0.5% (w/v) fatty-acid-free BSA, made in Percoll, pH 7.2 with HCl.
   (a) For 15, 21, 24, 40, and 50 % (vol/vol) Percoll, dilute 100 % Percoll solution with appropriate volume of IM.

2.2 Mitochondrial Sample Preparation for Mass Spectrometry

2.2.1 Reagents

1. LC-MS/MS grade water.
2. Sodium dodecyl sulfate (SDS).
3. Dithiothreitol (DTT).
4. LC-MS/MS grade 0.1 % formic acid (also need in Sect. 2.3).

2.2.2 Equipment

1. Oasis MCX 1 cc column (Waters).
2. Centrifugal vacuum concentrator with rotor for 1.5–2.0 mL microcentrifuge tubes, e.g., Savant SpeedVac Concentrator (Thermo Scientific).
3. NanoDrop 2000 (Thermo Scientific).
4. 11 mm Snap-It Cap for Autosampler Vials, 6 mm hole (Thermo Scientific).
5. National Limited Volume Wide Opening Plastic Crimp Top Autosampler Vials, 450 μL capacity (Thermo Scientific).

(Optional)

1. Mitochondrial Isolation Kit for Cultured Cells (Mitosciences).
2. Mitochondria Isolation Kit (Miltenyi Biotec).
3. Agilent 3100 OFFGEL Fractionator.
4. Pierce C-18 PepClean Spin Columns.
5. HRM calibration kit standard peptides (Biognosys).

2.2.3 Solutions to Prepare

1. *Mitochondrial storage buffer*: 250 mM sucrose, 10 mM Tris-base, made in LC-MS/MS grade water, pH to 7.2 with HCl.
2. *Lysis buffer*: 4 % (w/v) SDS, 100 mM Tris-base, 0.1 M DTT, made in LC-MS/MS grade water, pH to 7.6 with HCl.

2.3 Mass Spectrometry

2.3.1 Reagents

1. LC-MS/MS grade acetonitrile.

2.3.2 Equipment

1. cHiPLC-nanoflex system (Eksigent).
2. Nano-cHiPLC analytical column 200 μm × 15 cm ChromXP C18-CL 3 μm 120 Å (Eksigent)—for super-SILAC.
3. Nano-cHiPLC analytical column 75 μm × 15 cm ChromXP C18-CL 3 μm 300 Å (Eksigent)—SWATH-MS.
4. Nano-cHiPLC trap column 200 μm × 6 mm ChromXP C18-CL 3 μm 120 Å (Eksigent)—for super-SILAC.
5. Nano-cHiPLC trap column 75 μm × 0.5 mm ChromXP C18-CL 3 μm 120 Å (Eksigent)—for SWATH-MS.
6. Trap-elute jumper chip (Eksigent).
7. TripleTOF 5600 mass spectrometer equipped with a NanoSpray II Ion Source (AB SCIEX).

2.3.3 Solutions to Prepare

1. Solution A: 0 % aqueous solution with 0.1 % formic acid.
2. Solution B: 100 % acetonitrile with 0.1 % formic acid.

2.4 Data Processing: Super-SILAC and Generating the SWATH-MS Reference Spectral Library

2.4.1 Equipment

1. 8-Core computer with ProteinPilot installed and licensed (AB SCIEX).

2.5 Data Processing: Targeted Data Extraction for SWATH-MS

2.5.1 Equipment

1. Computer with PeakView v. 2.1 software installed and licensed (AB SCIEX) and the add-on, Protein Quantitation 1.0 MicroApp, installed and licensed.

3 Methods

3.1 Isolation of Neuronal Mitochondria

3.1.1 Isolation of Presynaptic Mitochondria from Nerve Terminals

The following protocol for the isolation of presynaptic mitochondria has been modified from previously published methods [14]. All steps should be performed on ice or at 4 °C.

Brain Tissue Homogenization

1. Obtain fresh brain tissue from "experimental" and "control" samples of interest.
2. Place brain tissue into petri dish on ice containing ice-cold IM.
3. Chop the brain tissue into small pieces and rinse once with ice-cold IM. Resuspend the tissue pieces in IM (2 mL for every 400 mg).

4. Transfer the tissue suspension into a 2 mL Dounce homogenizer.
5. Homogenize the sample using 10 up and down strokes (see **Note 1**).

Brain Homogenate Subfractionation

1. Transfer the homogenate into a 15 mL polypropylene tube and centrifuge the homogenate at 1300 × *g* for 3 min at 4 °C.
2. Transfer the supernatant (see **Note 2**) into a 35 mL Nalgene conical tube and place on ice.
3. Resuspend the pellet with 5 mL of IM and centrifuge at 1300 × *g* for 3 min at 4 °C.
4. Combine this supernatant (see **Note 2**) with the previously collected supernatant in the 35 mL Nalgene conical tube and centrifuge at 21,000 × *g* for 10 min at 4 °C.
5. Prepare the lower two layers of the Percoll gradient by adding 4 mL of 24 % Percoll into a Ultra-Clear centrifuge tube (14 × 95 mm) then slowly introduce 4 mL of 40 % Percoll to the bottom of the tube under the 24 % Percoll creating a discontinuous gradient of 24 % Percoll on top of the 40 % Percoll.
6. Discard the supernatant from the previous centrifugation, resuspend the pellet in 4 mL of 15 % Percoll, and layer this material slowly on top of the 24 % Percoll.
7. Centrifuge at 30,700 × *g* for 10 min at 4 °C using slow acceleration and deceleration (see **Note 3**).
8. The material banding near the interface of the upper two layers of the gradient contains synaptosomes (see **Note 4**), collect this material with an adjustable 1 mL volumetric pipette and transfer it to a 15 mL polypropylene tube containing 2 mL of IM to yield a suspension of synaptosomes in about 12% Percoll.

Releasing Presynaptic Mitochondria from Synaptosomes

1. Transfer the synaptosomal suspension into a precooled nitrogen cavitation vessel containing a stir bar and attach the nitrogen tank.
2. Increase the pressure inside the vessel slowly to 1500 psi by letting the nitrogen gas enter the vessel, then stir the suspension inside the vessel by magnetic stirrer for 15 min (see **Note 5**).
3. Connect outflow tubing to the valve located at the bottom of the vessel, release and collect the material. Layer this material on the top of 8 mL of 24% Percoll in a Ultra-Clear centrifuge tube (14 × 95 mm) and centrifuge at 30,700 × *g* for 10 min at 4 °C using slow acceleration and deceleration.
4. Collect the loose pellet at the bottom of the tube containing the free presynaptic mitochondria and add it to a 15 mL polypropylene tube containing 6 mL of IM, then centrifuge at 16,700 × *g* for 10 min at 4 °C.

5. Remove the supernatant, resuspend the mitochondrial pellet using 4.5 mL of IM, and centrifuge at 8000 × *g* for 10 min at 4 °C.
6. Remove the supernatant and keep the pellet, which contains the presynaptic mitochondria for proteomic analysis. Proceed directly to mitochondrial lysate preparation (Sect. 3.2.1).

3.1.2 Isolating Mitochondria from Primary Neurons

Primary cultures of neurons should be prepared from the brain region and animal model of interest using established protocols [15]. At least 1 × 10^8 cells will be required for the mitochondrial isolation and subsequent proteomic analysis. Mitochondria are isolated from primary neurons using a modified version of a previously described method [16].

Harvesting Primary Neurons

1. Remove media and wash (once) the primary neurons gently with IM being careful not to detach the cells.
2. Collect the cells by scraping off the dish/flask in IM.

Purification of Mitochondria

1. Transfer cell suspension to a precooled nitrogen cavitation vessel containing a magnetic stir bar.
2. Pressurize the vessel to 1500 psi and stir for 15 min (see **Note 5**).
3. Release the cell suspension through outflow tubing attached to the valve at the bottom of the vessel.
4. Centrifuge the collected material at 1500 × *g* for 3 min at 4 °C to pellet the cell debris.
5. Transfer the supernatant to a 35 mL Nalgene conical tube and centrifuge at 20,000 × *g* for 10 min at 4 °C.
6. Resuspend the resulting pellet (mitochondrial fraction) in 10 mL of 15 % Percoll.
7. Layer the mitochondrial suspension on top of a preformed discontinuous gradient of 4 mL of 21 % Percoll layered over 4 mL of 50 % Percoll in a 14 mL Ultra-Clear tube.
8. Centrifuge the gradient at 30,700 × *g* for 6 min and collect the material banding at the lower interface (between the 21 and 50 % Percoll layers), which contains mitochondria.
9. Dilute the mitochondrial fraction with IM (1:8) and centrifuge at 17,000 × *g* for 10 min at 4 °C.
10. Resuspend the pellet in 5 mL of IM and centrifuge at 8000 × *g* for 10 min at 4 °C. The resulting pellet contains purified neuronal mitochondria. Proceed directly to mitochondrial lysate preparation (Sect. 3.2.1).

3.2 Mitochondrial Sample Preparation for Mass Spectrometry

The following protocol for preparing the mitochondria for mass spectrometry utilizes the filter-aided sample preparation (FASP) method [17].

3.2.1 Preparation of Mitochondrial Lysate

1. Wash the mitochondrial pellets using 1 mL storage buffer (twice) and centrifugation at 8000 × *g* for 10 min at 4 °C (see **Note 6**).
2. Sonicate (speed 2 for about 2 s) the mitochondrial pellets in lysis buffer at room temperature (see **Note 7**).
3. Heat samples at 95 °C for 5 min.
4. Centrifuge at 8000 × *g* for 5 min.
5. Collect supernatant and determine protein concentration using Pierce 660 nm Protein Assay. These samples can be stored at −80 °C until sample processing for mass spectrometry.

3.2.2 Sample Processing for Mass Spectrometry

1. Trypsin digest the mitochondrial lysates (we recommend using 50 μg of protein) using the FASP method, which is compatible with SDS and has several benefits over in-gel or in-solution approaches [17] (see **Note 8**).
   (a) (Required for super-SILAC) Mitochondrial super-SILAC mix: For super-SILAC experiments, mitochondria should be isolated from the SILAC-labeled cell lines (see **Note 9**) using the Mitochondrial Isolation Kit for Cultured Cells (Mitosciences) which can be followed by anti-TOM22 immunomagnetic isolation (Miltenyi Biotec Mitochondria Isolation Kit) if high purity is desired [18]. The isolated heavy labeled mitochondrial lysates should be prepared as described in Sect. 3.2.1. The mitochondrial super-SILAC mix should be prepared by mixing equal amounts of the heavy mitochondrial lysate from each of the individual cell lines. This mitochondrial super-SILAC mix needs to be combined with the unlabeled lysate from either the presynaptic or cultured primary neuronal mitochondria in a 1:1 protein ratio prior to FASP.
2. Use Oasis mixed mode cation exchange cartridges following the manufacturer's recommendations to desalt the resultant peptides.
3. Dehydrate the desalted peptides using a vacuum concentrator.
4. Resuspend the peptides in a minimal volume of 0.1 % formic acid in LC-MS/MS grade water (see **Note 10**). Quantify the peptides based on peptide bond spectral absorbance at 205 nm [(peptide concentration) (mg/mL) = 31 (absorbance at 205 nm)] using a NanoDrop 2000 [19]. Transfer an aliquot (6 μg for super-SILAC or 2 μg for SWATH-MS) of the peptides from each sample to a clean autosampler vial (see **Note 11**). If the final volume is <6 μL, bring the solution to 6 μL using 0.1 %

formic acid in LC-MS/MS grade water. If the final volume is > 6 μL, dessicate the sample and resuspend in 6 μL 0.1 % formic acid in LC-MS/MS grade water.

5. (Optional for super-SILAC and SWATH-MS reference spectral library samples) Isoelectric focusing: To overcome the challenge of measuring the presence and abundance of individual peptides within the complex mixture present in the sample, isoelectric focusing of the resultant peptides can be performed using an Agilent 3100 OFFGEL Fractionator to yield 12 fractions from pH 3–10. The peptides from each fraction are prepared for mass spectrometry using Pierce C-18 PepClean Spin Columns according to the manufacturer's protocol followed by dehydration and resuspension in the appropriate volume of 0.1 % formic acid in LC-MS/MS grade water.
6. (Optional for SWATH-MS DIA) Spiking in peptides: Artificial peptides that have various predicted elution times can be added to the database and to each experimental sample. The difference in experimental elution times from predicted elution times can then be determined allowing for correction of the elution profile to enable better matching of the SWATH library reference spectra to the experimentally obtained DIA spectra. For this, add 1 μL of artificial peptides from the HRM calibration kit to each DIA sample in place of 1 μL of 0.1 % formic acid in LC-MS/MS grade water to maintain the final volume of 6 μL.

3.3 Mass Spectrometry

1. The Eksigent cHiPLC system requires three chips: the trap column that is used during sample loading, the cHiPLC column that is used for elution, and a trap-and-elute jumper chip. Replace nano cHiPLC columns if necessary (see **Note 12**).
2. The LC method for LC-MS/MS analyses of the digested peptides is provided in Table 1 (see **Note 13**). The cHiPLC columns should be equilibrated. If using the Eksigent LC system, insert pre-run flush for 0.1 min using 100 % initial flow rate into the LC in the second tab of Eksigent method.
3. The mass spectrometer should be operated in high sensitivity mode for data acquisition (data-dependent acquisition [DDA] is used for super-SILAC experimental samples and SWATH-MS reference spectral library samples; while data-independent acquisition [DIA] is used for SWATH-MS experimental samples). See Table 2 for mass spectrometry data acquisition methods. For information on samples for generation of the super-SILAC mix and SWATH-MS reference spectral library (see **Notes 9** and **14**, respectively).
4. Transfer the autosampler vial(s) prepared in step 4 of Sect. 3.2.2 to the autosampler and assign the samples to queue

Table 1
LC method

Sample loading		
Flow mode	Independent	
Time (min)	Flow soln. A (μL/min)	Flow soln. B (μL/min)
0	10	0
8.5	10	0
9	2	0
10	2	0
Elution		
Flow mode	Conserved total flow rate = 1 μL/min (super-SILAC) = 300 nL/min (SWATH-MS)	
Time (min) super-SILAC/SWATH-MS	% Soln. A	% Soln. B
0/0	95	5
0/0	95	5
1/1	95	5
180/60	65	35
182/62	10	90
192/72	10	90
193/73	95	5
200/90	95	5

accordingly (see **Note 15**). To avoid breaking the autosampler needle or unequal sample uptake, ensure that the autosampler vial lids are on flush and the tubes are not crooked.

5. Start the queue and monitor sample elution during the run using the Analyst program to visualize the total ion current (TIC) chromatogram (see **Note 16**).
6. When the mass spectrometry runs are complete, the TIC's can be overlaid in PeakView software using the open multiple WIFF tool for chromatogram comparison and evaluation of any differences between the samples (see **Note 16**).
7. Transfer all files to the hard drive of the computer that will be used to process the collected data (see **Note 17**).

3.4 Data Processing: Super-SILAC

1. Compile the FASTA file that will be used for database searching (see **Note 18**). For this step, we have used the UniProt-SwissProt (www.uniprot.org) database to export the reference proteomes for several species. A word processor, such as

Table 2
Mass spectrometry methods

DDA (super-SILAC and SWATH-MS library samples)	
Charge state	From +2 to +5
Intensity threshold	>100 counts
Switch after	50 Spectra
Advanced settings	True
Always exclude	True
Exclude for	15 s
Mass tolerance units	mDa
Mass tolerance	50
Use inclusion list	False
Use exclusion list	False
Ignore peaks within	6 Da
Real time	None
Dynamic collision energy	True
Fragment intensity multiplier	2
Maximum accumulation	2 s
Allow standard filters for smart IDA	True
Number of cycles	7615
Polarity	Positive
Period cycle time	2798 ms
Pulser frequency	14.170 kHz
ISVF (ion spray voltage floating)	2400 V
Pulser frequency	14.170 kHz
Precursor scan (MS1) experiment type	TOF MS
MS1 accumulation time	250.0 ms
MS1 start mass	400.0 Da
MS1 end mass	1800.0 Da
Precursor fragmentation (MS2) experiment type (50 selections)	TOF MS^2
MS2 accumulation time	50.0 ms
MS2 start mass	100.0 Da
MS2 end mass	1800.0 Da

(continued)

Table 2 (continued)

DIA (SWATH-MS experimental samples)	
Number of cycles	3555
Polarity	Positive
Period cycle time	3363 ms
Pulser frequency	14.170 kHz
ISVF (ion spray voltage frequency)	2400 V
Pulser frequency	14.170 kHz
Precursor scan (MS1) experiment type	TOF MS
MS1 accumulation time	50.0 ms
MS1 start mass	400.0 Da
MS1 end mass	1250.0 Da
SWATH-MS (MS2) experiment type	TOF MS^2
MS2 accumulation time	96.0 ms
MS2 start mass	100.0 Da
MS2 end mass	1800.0 Da
SWATH-MS experiment mass window	25 Da + 1 Da overlap
Fragment conditions	Rolling collision energy, charge state +2, Collision energy spread of 15 V

Notepad or Sublime Text, can be used to merge the files and add any additional FASTA sequences, such as the list of common laboratory contaminants provided by AB SCIEX. Transfer the FASTA file to the databases folder within the AB SCIEX ProteinPilot Application folder.

2. Launch ProteinPilot. In the workflow tasks panel, under the "Identify Proteins" tab select "LC." Use the "Add" button to add DDA samples to the search file (see **Note 19**). Process the file using a new Paragon method (see Table 3) and save the method using the "Save As" button. Back in the "Identify Proteins" dialog box, save the results file and assign its location using the "Save As" button. Click the "Process" button to begin the search. The file that is generated is a .group file.
3. Export the Peptide and Protein data as .txt files.
4. Open the .txt files using Excel. Remove proteins from the analysis identified as being in the reverse database and contaminants identified through the database search. The additional cutoff values of Unused ProtScore $\geq$1.3 and number of unique peptides $\geq$2 should also be applied to the data.

Table 3
Paragon method

Describe sample	Super-SILAC/SWATH-MS
Sample type	SILAC (Lys+6, Arg+10)/Identification
Cys alkylation	Iodoacetamide
Digestion	Trypsin
Instrument	TripleTOF 5600
Special factors	None selected
Species	Select particular species of interest
Specify processing	
Quantitate, bias correction, background correction	Quantitate/not able to be selected
ID focus	Biological modifications
Database	FASTA database compiled in step b
Search effort	Thorough ID
Results quality	Detected protein threshold >0.05 (10 %)
Run false discovery rate analysis	Checked

5. Protein ratios are determined using the heavy super-SILAC mix as an internal standard. The protein ratios are normalized to the mix and expressed as light-to-heavy (L/H, sample/super-SILAC internal standards).
6. Proceed to Sect. 3.7.

3.5 Data Processing: Generating the SWATH-MS Reference Spectral Library

1. Compile and transfer the FASTA file that will be used for database searching (see **Note 19**) to the databases folder within the AB SCIEX ProteinPilot Application folder.
2. Include artificial peptides in the FASTA database (optional for DIA, see **Note 20**). Open the FASTA database in Notepad or Sublime Text and provide an entry (the sequence and name) for each artificial peptide (the spiked-in peptides).
3. Launch ProteinPilot. In the workflow tasks panel, under the "Identify Proteins" tab select "LC." Use the "Add" button to add DDA samples to the search file. Process the file using a new Paragon method (see Table 3) and save the method using the "Save As" button. Back in the "Identify Proteins" dialog box, save the results file and assign its location using the "Save As" button. Click the "Process" button to begin the search. The file that is generated is a .group file.

3.6 Data Processing: Targeted Data Extraction for SWATH-MS

1. For targeted data extraction using the DIA-generated files, the computer must have PeakView installed and licensed with the Protein Quantitation MicroApp installed.
2. Launch PeakView. Under the "Quantitation" menu, click "Import Ion Library." Select the .group file that was generated in step 1 of Sect. 3.6 (see **Note 21**). After the library has been uploaded, a dialog box will automatically appear and request selection of the SWATH-MS files that will be used for targeted data extraction. Select all the files that will be used for export.
3. (Optional) If using spiked-in peptides, select the spiked-in peptides that will be used for retention time (RT) correction. In PeakView, search for the protein of interest and click the peptides that will be used for correction so that a check mark is apparent next to the peptide sequence. Next, click the "Add RT-Cal" button to add selected peptides to the RT calibration set. To edit the set of peptides used for RT calibration, click the "Edit RT-Cal" button and select peptides for deletion. Use the "Edit-RT-Cal" tool to calculate RT fit and apply RT modifications.
4. Click the processing settings button under the SWATH Processing dialog box (after RT correction, if applied). Set the processing settings. We recommend using five peptides, five transitions, 95 % peptide confidence threshold, 1 % false discovery rate, exclude shared peptides, XIC window of 10 min, and XIC width of 50 ppm. These settings will require optimization dependent on the samples (see **Note 22**).
5. Click "Process" to perform targeted data extraction. Following processing, export all information using the Quantitation menu, click "SWATH Processing", select export and all to generate an .txt file that can be opened with Excel or an .mrkvw file that can be opened with MarkerView.
6. (Optional) Normalize in MarkerView. Often when analyzing proteomics data, it may be necessary to normalize to correct for preparation errors. We recommend normalization using MarkerView, which offers a wide variety of normalization parameters that can be chosen depending on the experimental design. To perform the normalization open the extracted ion chromatogram (XIC) file in MarkerView. Then, choose the best normalization method based upon the experimental design (see **Note 23**).
7. Proceed to Sect. 3.7.

3.7 Statistical Analysis and Data Interpretation

1. Statistical analysis of proteomics data enables the evaluation of biologically significant changes; however, common multiple testing corrections (e.g., Bonferroni) tend to be too stringent

when comparing multiple proteins across multiple samples. To overcome this issue, Bayesian analysis, which can analyze high-dimensional data by demonstrating that the data must follow the rules of probability according to the Bayes' theorem [20], followed by multiple testing correction is a viable method for complex proteomics data.

2. CyberT (http://cybert.ics.uci.edu), an online Bayesian analysis calculator, can be used to analyze the statistical significance of protein expression changes in high-dimensional mass spectrometry data [21].
3. Format the data for CyberT upload according to the online calculator recommendations. Select the correct Bayesian analysis parameters following CyberT instructions for sliding window size and Bayesian confidence value. Select "Compute multiple test corrections" under "Standard multiple hypothesis testing corrections" and compute the Posterior Probability of Differential Expression (PPDE).
4. Proteins with a p-value <0.05 and a Cumulative PPDE >0.95 are considered to be significantly altered between samples. Cumulative PPDE corrects the PPDE to a false discovery rate of 0.05.
5. Export the data as a .txt file that can be opened using Excel.

3.8 Limitations of Mass Spectrometry-Based Quantitative Proteomics

Despite the continuous improvement of mass spectrometry-based proteomic approaches, quantification relies on the ability to detect small changes in protein and peptide abundance in response to cellular alterations. Thus, the development of methods for accurate protein quantitation remains challenging. The measures of relative and absolute abundances of proteins using stable isotope labeling and label-free methods have their own strengths and limitations, which must be considered when designing each study. Additionally, to study the mitochondrial proteome, enrichment and fractionation of mitochondrial proteins may prove beneficial for shotgun proteomic experiments. However, it is important to note that the isolation of mitochondria disrupts the intricate network that exists in the cell and results in the loss of the cellular context, which must be considered when interpreting the data. Further, the purity of the mitochondria must be assessed to ensure the sample being studied is indeed mitochondrial in nature. In addition to these caveats, adequate biological replicate numbers, proper controls, consistency in sample preparation, processing, and storage must be taken in to consideration when designing proteomics experiments and interpreting the results.

4 Notes

1. Perform the up and down strokes gently, avoid creating negative pressure under the pestle and between the pestle and the sample during the up stroke. The use of 10 strokes/sample allows for proper homogenization and maintenance of functional mitochondria.
2. Avoid collecting the fluffy, loose material from the top of the pellet by carefully transferring the supernatant.
3. Employ slow acceleration and deceleration speeds to avoid disturbing the Percoll gradient. No brake may be necessary for the deceleration step.
4. The material accumulating on the top of the gradient contains predominantly myelin and can be discarded. The material banding near the interface of the upper two layers of the gradient containing the synaptosomes should be collected. The material at the interface of the lower two layers of the gradient contains the nonsynaptic mitochondria, which can be discarded or collected for additional experiments if necessary.
5. As the nitrogen enters the solution the pressure will decrease and equilibrate at around 900 psi, stirring for 15 min facilitates this process. Depressurization of the solution to atmospheric pressure causes bubbles to form and expand, disrupting the synaptosomes membranes and releasing the mitochondria.
6. These washes are necessary to remove excess BSA prior to the proteomic analysis as this is considered a contaminant of the isolation procedure. However, the presence of BSA during the mitochondrial isolation procedure is preferable for the preservation of mitochondrial integrity and protection of mitochondrial proteins.
7. Resuspend the mitochondrial pellets using ~150 μL lysis buffer for synaptic mitochondrial pellets obtained from 400 mg tissue and for neuronal mitochondrial pellets obtained from 1×10^8 cells.
8. For trypsin digestion, we recommend using a protein–enzyme ratio of $< 1{:}50$. In addition to trypsin, other enzymes may also be used, such as LysC; however, these enzymes will require additional optimization.
9. The cells chosen for the super-SILAC mix should be SILAC-labeled by culturing in Advanced DMEM/F-12 media in which the natural lysine and arginine are replaced by heavy isotope-labeled amino acids, (U-$^{13}C_6{}^{15}N_4$)-l-arginine (Arg-10) and (U-$^{13}C_6$)-l-lysine (Lys-6) supplemented with 10 % dialyzed fetal bovine serum, SILAC glucose solution, L-glutamine, SILAC phenol red solution, and penicillin-streptomycin.

Cells should be cultured in the SILAC media until fully labeled as assessed by quantitative mass spectrometry (in our experience seven generations appears to be the minimum for most mouse and rat brain derived cell lines). The cell lines (we recommend a minimum of four different cell lines) should be chosen with the goal of adequate representation of the proteome of the experimental tissue of interest. For example, for proteomic studies of presynaptic mitochondria isolated from mouse brain, the use of mitochondria isolated from the SILAC-labeled mouse cell lines Neuro-2a, CATH.a, NB41A3, and C8-D1A were successfully used [6, 10].

10. In our experience, peptide quantity is usually 20–50% of the starting amount of protein measured in step 5 of Sect. 3.2.1. Thus, for a starting protein concentration of 50 μg, we recommend resuspending the sample in 25 μL to allow for accurate peptide quantification.
11. The quantity of peptide loaded on the LC column is dependent on the type of cHiPLC columns in use and will require optimization. We recommend that the same cHiPLC columns be used for the duration of the project. The provided peptide quantities have been optimized for the cHiPLC columns suggested here (see **Note 12**).
12. If using our recommended peptide quantities and LC methods, the cHiPLC columns suggested in the Methods Equipment Sect. 2.3.2 have been optimized depending on the mass spectrometry approach, super-SILAC or SWATH-MS.
13. Table 1 provides a suggested LC protocol for super-SILAC and SWATH-MS DDA runs; however, optimization is necessary, and depending on the experimental setup, the gradient may be shortened or lengthened. Keep in mind that it is important that all DDA and DIA runs for an individual experiment should be performed with the same gradient.
14. The generation of the SWATH-MS reference spectral library in DDA mode is an important component of the experiment. Several methods are available to generate a library: (a) use a preconstructed library, (b) generate a library from experimental samples, or (c) generate a library from a variety of cell lines or other suitable samples containing proteins covering the range of those found in the experimental samples. Until issues of elution time alignment are resolved, we do not recommend using a preconstructed library. Further, if sample availability is a limitation, the generation of a library using cell lines might be beneficial. We have previously performed such an experiment for the analysis of mitochondria isolated from rat brain [9, 13].

15. The samples should be separated into SWATH-MS reference spectral library or SWATH-MS or super-SILAC runs. Samples are then randomized within each group so that the mass spectrometry methods only need to be changed between groups of samples, and not within groups of samples. This is only a problem if running samples from multiple experiment types (super-SILAC and SWATH-MS) or if running SWATH-MS reference spectral library and SWATH-MS experimental samples together.
16. We recommend using a sample that is comparable to the experimental samples to test all the procedures and to optimize the elution gradient. Elution times will be influenced by the type of reverse phase HPLC resin as well as the sample composition, although we observe that most peptides begin eluting at ~30 min. Slight differences between samples are expected; however, TICs that are markedly dissimilar may indicate unequal loading, sample impurities, cHiPLC system issues, or problems with the mass spectrometry methods.
17. We recommend using a different computer to collect and process the collected data due to the processor intensive demands for searching and targeted data analysis.
18. The SwissProt section of UniProtKB provides a list of proteins that are manually curated to remove redundant sequences. The TrEMBL section contains computationally analyzed records that are obtained from the translation of annotated coding sequences of the EMBL-bank/GenBank/DDBJ nucleotide databases. Thus, TrEMBL contains redundant sequences and is limited in experimental validation of sequences. Alternatively, NCBI (www.ncbi.nlm.nih.gov) provides the RefSeq: NCBI Reference Sequence Database, which is also a well-annotated, nonredundant set of reference sequences.
19. If OFFGEL fractionation of the super-SILAC experimental samples was performed, upload the data files for all 12 fractions for each individual sample simultaneously for joint analysis in ProteinPilot.
20. The website http://www.biognosys.ch provides FASTA files for artificial peptides contained in the HRM calibration kit.
21. The proteins with lower confidence that are being imported into PeakView will be filtered out automatically during targeted data extraction. Additionally, the proteins can be filtered out manually after export.
22. Targeted data extraction should only be performed for samples that will be directly compared to one another to avoid unwanted impacts of low confidence spectral assignments in samples that may impact the quality of the export of all samples exported in unison. The processing settings for targeted data

extraction in PeakView will need to be optimized for each individual experiment. Particularly, the extraction window will effect the assignment of spectra; however, using the RT correction tool tends to improve RT variability and allows for narrowing of the extraction window. Although setting a stringent FDR threshold (e.g., 1 %) improves data quality, this comes with the cost of decreasing the total number of peptides used for quantification and the number of proteins exported.

23. Information about the four normalization methods: (a) selected peak, (b) total peak intensity, (c) median peak intensity, and (d) manual scale factor provided by MarkerView can be obtained in the MarkerView program.

References

1. Schapira AH (2006) Mitochondrial disease. Lancet 368(9529):70–82. doi:10.1016/S0140-6736(06)68970-8
2. Chan DC (2012) Fusion and fission: interlinked processes critical for mitochondrial health. Annu Rev Genet 46:265–287. doi:10.1146/annurev-genet-110410-132529
3. Picard M, Turnbull DM (2013) Linking the metabolic state and mitochondrial DNA in chronic disease, health, and aging. Diabetes 62(3):672–678. doi:10.2337/db12-1203
4. Jiang Y, Wang X (2012) Comparative mitochondrial proteomics: perspective in human diseases. J Hematol Oncol 5:11. doi:10.1186/1756-8722-5-11
5. Forner F, Kumar C, Luber CA, Fromme T, Klingenspor M, Mann M (2009) Proteome differences between brown and white fat mitochondria reveal specialized metabolic functions. Cell Metab 10(4):324–335. doi:10.1016/j.cmet.2009.08.014
6. Stauch KL, Purnell PR, Fox HS (2014) Quantitative proteomics of synaptic and nonsynaptic mitochondria: insights for synaptic mitochondrial vulnerability. J Proteome Res 13(5):2620–2636. doi:10.1021/pr500295n
7. Geiger T, Cox J, Ostasiewicz P, Wisniewski JR, Mann M (2010) Super-SILAC mix for quantitative proteomics of human tumor tissue. Nat Methods 7(5):383–385. doi:10.1038/nmeth.1446
8. Gillet LC, Navarro P, Tate S, Rost H, Selevsek N, Reiter L, Bonner R, Aebersold R (2012) Targeted data extraction of the MS/MS spectra generated by data-independent acquisition: a new concept for consistent and accurate proteome analysis. Mol Cell Proteomics 11(6):O111 016717. doi:10.1074/mcp.O111.016717
9. Villeneuve LM, Stauch KL, Fox HS (2014) Proteomic analysis of the mitochondria from embryonic and postnatal rat brains reveals response to developmental changes in energy demands. J Proteomics. doi:10.1016/j.jprot.2014.07.011
10. Stauch KL, Purnell PR, Fox HS (2014) Aging synaptic mitochondria exhibit dynamic proteomic changes while maintaining bioenergetic function. Aging 6(4):320–334
11. Deeb SJ, D'Souza RC, Cox J, Schmidt-Supprian M, Mann M (2012) Super-SILAC allows classification of diffuse large B-cell lymphoma subtypes by their protein expression profiles. Mol Cell Proteomics 11(5):77–89. doi:10.1074/mcp.M111.015362
12. Liu Y, Huttenhain R, Surinova S, Gillet LC, Mouritsen J, Brunner R, Navarro P, Aebersold R (2013) Quantitative measurements of N-linked glycoproteins in human plasma by SWATH-MS. Proteomics. doi:10.1002/pmic.201200417
13. Villeneuve LM, Purnell PR, Boska MD, Fox HS (2014) Early expression of Parkinson's disease-related mitochondrial abnormalities in PINK1 knockout rats. Mol Neurobiol. doi:10.1007/s12035-014-8927-y
14. Kristian T (2010) Isolation of mitochondria from the CNS. Current protocols in neuroscience / editorial board, Jacqueline N Crawley [et al.] Chapter 7:Unit 7 22. doi:10.1002/0471142301.ns0722s52
15. Beaudoin GM 3rd, Lee SH, Singh D, Yuan Y, Ng YG, Reichardt LF, Arikkath J (2012) Culturing pyramidal neurons from the early postnatal mouse hippocampus and cortex. Nat Protoc 7(9):1741–1754. doi:10.1038/nprot.2012.099

16. Gottlieb RA, Adachi S (2000) Nitrogen cavitation for cell disruption to obtain mitochondria from cultured cells. Methods Enzymol 322: 213–221
17. Wisniewski JR, Zougman A, Nagaraj N, Mann M (2009) Universal sample preparation method for proteome analysis. Nat Methods 6(5):359–362
18. Franko A, Baris OR, Bergschneider E, von Toerne C, Hauck SM, Aichler M, Walch AK, Wurst W, Wiesner RJ, Johnston IC, de Angelis MH (2013) Efficient isolation of pure and functional mitochondria from mouse tissues using automated tissue disruption and enrichment with anti-TOM22 magnetic beads. PLoS One 8(12):e82392. doi:10.1371/journal.pone.0082392
19. Scopes RK (1974) Measurement of protein by spectrophotometry at 205 nm. Anal Biochem 59(1):277–282
20. Baldi P, Long AD (2001) A Bayesian framework for the analysis of microarray expression data: regularized t-test and statistical inferences of gene changes. Bioinformatics 17(6): 509–519
21. Kayala MA, Baldi P (2012) Cyber-T web server: differential analysis of high-throughput data. Nucleic Acids Res 40(Web Server Issue):W553–W559. doi:10.1093/nar/gks420

Chapter 16

Measuring Mitochondrial Pyruvate Oxidation

Lawrence R. Gray, Alix A.J. Rouault, Lalita Oonthonpan, Adam J. Rauckhorst, Julien A. Sebag, and Eric B. Taylor

Abstract

Pyruvate is a central metabolic intermediate and plays a prominent role in nervous system function. Neurons are highly reliant on pyruvate oxidation for maintenance of cellular energetics. Disorders in pyruvate metabolism may result in severe neurological defects. Pyruvate is imported into the mitochondrial matrix by the mitochondrial pyruvate carrier (MPC) for oxidation by the TCA cycle to support oxidative phosphorylation. Understanding the function of the MPC requires specialized methods for investigating mitochondrial pyruvate metabolism. Multiple instruments are available for measuring respiration of animal tissues, intact cells, permeabilized cells and tissue, and isolated mitochondria. Compared to other platforms, the comparative advantage of the Seahorse extracellular flux analyzer is the ability to sequentially administer treatment compounds and observe the effects on cellular respiration in a multiplexed 96-well format. Here we describe the methods and procedures for: (1) assessing mitochondrial pyruvate oxidation by intact cultured neuronal cells; (2) by mitochondria isolated from HEK293T cells, and (3) by mitochondria isolated from mouse liver, which are useful as a general model of mitochondrial function.

Key words Mitochondria, Pyruvate, Oxidation, Respiration, Metabolism

1 Introduction

Pyruvate is a central metabolic intermediate lying at the intersection of cytosolic glycolytic and mitochondrial oxidative metabolism [1]. Mitochondrial pyruvate metabolism is important for multiple biosynthetic pathways and ATP production by oxidative phosphorylation. Mitochondrial pyruvate oxidation plays an especially prominent role in the nervous system because of the general absence of fatty-acid oxidation. Neurons are highly reliant on pyruvate oxidation for maintenance of cellular energetics and inborn errors in pyruvate metabolism may result in severe neurological dysfunction.

Pyruvate may be derived from multiple sources in the cytoplasm, including glycolysis, amino acids, and imported lactate produced by other cells. In the nervous system, the latter is a key feature of the astrocyte-neuron lactate exchange shuttle, where astrocytes

Stefan Strack and Yuriy M. Usachev (eds.), *Techniques to Investigate Mitochondrial Function in Neurons*, Neuromethods, vol. 123, DOI 10.1007/978-1-4939-6890-9_16, © Springer Science+Business Media LLC 2017

support neuronal metabolism by providing a steady supply of glycolytically produced lactate [2]. Lactate imported by neurons is oxidized to pyruvate in the cytoplasm by the enzyme lactate dehydrogenase. The resulting pyruvate is then imported into the mitochondrial matrix by the mitochondrial pyruvate carrier (MPC), where it is oxidized to CO_2 by the TCA cycle to drive oxidative phosphorylation and support cellular energetics. The importance of the enzymes pyruvate carboxylase and pyruvate dehydrogenase for mitochondrial pyruvate metabolism has been well-demonstrated. In contrast, the molecular function and in vivo importance of the MPC, which supplies pyruvate to both of these enzymes, has only recently begun to be addressed.

The recent discovery of the genes encoding the MPC now enables targeted molecular-genetic studies on the function of the MPC in the regulation of cellular pyruvate metabolism [3, 4]. To date there is one published case report of a human patient with loss of MPC activity, who was born with severe neurological abnormalities and died at age 19 months [5]. The causative mutation was later identified to reside in the *MPC1* gene, changing a highly conserved arginine residue in the MPC1 protein to tryptophan [3]. Recent studies have begun to define how altered MPC function contributes to diseases like cancer and type 2 diabetes [6–10]. Understanding the expectedly important function of the MPC in the nervous system will require specialized methods for investigating neuronal pyruvate metabolism. Here we present procedures for assessing mitochondrial pyruvate oxidation (1) by intact cultured mouse hypothalamic neuronal GT1-1 cells [11]; (2) by mitochondria isolated from HEK293T cells, which exhibit some neuron-like characteristics and have been utilized as a model of synaptogenesis [12, 13]; and (3) by mitochondria isolated from mouse liver, which, because they respire robustly and are relatively straightforward to isolate, may be useful as a general model of mitochondrial function. We also recommend liver mitochondria as a reliable positive control during experiments designed to measure respiration by mitochondria originating from other tissues.

Multiple hardware platforms are available for measuring respiration of animal tissues, intact cells, permeabilized cells and tissue, and isolated mitochondria. These include traditional Clark-type electrodes [14, 15] and more recently developed instruments like the Oroboros Oxygraph [16] and the Seahorse extracellular flux analyzers [17]. The procedures presented here are for use with the Seahorse XF 96 flux analyzer instrument. Nonetheless, we expect that they are compatible with other platforms after accounting for volume differences among the respective measurement chambers and resultant changes in sample mass requirements.

The key advantages of utilizing the Seahorse extracellular flux analyzer are (1) the ability to sequentially administer up to four different compounds and observe effects on cellular respiration;

(2) higher throughput because of a plate-based, multi-well format; and (3) a relative decrease in sample mass requirements. A standard initial test of mitochondrial function involves measurement of baseline respiration rates, of uncoupled respiration by blocking the ATP synthase, of maximum respiration induced by chemical uncoupling, and of non-mitochondrial oxygen consumption after application of electron transport chain poisons.

First, initial oxygen consumption measurements are performed to provide baseline respiration values. Second, oligomycin is used to inhibit Complex V, the ATP synthase, blocking coupled oxygen consumption. Remaining respiration is attributed to oxygen consumption that is uncoupled from mitochondrial oxidative phosphorylation. Third, FCCP, a proton ionophore, is used to artificially uncouple mitochondrial respiration and thus bypass inhibition of respiration by oligomycin. Treatment with FCCP increases oxygen consumption because it evokes homeostatic increases in mitochondrial metabolism aimed at protecting cellular ATP levels. Of the mitochondrial poisons described here, the optimization of FCCP concentration is of special importance. If too little FCCP is used, the stimulated oxygen consumption rate will not be maximal. Conversely, if too much FCCP is used, mitochondria become completely depolarized and oxygen consumption is lost. Finally, the electron transport chain Complex I inhibitor rotenone, sometimes in combination with the Complex III inhibitor antimycin A, depending on the respiratory substrate, is used to inhibit overall mitochondrial respiration. Relevant to this chapter, pyruvate-driven respiration is Complex I-dependent and rotenone alone is usually sufficient to essentially stop pyruvate-driven respiration.

2 Materials, Reagents, and Buffers

Buffers:

Cellular respiration buffer (RB): 1 bottle of powdered DMEM and 1.85 g NaCl. Bring to 1 L and pH adjust to 7.4 at 37 °C. Filter-sterilize and place at 4 °C. Media is good for up to 6 weeks.

Mitochondria respiration buffer (MSB): 70 mM sucrose, 220 mM d-mannitol, 10 mM KH_2PO_4, 5 mM $MgCL_2$, 5 mM HEPES pH 7.4, 1 mM EGTA, and 0.2% fatty acid free BSA.

Liver mitochondria isolation buffer (LMIB): 70 mM sucrose, 210 mM d-mannitol, 5 mM HEPES pH 7.4, 1 mM EGTA, and 0.5% fatty acid free BSA.

293T Hypotonic mitochondrial isolation buffer (293THypo): 20 mM HEPES pH 7.8, 5 mM KCl, 1.5 mM $MgCL_2$, 1 mg/mL fatty acid free BSA, 1 mM PMSF, 2 mM DTT, and 1 Roche protease inhibitor tablet per 50 mL buffer (add PMSF, DTT, and protease inhibitors fresh).

293T Hypertonic mitochondrial isolation buffer (293THyper): 525 mM mannitol, 175 mM sucrose, 50 mM HEPES pH 7.8, 1 mg/mL fatty acid free BSA, 1 mM PMSF, 2 mM DTT, and 1 Roche protease inhibitor tablet per 50 mL buffer (add PMSF, DTT, and protease inhibitors fresh).

3 Procedures

3.1 Oxygen Consumption by Cultured GT1-1 Mouse Hypothalamic Neuronal Cells

Measuring oxygen consumption in intact, adherent cells is a 2-day experiment. The first day involves seeding the cell culture plate with cells, incubating the respiration cartridge, and preparing base buffers. The second day involves preparing media with treatment compounds and running the respiration assay. Critical experimental variables for prior consideration include the number of cells to be seeded, substrates and concentrations, treatment compounds and concentrations, order of treatment, and plate layout including wells to be a part of the same experimental group. These variables are discussed in the following sections. Finally, optimal experimental conditions need to be determined by pilot experiments, often with wild-type cultured cells or mitochondria. Empirical determination of optimal conditions during pilot experiments will markedly increase the likelihood of experimental success when working with limited or expensive samples. Seahorse Bioscience curates an extensive library of publications that utilize their instruments and is a useful resource to find initial conditions for specific cultured cells or mitochondria sources.

Day #1

Any cell line that successfully adheres to the Seahorse assay plate may be used for respirometry experiments. Indeed, lines ranging from C2C12 mouse myoblasts to primary rat hepatocytes have been successfully analyzed. Separate protocols are available for working with cells in suspension. The Seahorse assay plate is amenable to a variety of surface coatings including poly-lysine, collagen, and fibronectin. For the experiments described here, the GT1-1 cell line was used. The GT1-1 cell line is an immortalized mouse hypothalamic neuronal cell line that was generated by Richard Weiner's lab. This was achieved by expressing the SV-40 T-antigen under the control of the GnRH promoter in mice [11]. GT1-1 cells were cultured in DMEM-F12 supplemented with 5% FBS and 1% penicillin/streptomycin at 37 °C in a humidified 5% CO_2 incubator.

For a typical respiration assay, two culture plates are seeded with cells. The first is the 96-well Seahorse culture plate. Cell seeding parameters are dependent upon the specific assay being performed (see Sect. 4.1). However, the four corner wells (A1, A12, H1, and H12) must remain free of cells to serve as background controls. Media is still placed in these wells, however. The desired number of cells is seeded in 150–200 μL growth media. GT1-1 cells were seeded at 12,000 cells per well and reached 75–80% confluency by the next day when experiments were conducted. See Sect. 4.2 for additional considerations when planning and seeding assay plates. The second plate is a standard tissue culture dish, such as a 24-well dish, and functions as a mirror plate for normalizing respiration data. The normalization plate is seeded, in triplicate, from the same cell stock used to seed the Seahorse plate and at an identical cell number-to-surface area ratio. Both plates are incubated overnight. See Sect. 4.3 for additional information.

A necessary component of the respiration assay is the Seahorse cartridge. This cartridge houses the pH and oxygen probes and needs to be hydrated prior to use. The Seahorse cartridge is hydrated with 200 μL of manufacturer supplied calibrant per well. The cartridge is equilibrated overnight in a 37 °C non-CO_2 incubator. If sealed tightly with Parafilm, the cartridge is viable up to 2–3 days.

The base Seahorse media is also prepared in advance. The formulation is given above. This media contains all of the primary components of DMEM without glutamine, pyruvate, phenol red, glucose, and, importantly, bicarbonate. The Seahorse flux analyzer is not a CO_2 incubator, which would confound measurements of proton production as an index of glycolytic rate. Therefore, bicarbonate-based media must not be used, to avoid alkalinization of media during the experiment.

Day #2

A series of media must be prepared prior to the start of the respirometry experiment. The first are the substrate media. After bringing the base Seahorse media to room temperature, substrates, either alone or in combination, are added. Commonly utilized substrates include pyruvate, glutamine, and glucose. The concentration of substrate supporting the highest rates of respiration must be experimentally determined and optimized for each cell line. A recommended starting concentration is 10 mM with either half-log dilutions (10 mM, 3.2 mM, and 1 mM) or twofold dilutions (10 mM, 5 mM, 2.5 mM, and 1 mM). For the workflow outlined here, 40 mL of each substrate buffer is prepared. Powdered substrates are dissolved directly into base media, adjusting pH as necessary.

The second set is the injection media. These media contain the metabolic poisons or other test compounds. Up to four compounds can be tested per experiment. A typical initial mitochondria metabolism profiling experiment utilizes, in the following order, 1 μM Oligomycin, 0.1–1.0 μM FCCP, and 1 μM Rotenone with or without 1 μM Antimycin A depending on the respiratory substrate. While the concentrations of these reagents may all be optimized,

Table 1
Reagents and stocks required for the assays described herein

Reagent	Supplier	Cat #	Stock	Composition	Storage
ADP	Sigma-Aldrich	A2754-1G	500 mM	1.00 g in 4.68 mL H_2O pH to 7.4	−20 °C
FCCP	Sigma-Aldrich	C2920-50MG	10 mM	12.7 mg in 5.00 mL DMSO or EtOH	−20 °C
Rotenone	Sigma-Aldrich	R8875-1G	10 mM	19.7 mg in 5.00 mL DMSO or EtOH	−20 °C
Oligomycin	Sigma-Aldrich	75351-5MG	10 mM	5.00 mg in 0.63 mL DMSO	−20 °C
CHC	Sigma-Aldrich	476870-2G	100 mM	189.1 mg in 10 mL H_2O pH 7.4	−20 °C
DMEM	Sigma-Aldrich	D5030-10X1L	–	–	–
Fatty acid free BSA	Sigma-Aldrich	A3803-10G	–	–	–
Sodium Pyruvate	Sigma-Aldrich	P5280-25G	–	–	–
Sodium Glutamate	Sigma-Aldrich	G3126-100G	–	–	–
Disodium Malic Acid	Sigma-Aldrich	M9138-25G	–	–	–
Seahorse XF^{e}96 FluxPak	Seahorse Bioscience	102416-100	–	–	–
Protein Assay Dye Reagent	Bio-Rad	500-0006	–	–	–

the concentration of FCCP is most critical, for reasons detailed above. The injection media are prepared by diluting concentrated reagents stocks (see Table 1) into the substrate-containing buffers described above. For example, if Oligomycin, FCCP, and Rotenone/Antimycin A are being used to stress cells respiring on 10 mM glucose, these compounds are diluted into media containing 10 mM glucose. Furthermore, each progressive injection media must be more concentrated than the one preceding it. By way of illustration, assume that the starting volume of media above the cells is 180 μL. Each injection is 20 μL. Therefore the first injection is 10× (180 + 20 = 200 μL). The second injection must be 11× (200 μL + 20 μL = 220 μL). And the third injection is 12× (220 + 20 = 240 μL). The maximum volume the wells will hold is approximately 275 μL. Care must be taken not to go over this volume or there is risk of cross-contaminating the probes. The injection media are then dispensed into 96-well plates in the layout required for the planned assay. For this step, pipet 40 μL into each well and seal with a plate sealer. A separate, standard (not Seahorse) 96-well plate is needed for each unique injection media. This simplifies the loading of the injection ports on the Seahorse cartridge by allowing for the use of a multichannel pipet.

After preparation, the injection media are loaded into the prehydrated Seahorse cartridge. We recommend loading the cartridge at the location of the Seahorse Flux Analyzer to prevent spills that may occur during transport. Loading guides supplied by Seahorse Bioscience may be used to assist the loading process. Ports are arranged with Port A in the upper left, Port B in the upper right, Port C in the lower left, and Port D in the lower right. The central hole is where the metallic probe is positioned. Because injections are performed pneumatically, if a given port is to be used, injection media must be loaded into that port for every single well, including wells that do not contain cells. Wells that lack injection media will constitute a pressure leak and result in a pressure gradient and uneven injections across the entire plate. In contrast, only the ports to be used in the experiment need to be filled (i.e., if only three injections are planned, no media needs to be loaded into Port D). Using the protocol supplied here, each injection will have a volume of 20 μL.

After loading the Seahorse cartridge, the experimental protocol is programmed into the Seahorse Analyzer. It is beyond the scope of this chapter to provide a detailed walkthrough on how to design Seahorse protocols or on how to operate the flux analyzer itself. Trained personnel should be available to assist with these steps. However, a standard operating protocol is provided here. A measurement cycle is divided into three phases; a 1 min mix step, a 1 min wait step, and a 3 min measurement step. This cycle is repeated three times to obtain basal measurements. Port A is injected and the three measurement cycles may be performed again with Ports B, C, and D, giving up to a total of 15 measurements if all ports are used. This protocol is a general guide and can be modified as needed. Using this protocol, a typical experiment will take approximately 90 min to complete.

With the experimental protocol programmed, the experiment is started. The first step is to calibrate the Seahorse cartridge. Upon initiation of the protocol, the cartridge is loaded into the flux analyzer. Only the green cartridge and the lower calibrant plate are loaded. The injection port loading guides and cartridge cover are discarded and should not be loaded into the analyzer. Over the course of approximately 20 min, the flux analyzer calibrates each cartridge probe.

During the 20 min window when the Seahorse Analyzer calibrates the cartridge, the media in the cell culture plate is changed. Great care must be taken during this step to ensure that the cells are not disrupted or sheared off the plate. Using a Pasteur pipet or a 200 μL pipet tip attached to a vacuum line, the media within each well is gently aspirated away. The pipet tip is inserted approximately a third to half of the way into the well. We recommend leaving approximately 15 μL of media in each well. Removal of all of the media will almost always result in loss of cells. 180 μL of substrate-containing Seahorse media is then gently and slowly applied to the side of the wells. The media is aspirated a second time and a final 180 μL is applied. We recommend changing the media for no more than two rows of cells at a time because delays may result in drying of cells. This process will take practice to master. Once media is changed for all of the wells, including the control wells, the culture plate is transported to the location of the Seahorse flux analyzer. We recommend transporting the plate inside a Styrofoam box as a secondary containment, to safeguard against spills and to insulate the plate.

Washing the cell culture plate will take 20 min or longer, depending upon the complexity of the assay. Calibration of the cartridge will be at or near completion by this time. The Seahorse analyzer will pause once cartridge calibration has finished. Interact with the analyzer to eject the calibrant plate. The green cartridge will be retained inside. Replace the calibrant plate with the cell culture plate and continue with the assay. Once accepted by the flux analyzer, no further input from the researcher is required. When the assay has finished, the expired Seahorse cartridge and cell culture plate may be discarded. The culture plate may also be retained to inspect wells for consistent cell adherence to the end of the assay. While the experiment is being performed, continue with the next step.

The Seahorse experiment will take over 1 h to perform. During this time, the normalization assay can be performed. The method of normalization is up to the researcher. The cells in the mirror plate may be washed with PBS and solubilized in buffer for use in a protein assay. Alternatively, cells may be counted with a hemocytometer and the number of cells in each well calculated. The values obtained are used as the denominator for subsequent normalization of data obtained during the Seahorse assay.

After the Seahorse experiment has been completed, analysis of the results may start. The Seahorse analyzer is supplied with a suit of analysis software that allows the researcher to view the data from experimental groups down to a single well. Publication quality images can be obtained using the Seahorse software. Alternatively, the raw oxygen consumption rates (OCR) and extracellular acidification rate (ECAR) data may be analyzed separately using a data analysis software such as Microsoft Excel. Wells producing unreliable data may be empirically identified by checking every well to make sure that all injections were successful and that main responses were consistent with other wells in the same experimental group. Statistical tests such as the Grubbs' test may also be employed to

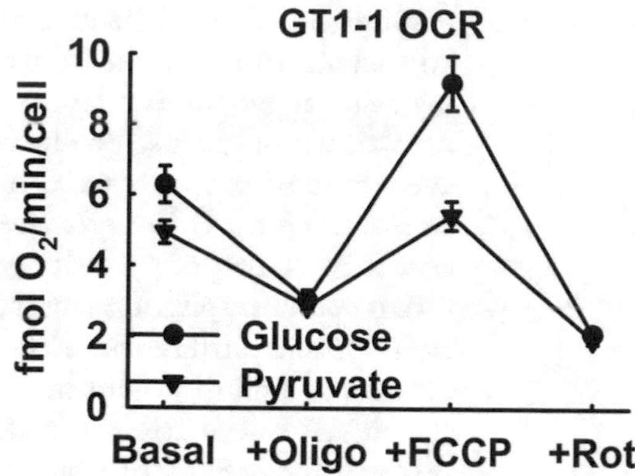

Fig. 1 OCR measurements utilizing GT1-1 cells on either (*solid circles*) glucose or (*solid triangles*) pyruvate ($n = 12$)

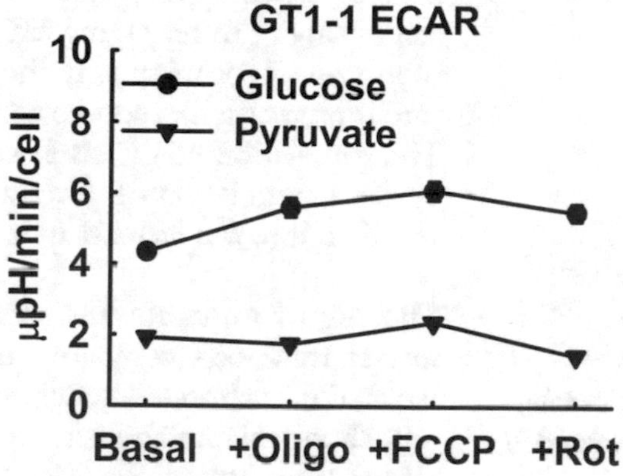

Fig. 2 ECAR measurements utilizing GT1-1 cells on either (*solid circles*) glucose or (*solid triangles*) pyruvate ($n = 12$)

identify outliers. Average the three replicate measurements under each treatment condition. Then average all wells under the same experimental groups. These groups may be compared by analysis of variance or other appropriate statistical methods, depending on the overall experimental design. For each experimental group, up to five values will be calculated, which include the basal condition and up to four treatment conditions. Graphing software, such as Microsoft Excel or SigmaPlot, can be used to prepare figures.

Figure 1 shows OCR measured using GT1-1 cells respiring on either glucose or pyruvate, under the typical mitochondria stress sequence of Oligomycin, FCCP, and Rotenone/Antimycin A. Compared to pyruvate-driven respiration, glucose-driven respiration shows slightly higher basal and FCCP-stimulated respiration. Figure 2 shows the ECAR of the same cells. ECAR does not change significantly under pyruvate-driven respiration because pyruvate is not a substrate for glycolysis. Glucose-driven respiration is stimulated upon Oligomycin treatment and is consistent through the remaining treatments. Notice that the behavior of pyruvate and glucose are not identical in this assay. The breakdown of glucose produces two molecules of pyruvate. Thus, even though the cells were treated equimolar concentrations of substrate (3 mM) the *effective* substrate concentration with glucose is twice that of pyruvate, which in some cases may explain the difference in respiration observed under FCCP-stimulated conditions. Other differences may arise because of the versatility of glucose as a multi-potent metabolic substrate.

ECAR values arise from the conversion of pyruvate to lactate through the activity of lactate dehydrogenase and the subsequent export of lactate into the extracellular milieu by the monocarboxylate transporters (MCTs), a process which is proton-linked. While this process occurs when both pyruvate and glucose are used as fuel sources, when applied extracellularly, pyruvate is taken up by the cell through the very same MCTs leading to a net zero movement of protons, explaining why the measured ECAR value when pyruvate is used as a fuel source changes little through the experiment. The stimulation of ECAR after oligomycin treatment is explained by the upregulation of glycolysis in order to maintain ATP levels, and therefore the increase in lactate production and export.

3.2 Oxygen Consumption Measurements Using Isolated Mitochondria

The use of mitochondria in respirometry experiments involves several trade-offs compared to the use of intact cultured cells. Isolated mitochondria generate greater respiratory signal and allow mitochondrial metabolism to be assayed more directly, without effects from interaction with non-mitochondrial metabolism and the broader intracellular milieu. Furthermore, in the absence of the plasma membrane barrier, a wider range of compounds may be tested that would otherwise be non-permeating or exhibit undesired off-target effects. For example, the MPC inhibitor α-cyano-4-hydroxycinnamate (CHC) not only inhibits the MPC in the inner mitochondria membrane but also MCTs present at the plasma membrane. Thus, if CHC were used to treat whole cells, both the MPC and MCTs would be inhibited, confounding analysis. Drawbacks of working with isolated mitochondria include the potentially confounding effects of disrupting the larger mitochondrial ultrastructure and the requirement that active, respiration-competent mitochondria are obtained following isolation. Control experiments, such as a Clark electrode respiration experiments, may be performed to ensure that quality mitochondria are obtained.

The procedures applied to isolated mitochondrial below are similar to those described above with intact cells, with specific differences described below. Mitochondria isolation procedures will vary by source and tissue. Included below are protocols for the isolation of mitochondria from HEK 293T cells and mouse liver.

3.2.1 Isolation of Mitochondria from 293T Cells

Mitochondria from 293T cells are isolated essentially as described by Gerhold et al. [18]. 5–10 × 10 cm^2 tissue culture plates at approximately 80–100% confluency are required. Cells are detached using a cell scrapper in 5 mL ice-cold PBS and collected into a pre-weighed 50 mL tube. To increase amount of cells collected, the scraped plates are washed with an additional 5 mL ice-cold PBS. Cells are pelleted at 200 × *g* for 5 min at 4 °C. The supernatant is removed and the cell pellet weighed. Cells are resuspended in 9 volumes of 293THypo buffer (1 g cells = 0.8 mL volume) and incubated on ice for 6–10 min (no longer). The cell suspension is homogenized using a tight fitting glass-on-glass Dounce homogenizer in 20–25 strokes. After homogenization, 2 mL of 293THyper buffer are added for every 3 mL homogenate. Centrifuge the homogenate at 1600 × *g* for 10 min at 4 °C. Transfer the supernatant into a clean tube and discard the pellet. Repeat the previous centrifugation step. The resulting supernatant is transferred to a clean centrifuge tube and centrifuged at 13,000 × *g* for 10 min at 4 °C. The pellet is resuspended in 1 mL of MSB, split into several 1.5 mL tubes, and centrifuged a final time at 13,000 × *g* for 10 min at 4 °C. Mitochondria may be stored as pellets on ice for several hours.

3.2.2 Isolation of Mitochondria from Mouse Liver

All experiments with mice need to be approved by institutional committees on the care and use of animals. Mouse liver mitochondria are isolated essentially as described by Rogers et al. [19]. Mice are anesthetized by isoflurane inhalation and sacrificed via cervical dislocation. Each liver is washed first in 10 mL in ice-cold PBS and second in ice-cold LMIB. Using scissors the livers are finely minced in 10 volumes of fresh LMIB and homogenized using a motor-rotated, Teflon-on-glass homogenizer during 4 × 400–500 RPM passes. The homogenate is centrifuged at 800 × *g* for 10 min at 4 °C in 50 mL tubes. The resulting supernatant is transferred to a clean tube and centrifuged at 8000 × *g* for 10 min. After this spin has completed the supernatant is discarded. The resulting pellet, containing isolated mitochondria, is resuspended in 10 volumes of fresh LMIB. Multiple 0.5 mL aliquots of resuspended mitochondria are transferred into labeled 1.5 mL tubes and re-pelleted by centrifugation at 9000 × *g* for 10 min at 4 °C. The supernatant is discarded and the pellet is kept on ice until needed. Isolated mitochondria, kept as a pellet, are suitable for Seahorse analysis up to 4 h after isolation.

3.2.3 Preparation of Isolated Mitochondria for Seahorse Assays

Each mitochondrial pellet, regardless of source, is resuspended in 100–200 μL MSB to create a concentrated mitochondrial suspension. A portion of the concentrated suspensions are further diluted from 1:20 to 1:40 in water and a protein quantification assay such as a Bradford assay using the Bio-Rad Protein Assay Dye Reagent is performed according to the manufacturer's protocols [20].

A buffer blank control is included because MSB contains 2 mg/mL BSA. After determination of protein content, the remaining concentrated mitochondrial suspension is diluted to 0.25 mg/mL mitochondrial protein in MSB to create a working stock. 20 μL of the mitochondrial working stock is loaded per well on the Seahorse assay plate, for a total of 5 μg mitochondrial protein per well. If necessary, a secondary post-seahorse assay protein quantification measurement can be performed on unused diluted mitochondrial samples. In this case we recommend at least quadruplicates of the same diluted sample be measured as the signal to noise ratio will be quite low (0.25 mg/mL mitochondrial protein in a background of 2 mg/mL BSA).

Day #1

Equilibrate Seahorse Cartridge with supplied calibrant as in Sect. 3.1. The mitochondria Seahorse buffers are prepared in a similar manner as in Sect. 3.1. These buffers are described by Rogers et al. [19]. The base buffer is MSB, described above. 100 mM substrate stocks can be made in advance and kept at −20 °C. Common stocks to prepare are pyruvate, glutamate, and malate. See Sect. 4.4 for additional information.

Day #2

Substrate-containing media are prepared in a similar fashion as described in Sect. 3.1. Common substrate concentrations range from 0.5 to 5.0 mM. Include malate at 1/5th or 1/10th the concentration of the primary substrate when examining pyruvate or glutamate-driven respiration. Prepare buffers at two concentrations: 1.5× and 1.125×. The 1.5× substrate buffer will be used to prepare the injection media. The 1.125× will be used to bring the assay plate up to the final volume. For example, if 5 mM pyruvate is the substrate being examined, the 1.5× buffer should contain 7.5 mM pyruvate and 1.50–0.75 mM malate. The 1.125× buffer will then contain 5.625 mM pyruvate and 0.523–1.125 mM malate.

Prepare injection media as in Sect. 3.1. A sequence of Oligomycin, ADP/FCCP, and Rotenone/Antimycin A may be utilized as with cultured cells. Other compounds may be tested instead. The 1.5× substrate buffer is used to prepare these buffers. Add appropriate reagents and bring up to 1.0× with MSB. Pipet into 96-well plates, secure with a plate sealer, and place at 4 °C for up to 6 h or until needed. See Sect. 4.5 for additional information.

Once all of the buffers have been prepared, mitochondria are isolated. This chapter describes protocols and procedures for the isolation of mitochondria from cultured HEK 293T cells (Sect. 3.2.1) and mouse liver (Sect. 3.2.1). However, mitochondria can be isolated from a wide range of tissues and cultured cells and the specific isolation procedure will vary depending upon the tissue being examined. We expect the assay procedure outlined here will be generally compatible with mitochondria isolated from other tissues but that it in some cases it may need to be modified to account for tissue-specific mitochondrial differences. These procedures will take 2–3 h to complete. Store mitochondria as pellets on ice until needed for the protein assay and dilution.

After isolation, resuspension, and dilution, the mitochondria are loaded onto a Seahorse cell culture plate. The same plates used for cell-based respirometry experiments are also used for mitochondria assays. For loading, and as noted above, mitochondria are diluted to 0.25 mg/mL mitochondrial protein and 20 μL is placed into each well, for 5 μg total mitochondrial protein. This process is done on ice. 20 μL MSB is added to control and unused wells. The Seahorse plate is centrifuged for 20 min at 2000 × *g* at 4 °C. Mitochondria will attach to the bottom of the plate. During this spin, continue with the next three steps outlined below. When the spin is complete, 160 μL of 1.125× substrate-containing buffer is added to every well to yield a final volume of 180 μL. The plate is then ready for analysis. See Sect. 4.6 for additional information.

During the 20 min centrifugation to plate mitochondria, several additional steps can be performed. As in Sect. 3.1, the Seahorse cartridge is loaded with injection media, the experimental protocol is programmed, and the Seahorse cartridge is calibrated. Once these steps and the centrifugation step are completed, the Seahorse experiment may be performed. Again, refer to Sect. 3.1. Unlike when using cultured cells, no additional normalization plate is required, because mitochondrial content is measured before addition to the assay plate. At the conclusion of the experiment, the results may be analyzed as in Sect. 3.1. However, only the OCR measurements will yield interpretable data when analyzing isolated mitochondria. ECAR measurements lack context because of the absence of glycolysis.

Examples of oxygen consumption rates (OCR) measurements using isolated mitochondria are shown in Fig. 3 (293T) and Fig. 4 (mouse liver). HEK 293T mitochondria respired on 2.5 mM pyruvate + 0.25 mM malate or 2.5 mM glutamate + 0.25 mM malate. After basal measurements, respiration was stimulated by the addition of 4 mM ADP + 1 μM FCCP. The second injection was 1 mM CHC, a specific inhibitor of the MPC, which diminishes pyruvate-driven respiration, but not glutamate-driven respiration. Addition of 1 μM Rotenone completely inhibited respiration. A similar result is shown in Fig. 4 using mouse liver mitochondria respiring on 10 mM pyruvate + 2 mM malate or 10 mM glutamate + 2 mM malate. The use of the MPC inhibitor CHC allows for careful dissection of the role the MPC plays in cellular metabolism. Indeed, inhibition of the MPC results in a drastic decrease in mitochondrial respiration. In contrast, glutamate-driven respiration is unaffected by CHC. This result is all the more informative because both pyruvate

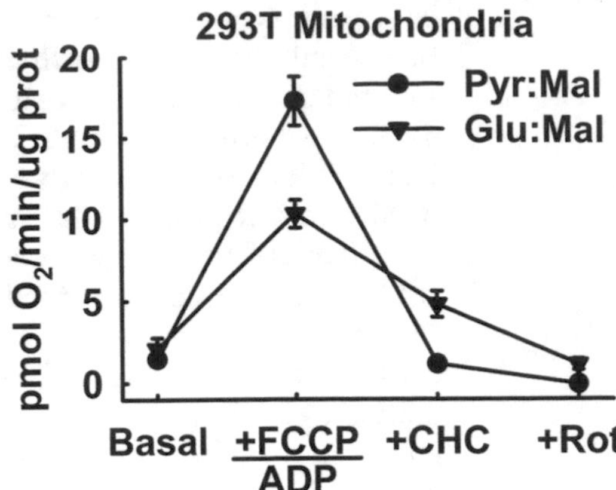

Fig. 3 OCR measurements utilizing isolated 293T mitochondria respiring on either (*solid circles*) pyruvate–malate or (*solid triangles*) glutamate–malate ($n = 6$)

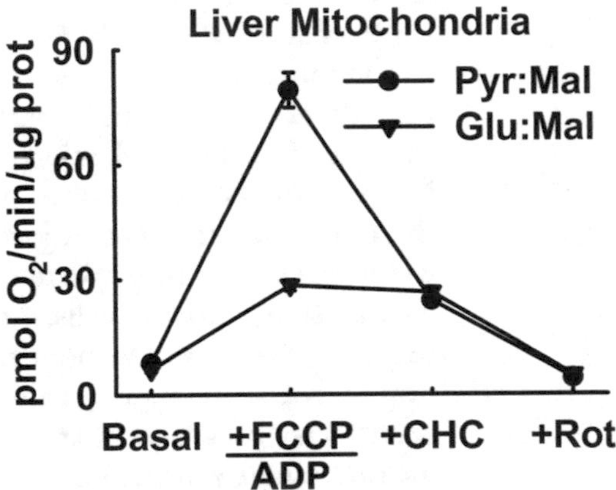

Fig. 4 OCR measurements utilizing isolated liver mitochondria respiring on either (*solid circles*) pyruvate–malate or (*solid triangles*) glutamate–malate ($n = 6$)

and glutamate are Complex I substrates, and thus, utilize much of the same cellular machinery to drive respiration. The lower OCRs observed with glutamate-driven respiration compared to pyruvate-driven respiration are routinely observed.

4 Troubleshooting

4.1 Estimating Proper Cell Density

Sufficient signal is a key requirement for obtaining high quality respirometry data from Seahorse experiments. When using cultured cells, this requirement is met through sufficient cell density and even plating. The appropriate cell density for each cell line must be determined prior to the optimization of other variables.

An inexpensive approach is to seed a Seahorse cell culture plate with a wide range of cell densities. After overnight incubation, the wells are examined using a standard microscope to determine what cell density gave the optimal 80–95% confluency needed. Each FluxPak contains 20 cell culture plates and 18 assay cartridges. Thus, a "spare" culture plate may be used to test plating density. Growing cells to 100% confluency should be avoided as cellular metabolism is often altered as cells leave log-phase growth because of contact inhibition.

For experiments requiring longer culture or cellular differentiation, the Seahorse cell plate may be treated like a standard tissue culture plate. Cells may be grown for several days, differentiated, and pretreated with additional compounds. Like other cultures that are grown for long periods of time, the culture media is changed every 1–3 days.

4.2 Assay Plate Considerations

All 96-well plates will, to some extent, display edge effects. The edge effects may manifest in several different ways. One of the most important for the context of these assays is increased evaporation and therefore concentration of media components in edge wells. This may lead to altered culture conditions for cells grown in these wells. Thus, the wells that provide the most consistent data are the interior 60. Nonetheless, the 36 edge wells may be used to increase sample group numbers. However, we recommend that no single condition is allocated exclusively to edge wells and that a minimum of six wells per experimental condition are utilized. Furthermore, data obtained from edge wells must be examined to ensure patterns of differences between groups are consistent with those from interior wells.

4.3 Normalization Plate Information

Every cell line has distinct properties with growth rate being a key parameter to consider during experimental design. Indeed, cell lines that have been manipulated in some way (i.e., knockdown or overexpression of a protein of interest) will often proliferate at different rates. This characteristic makes achieving identical cell densities at time the assay is performed challenging. An important consideration, therefore, is how to normalize the data generated by the Seahorse experiment so that reliable conclusions can be made. To this end a normalization plate is required. This plate, typically a 12 or 24-well culture dish, is seeded from the same cell suspension used to seed the Seahorse cell culture plate. To faithfully recapitulate the growth conditions in the Seahorse cell culture plate, the amount of cell suspension added to the normalization plate is scaled by surface area. For example, a 24-well culture dish has a well surface area of 1.90 cm^2, or 18.6× the surface area of a XF^e96 Seahorse cell culture plate (0.102 cm^2). Therefore, if 0.15 mL is used to seed the Seahorse plate 2.80 mL (0.15 mL × 18.6) is used to seed each well of a 24-well culture dish. If possible the normalization plate is seeded in triplicate so that an average measurement can be made. The method of normalization (i.e., cell count or protein assay) is up to the researcher.

4.4 Mitochondria Substrates

The specific use of different mitochondrial substrates allows researchers to test how specific aspects of metabolism are altered in various disease states. For example, glutamate and pyruvate are used to investigate respiration dependent upon Complex I activity while succinate is used to investigate Complex II activity. Traditional substrate cocktails include pyruvate + malate, glutamate + malate, pyruvate + glutamate + malate, and succinate + rotenone. Malate is required for respiration of pyruvate and glutamate in isolated mitochondria. Failure to include malate will result in little to no measurable respiration. Glucose cannot be utilized by isolated mitochondria as a fuel because the enzymes required to metabolize glucose to pyruvate are cytosolic and are lost during the mitochondrial isolation. Pyruvate should be limited to 5 mM or below. At concentrations greater than 5 mM, pyruvate uptake into mitochondria may become non-carrier mediated [21].

4.5 Mitochondria Injection Buffers

ADP and FCCP are the most common compounds used to stimulate respiration in isolated mitochondria. FCCP acts by uncoupling the mitochondria by shuttling protons across the inner mitochondria membrane. ADP stimulates ATP synthase thereby driving respiration forward. However, depending upon the substrate being examined, ADP and FCCP elicit different responses. In our experience ADP, is more effective in stimulating glutamate-driven respiration than FCCP, while the reverse is true for pyruvate-driven respiration (FCCP > ADP). While ADP and FCCP may be utilized separately, a mixture of the two is effective for both pyruvate and glutamate. 4 mM ADP and 1 μM FCCP (final concentration) is sufficient to drive maximal respiration in isolated mitochondria. FCCP, not ADP, is used to stimulate respiration in cultured cells. Cellular uptake of ADP is either too slow or insubstantial to stimulate respiration at the timescale of the respiration experiments.

Two different concentrations of substrate-containing buffers are required for a typical experiment. The 1.5× buffer is needed to prepare the injection buffers. The buffers are 1.5× in order to account for the potentially large volume of reagents (FCCP, ADP, etc.) needed. The 1.125× is needed to bring the volume in the Seahorse assay plate up to 180 μL. This buffer is concentrated to take into account for the dilution that occurs when the 160 μL buffer is added to the 20 μL used to apply mitochondria to the plate. Resuspending mitochondria in a common buffer, and adding substrates after centrifugation, allows for multiple substrates to be tested on the same working stock of mitochondria.

4.6 Timing of Seahorse Run

Mitochondria assays are more time-sensitive when compared to assays using cultured cells. A minimum amount of time should elapse between resuspending the mitochondria and starting the experiment. A concentrated pellet of mitochondria may be stable for up to 4 h after isolation. In contrast, once resuspended, mitochondria rapidly expire and lose the ability to respire. Resuspended mitochondria are viable for up to an hour. If resuspended mitochondria are over an hour old, perhaps due to an unexpected delay, we recommend using a different pellet of the same mitochondrial sample. For this reason, we recommend preparing multiple pellets of mitochondria from a given sample to match or exceed the number of temporally separated resuspensions that will be required. A key time-saving step is to load the injection media into the cartridge, program the experiment, and start the cartridge calibration while the mitochondria are attaching to the plate. Typically, calibration of the Seahorse cartridge is nearly complete by the time the assay plate is ready for analysis.

Acknowledgments

We would like to thank Brett Wagner and the Radiation and Free Radical Research Core Facility for assistance performing the experiments utilizing the Seahorse Bioscience XF96. This research was supported by NIH grants R01 DK104998 (E.B.T.); R00 AR059190 (E.B.T.); F32 DK101183 (L.R.G.); T32 HL007121 to Francois Abboud (L.R.G.); T32 HL007638 to Michael Welsh (A.J.R.); and P30CA086862 to George Weiner, which contributed to support of core facilities utilized for this research.

References

1. Gray LR, Tompkins SC, Taylor EB (2014) Regulation of pyruvate metabolism and human disease. Cell Mol Life Sci 71(14):2577–2604. doi:10.1007/s00018-013-1539-2
2. Pellerin L, Magistretti PJ (2012) Sweet sixteen for ANLS. J Cereb Blood Flow Metab 32(7): 1152–1166. doi:10.1038/jcbfm.2011.149
3. Bricker DK, Taylor EB, Schell JC, Orsak T, Boutron A, Chen YC, Cox JE, Cardon CM, Van Vranken JG, Dephoure N, Redin C, Boudina S, Gygi SP, Brivet M, Thummel CS, Rutter J (2012) A mitochondrial pyruvate carrier required for pyruvate uptake in yeast, Drosophila, and humans. Science 337(6090):96–100. doi:10.1126/science.1218099
4. Herzig S, Raemy E, Montessuit S, Veuthey JL, Zamboni N, Westermann B, Kunji ER, Martinou JC (2012) Identification and functional expression of the mitochondrial pyruvate carrier. Science 337(6090):93–96. doi:10.1126/science.1218530
5. Brivet M, Garcia-Cazorla A, Lyonnet S, Dumez Y, Nassogne MC, Slama A, Boutron A, Touati G, Legrand A, Saudubray JM (2003) Impaired mitochondrial pyruvate importation in a patient and a fetus at risk. Mol Genet Metab 78(3):186–192
6. Gray LR, Sultana MR, Rauckhorst AJ, Oonthonpan L, Tompkins SC, Sharma A, Fu X, Miao R, Pewa AD, Brown KS, Lane EE, Dohlman A, Zepeda-Orozco D, Xie J, Rutter J, Norris AW, Cox JE, Burgess SC, Potthoff MJ, Taylor EB (2015) Hepatic mitochondrial pyruvate carrier 1 is required for efficient regulation of gluconeogenesis and whole-body glucose homeostasis. Cell Metab 22(4):669–681. doi:10.1016/j.cmet.2015.07.027

7. McCommis KS, Chen Z, Fu X, McDonald WG, Colca JR, Kletzien RF, Burgess SC, Finck BN (2015) Loss of mitochondrial pyruvate carrier 2 in the liver leads to defects in gluconeogenesis and compensation via pyruvate-alanine cycling. Cell Metab 22(4):682–694. doi:10.1016/j.cmet.2015.07.028
8. Schell JC, Olson KA, Jiang L, Hawkins AJ, Van Vranken JG, Xie J, Egnatchik RA, Earl EG, DeBerardinis RJ, Rutter J (2014) A role for the mitochondrial pyruvate carrier as a repressor of the Warburg effect and colon cancer cell growth. Mol Cell 56(3):400–413. doi:10.1016/j.molcel.2014.09.026
9. Vigueira PA, McCommis KS, Schweitzer GG, Remedi MS, Chambers KT, Fu X, McDonald WG, Cole SL, Colca JR, Kletzien RF, Burgess SC, Finck BN (2014) Mitochondrial pyruvate carrier 2 hypomorphism in mice leads to defects in glucose-stimulated insulin secretion. Cell Rep 7(6):2042–2053. doi:10.1016/j.celrep.2014.05.017
10. Yang C, Ko B, Hensley CT, Jiang L, Wasti AT, Kim J, Sudderth J, Calvaruso MA, Lumata L, Mitsche M, Rutter J, Merritt ME, DeBerardinis RJ (2014) Glutamine oxidation maintains the TCA cycle and cell survival during impaired mitochondrial pyruvate transport. Mol Cell 56(3):414–424. doi:10.1016/j.molcel.2014.09.025
11. Mellon PL, Windle JJ, Goldsmith PC, Padula CA, Roberts JL, Weiner RI (1990) Immortalization of hypothalamic GnRH neurons by genetically targeted tumorigenesis. Neuron 5(1):1–10
12. Stepanenko AA, Dmitrenko VV (2015) HEK293 in cell biology and cancer research: phenotype, karyotype, tumorigenicity, and stress-induced genome-phenotype evolution. Gene 569(2):182–190. doi:10.1016/j.gene.2015.05.065
13. Fu Z, Washbourne P, Ortinski P, Vicini S (2003) Functional excitatory synapses in HEK293 cells expressing neuroligin and glutamate receptors. J Neurophysiol 90(6):3950–3957. doi:10.1152/jn.00647.2003
14. Li Z, Graham BH (2012) Measurement of mitochondrial oxygen consumption using a Clark electrode. Methods Mol Biol 837:63–72. doi:10.1007/978-1-61779-504-6_5
15. Silva AM, Oliveira PJ (2012) Evaluation of respiration with clark type electrode in isolated mitochondria and permeabilized animal cells. Methods Mol Biol 810:7–24. doi:10.1007/978-1-61779-382-0_2
16. Oroboros Instruments. http://wiki.oroboros.at/index.php/OROBOROS_home. Accessed 28 Dec 2015
17. Seahorse Bioscience. http://www.seahorse-bio.com/. Accessed 28 Dec 2015
18. Gerhold JM, Cansiz-Arda S, Lohmus M, Engberg O, Reyes A, van Rennes H, Sanz A, Holt IJ, Cooper HM, Spelbrink JN (2015) Human mitochondrial DNA-Protein complexes attach to a cholesterol-rich membrane structure. Sci Rep 5:15292. doi:10.1038/srep15292
19. Rogers GW, Brand MD, Petrosyan S, Ashok D, Elorza AA, Ferrick DA, Murphy AN (2011) High throughput microplate respiratory measurements using minimal quantities of isolated mitochondria. PLoS One 6(7):e21746. doi:10.1371/journal.pone.0021746
20. Bradford MM (1976) A rapid and sensitive method for the quantitation of microgram quantities of protein utilizing the principle of protein-dye binding. Anal Biochem 72 (1):248–254. doi: 10.1016/0003-2697(76)90527-3
21. Gnaiger E (2012) Mitochondrial pathways and respiratory control. An introduciton to OXPHOS analysis, 3rd edn. OROBOROS MiPNet Publications, Innsbruck. http://www.oroboros.at/?MiPNet-protocols

Index

A

B

C

Stefan Strack and Yuriy M. Usachev (eds.), *Techniques to Investigate Mitochondrial Function in Neurons*, Neuromethods, vol. 123, DOI 10.1007/978-1-4939-6890-9, © Springer Science+Business Media LLC 2017

H

I

J

K

L

M

Q

R

S

T

U

V

W

Y

Z

Printed in the United States
By Bookmasters